T0202367

# Big Science, Innovation, and Societal Contributions

# Big Science, Innovation, and Societal Contributions

## The Organisations and Collaborations in Big Science Experiments

*Edited by*

Shantha Liyanage
Markus Nordberg
and
Marilena Streit-Bianchi

OXFORD
UNIVERSITY PRESS

# OXFORD

### UNIVERSITY PRESS

Great Clarendon Street, Oxford, OX2 6DP,
United Kingdom

Oxford University Press is a department of the University of Oxford.
It furthers the University's objective of excellence in research, scholarship,
and education by publishing worldwide. Oxford is a registered trade mark of
Oxford University Press in the UK and in certain other countries

Published in the United States of America by Oxford University Press
198 Madison Avenue, New York, NY 10016, United States of America

British Library Cataloguing in Publication Data

Data available

Library of Congress Control Number: 2023946226

ISBN 9780198881193

DOI: 10.1093/oso/9780198881193.001.0001

Printed and bound by
CPI Group (UK) Ltd, Croydon, CR0 4YY

Links to third-party websites are provided by Oxford in good faith and
for information only. Oxford disclaims any responsibility for the materials
contained in any third party website referenced in this work.

# Foreword

It gives me great pleasure to introduce this book on Big Science to the interested reader. It combines an impressive group of authors from many disciplines, each contributing to this book by offering examples and sharing their own experiences. This I find of great interest and value. Many authors I know personally, as I have spent most of my own professional career in experimental high energy physics. Although some authors I am less familiar with, it is fascinating to read their own accounts, which resonate with and build upon my own experiences of the many aspects of designing, managing, and running Big Science experiments, and witnessing many of the fruits of our research being picked up by society.

I will leave it up to the reader to discover the wealth of insights offered by the following chapters. But I'd like to comment just on a few, those that resonated particularly well with me. First, the importance of the principles of open science and open innovation being recognised as one of the key ingredients in Big Science throughout the book.

In my view, this is the very essence of being successful in shaping and defining the burning research questions and then designing, constructing, and operating a Big Science facility. The second important element is the inclusion of serendipity. That is, designing the scientific apparatus with enough parameter space to allow for *unexpected* discoveries. Of course, scientific instruments are rigorously designed and constructed to minute detail, driven by the expectations offered by the scientific theories being tested. However, the very nature of fundamental research is that not all outcomes can be pre-determined ex-ante.

An element of (pleasant)surprise is thus always present. I find this notion particularly relevant these days, when large-scale basic research projects are almost expected to make all dreams of discovery true. Well, mostly they do—like the discovery of the Higgs or gravitational waves—but not always. And when the latter happens, it may be even more significant for our understanding of how nature works.

However understandably, this is hard to explain to our funding authorities, the ultimate financiers, and the taxpayers.

This brings me to my third point that a good part of this book explores: apart from the scientific discovery potential that Big Science offers, what is in it then for society at large? Starting with the obvious: technologies developed for scientific instrumentation eventually find their way to good (and bad) use in society. Examples include the web, fast telecommunications, data storage ('the cloud'), medical applications and so on. However, as the chapters demonstrate, there are other aspects, the less obvious of which is the power of ideation and experimentation with new concepts on a larger scale, like addressing Sustainable Development Goals. That is, using Big Science labs as test-beds for activities other than their primary focus.

Our society desperately needs solid, evidence-based platforms while gathering crucial information on how to take its next steps. Neutral spaces, such as the CERN cafeteria, where scientists from all over the world can meet informally, are invaluable for exchanging ideas far beyond the scope of science alone. Other aspects include the education of a large number of next-generation scientists and innovators. Statistics show that less than half of the students, for example, will stay in high energy physics research. In large experiments at CERN alone, there are thousands of technical and PhD students being educated as we speak. Then there is the network of hundreds of universities and research laboratories all around the world that are connected, forming several interconnected layers of ecosystems of research and innovation that Big Science labs heavily draw upon. So, this is not 'just' about these big centres; it is about communities reaching up to many tens of thousands of active researchers and innovators.

My last point concerns the future of Big Science itself. Many chapters, I believe, demonstrate the point that future Big Science initiatives will require a stronger element of public engagement, such as outreach efforts, without jeopardising their basic research foundations. Examples of current efforts provided in this book point to determined efforts by many, and I believe that more can be done, particularly to engage funding agencies, industry, and government at an early stage as the size of these investments grows. The importance of openness in our science and its positive consequences for the longer-term prosperity and democratic principles of our society cannot be overemphasised.

The emerging Sustainability Science could benefit a wide range of organisations and connections, as shown by the Big Basic Sciences. We need in this time of frightening global challenges, the same spirit which prevailed at the origin of CERN after the Second World War, focused this time on Sustainability Science. Global challenges may present a unique opportunity to create a better world.

Although I find the term 'Big Science' to be somewhat pompous and runs the risk of unintentionally polarising what could then be categorised as 'Small Science', as they are intricately linked, this book nonetheless nicely captures what the drivers are of large-scale scientific enterprises, how they are built and managed bottom-up, and how they can also act as drivers of our society at large.

Dr Michel Spiro
*President of the International Union of Pure and Applied Physics (IUPAP)*
*Chair of the International Year of Basic Sciences for Sustainable Development in 2022 (IYBSSD 2022)*
*Chair of the Board of the CERN & Society Foundation and Former President (2010–2013) of the CERN Council*
*Paris, November 2022*

# Acknowledgements

Editors can only make a book like this possible with dedicated work by our esteemed authors, not one or two—34—of them, who have contributed to respective chapters. It was, of course, a challenge to bring diverse views into a coherent Big Science themes. We are grateful to Professor Peter Jenni, former Spokeperson for ATLAS, CERN whose relentless support and guidance in helping us achieve our goals. We sincerely thank many scholars and others from Big Science organisations, CERN, ESO, LIGO, and other experiments for supporting us in many ways to gather data and valuable insights. There are too many to list them individually, but we sincerely thank them for their support.

Editors would like to thank all our contributing authors who have generously given their time to produce high-quality content and shared their intellectual capital to shape the ideas of tackling complex Big Science topics like high energy physics and astrophysics and connecting science with society. We are thankful for their patience and continued support as we went through several editorial versions and clarifications of contents to ensure quality and consistency.

We are grateful to many colleagues, researchers, and experts in the field who agreed to take part in the interviews and contributed to various chapters. The editors are grateful to several people at CERN; Thierry Lagrange, Hans-Peter Beck, Jens Vigen, Giovanni Anelli, Jonathan Drakeford, Johanna McEntyre, and Lindsey Crosswell at EMBL-EBI and Rupert Lück, Jessica Klemeier at EMBL, and David Manset at United Nations ITU, for their important contributions and insights into Big Science operations.

We would like to thank the Oxford University Press staff, and our commissioning editor, Adam Swallow, who ably guided us in this book project from uncharted waters to a safe port. We are also grateful for all the production assistance from Phoebe Aldridge-Turner, OUP, Purushothaman Govindhasamy from Integra, and Martin Noble for copyediting the text. We sincerely thank our reviewers who provided valuable insights and useful criticism, which helped us focus on the fundamental objectives and goals of Big Science initiatives.

Our sincere gratitude goes to several critical friends who provided valuable comments on the content and guided us to focus on the key issues. We are indebted to our critical friends, Doris Forkel-Wirth, Pra Murthy, Stephen Hill, and Robert Stevens, for their valuable comments and suggestions. We gratefully acknowledge several people who helped us with the editorial work: Stig Johannessen, Sachini Marasinghe, Lochana Kulatunga, and Satis Arnold for their editorial assistance. We are especially grateful to Nalin Wasantha Raj Thomas whose critical eyes and editorial work immensely helped to improve quality and added value to linguistic expressions that aligned with the comments of our critical friends and reviewers. We are grateful to many OUP reviewers who provided us with valuable comments at various stages of the production of this book and their advice was very valuable in improving the content and bringing this book to the publishing quality required by Oxford University Press.

# Contents

# List of Figures

# List of Tables

# Abbreviations—ACRONYMS

| | |
|---|---|
| AAI | Architecture Adaptive Integrator |
| ACFA | Asian Committee for Future Accelerators |
| ACNS | Australian Centre for Neutron Scattering |
| ACT | Access to Covid-19 Tools Accelerator |
| AD | Antiproton Decelerator |
| ADMX | Axion Dark Matter eXperiment |
| ADP | Avalanche Photodiode |
| AGLAE | Accélérateur du Grand Louvre d'Analyse Élémentaire |
| AI | Artificial Intelligence |
| AISBL | Association International Sans But Lucratif (International not-for-profit Association) |
| ALEPH | Apparatus for LEP Physics |
| ALICE | A Large Ion Collider Experiment |
| ALMA | Atacama Large Millimeter/submillimeter Array |
| ALP | Axion-like Particle |
| ALPHA | Absolute Luminosity for ATLAS |
| ALS | Amyotrophic Lateral Sclerosis |
| ANSTO | Australian Nuclear Science and Technology Organization |
| ANTARES | Australian National Tandem Research Accelerator |
| APEX | Atacama Pathfinder EXperiment Telescope |
| ARC | Australian Research Council |
| ARC-DMPP | Australian Research Centre—Dark Matter Particles Physics |
| ArDM | Argon Dark Matter detector |
| AS | Australian Synchrotron |
| ASCOT | Apparatus with Super Conducting Toroids |
| ASKAP | Australian Square Kilometre Array Pathfinder |
| ATF | Accelerator Test Facility |
| ATLAS | A Toroidal LHC ApparatuS |
| ATTRACT | breAk Through innovaTion pRogrAmme for deteCtor infrastructure ecosysTem |
| AuthN/AuthZ | Authentication / Authorisation |
| BCH | Billions Swiss franc |
| BEPC | Beijing Electron Positron Collider |
| BGU | Ben Gurion University |
| BINP | Budker Institute of Nuclear Physics |
| BNCT | Boron Neutron Capture Therapy |
| BioDynaMo | Biology Dynamic Modeller |
| BSCCO | Bismuth Strontium Calcium Copper Oxide |
| BSM | Beyond Standard Model |
| CAD | Computer-Aided Design |
| CAST | CERN Axion Solar Telescope |

| | |
|---|---|
| CBA | Cost Benefit Analysis |
| CBI | Challenge Based Innovation |
| CBI | Computer-Based Instruction |
| CCC | CERN Control Centre |
| CCD | Charge-Coupled Device |
| CDM | Customer Data Management |
| CDR | Conceptual Design Report |
| CEA | Commissariat à l'Energie Atomique (et aux énergies alternatives) |
| CEO | Chief Executive Officer |
| CEPC | Circular Electron Positron Collider |
| CEPI | Coalition for Epidemic Preparedness Innovation |
| CERN | Conseil Européen pour la Recherche Nucléaire (European Organization for Nuclear Research) |
| CESP | CERN Entrepreneurship Student Programme |
| CH | Confoederatio Helvetica |
| CIRT | Carbon Ion RadioTherapy |
| CNAO | Centro Nazionale di Adroterapia Oncologica |
| COIF | Collaborative Open Innovation Framework |
| CSIRO | Commonwealth Scientific and Industrial Research Organization |
| CLIC | Compact LInear Collider |
| CLOUD | Cosmic Leaving Outdoor Droplets |
| CMOS | Complementary Metal-Oxide Semiconductor |
| CMS | Compact Muon Solenoid |
| CMB | Cosmic Microwave Background |
| CNS | Central Nervous System |
| Covid-19 | CoronaVirus Disease 2019 |
| CP | Charge Parity |
| CPU | Central Processing Unit |
| CRESST | Cryogenic Rare Event Search with Superconducting Thermometers |
| CSR | Corporate Social Responsibility |
| CT | Computer Tomography |
| CW | Continuous Wave |
| DAMIC-M | Dark Matter In CCDs in Modane |
| DCF | Discounted Cash Flow analysis |
| DELPHI | DEtector with Lepton, Photon and Hadron Identification |
| DESY | Deutsches Elektronen Synchrotron |
| DFGN | Design Factory Global Network |
| DIPG | Diffuse Intrinsic Pontine Glioma |
| DM | Data Management |
| DMP | Data Management Plan |
| DNA | DeoxyriboNucleic Acid |
| DOE | Department of Energy |
| DORA | San Francisco Declaration on Research Assessment |
| DUNE | Deep Underground Neutrino Experiment |
| EAGLES | Experiment for Accurate Gamma, Lepton and Energy Measurements |
| EBI-EMBL | European Bioinformatic Institute—European Molecular Biology Laboratory |

| | |
|---|---|
| EBRT | External Beam RadioTherapy |
| ECFA | European Committee for Future Accelerators |
| ECR | Engineering Change Request |
| EDELWEISS | Expérience pour DÉtecter Les WIMPs En Site Souterrain |
| EHT | Event Horizon Telescope |
| EIB | European Investment Bank |
| EIC | The Electron-Ion Collider |
| EIROforum | European Intergovernmental Research Organization forum |
| ELENA | Extra Low ENergy Antiproton ring |
| ELT | Extremely Large Telescope |
| EMBL | European Molecular Biology Laboratory |
| EMPA | Eidengenossische Materialprufungs und Forschungsanstalt (Swiss Federal Laboratories for Material Science and Technology) |
| EoIs | Expression of Interests |
| EOSC | European Open Science Cloud |
| EPOG | European Particles Physics Outreach Group |
| EPPCN | European Particles Physics Communication Network |
| ESPPPG | European Strategy Particles Physics Preparatory Group |
| EPPSU2020 | European Particles Physics Strategy Update 2020 |
| ESA | European Space Agency |
| ESADE | École Supérieure d'Administration et de Direction d'Entreprise |
| ESRF | European Synchrotron Radiation Facility |
| ESS | European Spallation Source |
| ESO | European Southern Observatory |
| EYETS | Extended Year-End Technical Stop |
| FAs | Funding Agencies |
| FAIR | Findability, Accessibility, Interoperability and Reusability principles |
| FAIR | FAcility for Iron and Antiproton |
| FALC | Funding Agencies for Large Colliders |
| FAST | Five hundred metre Aperture Spherical Telescope |
| FCC | Future Circular Collider |
| FCC-ee | Future Circular Collider-electrons |
| FCC-hh | Future Circular Collider hadrons |
| FCC-h | FCC-hadrons |
| F&R | Feet and Rails System |
| FST | Fleet Space Technologies |
| GaiaNIR | Global Astrometric Interferometer for Astrophysics Near Infrared Research |
| GAVI | Global Alliance for Vaccines and Immunisation |
| GCR | Galactic Cosmic Rays |
| GDE | Global Design Effort |
| GDP | Gross Domestic Product |
| GDPR | General Data Protection Regulation |
| GEAR | Gather Evaluate Accelerate Refine |
| GEMPix | Gas Electron Multipliers Pixel |
| GeV | Giga electron Volt |
| GHG | Greenhouse Gas |

| | |
|---|---|
| GM | Genetically Modified |
| GKGSB | Cheung Kong Graduate School of Business |
| GPS | Global Positioning System |
| GR | General Relativity |
| GRB | Gamma-Ray Burst |
| GSI | Gesellschaft für Schwerionenforschung (Society for Heavy Ions Research) |
| GWIC | Gravitational Wave International Committee |
| HEP | High Energy Physics |
| HERA | Hadron Elektron Ring Anlage |
| HGP | Human Genome Project |
| HIAF | Heavy Ion Analytical Facility |
| HL-LHC | High Luminosity-Large Hadron Collider |
| HPDT | High Precision Drift Tubes |
| HSC | Honeycomb Strip Chambers |
| HTS | High Temperature Superconducting |
| HVDC | High Voltage Direct Current |
| Hz | Hertz |
| IAEA | International Atomic Energy Agency |
| IAP | Institute of Applied Physics |
| IASS | Institute of Advanced Sustainability Study |
| IAXO | International AXion Observatory |
| IBA | Ion Beam Analysis |
| IBM | International Business Machines corporation |
| IBSE | Inquiry-Based Science Education |
| ICFA | International Committee for Future Accelerators |
| ICT | Information and Communication Technology |
| ICTP | International Centre for Theoretical Physics |
| ID | Inner Detector |
| IDT | International Development Team |
| IHEP | Institute for High Energy Physics |
| ILC | International Linear Collider |
| ILL | Institut Laue-Langevin |
| IMRT | Intensity Modulated Radiotherapy |
| INFN | Istituto Nazionale di Fisica Nucleare |
| IN2P3 | Institut National de Physique Nucléaire et de Physique des Particules |
| INTAS | International Association for the promotion and cooperation with scientists from the independent states of the former Soviet Union. |
| IPPOG | International Particles Physics Outreach Group |
| IPR | Intellectual Property Right |
| ISOLDE | Isotope Separator On Line DEvice |
| I-Space | Information Space |
| ISR | Intersecting Storage Ring |
| ISTC | International Science and Technology Center |
| ISCT | International Society for Computer Tomography |
| ITER | International Thermonuclear Experimental Reactor (in Latin *the way*) |

| | |
|---|---|
| JCDC | Jet Cell Drift Chambers |
| JET | Joint European Torus |
| JINR | Joint Institute for Nuclear Research (Russia) |
| JUNO | Jiangmen Underground Neutrino Observatory |
| IUPAP | International Union of Pure and Applied Physics |
| JWST | James Webb Space Telescope |
| KAGRA | Kamioka Gravitational Wave Detector |
| KAMLAND | KAMioka Liquid scintillator AntiNeutrino detector |
| KEK | Japanese abbreviation for High Energy Research Organisation |
| ΛCDM | Lambda-Cold Dark Matter |
| LEO | Low Earth Orbit |
| LEP | Large Electron–Positron collider |
| LET | Linear Energy Transfer |
| LHC | Large Hadron Collider |
| LHCb | Large Hadron Collider beauty |
| LHCC | Large Hadron Collider Committee |
| LIGHT | Linac for Image Guided Hadron Therapy |
| LIGO | Laser Interferometer Gravitational-wave Observatory |
| Linac | Linear Accelerator |
| LmADP | Linear mode Avalanche Photodiode |
| LMICs | Lower and Medium Income Countries |
| LBNF/DUNE | Long-Baseline Neutrino Facility/Deep Underground Neutrino Experiment |
| LoI | Letter of Intention |
| LOFAR | LOw Frequency ARray |
| LOSC | Ligo Open Science Centre |
| LRI | Large Research Infrastructures |
| LS1 | Long Shutdown 1 |
| LS2 | Long Shutdown 2 |
| LSC | Ligo Scientific Collaboration |
| LSS | Long Straight Sections |
| LUX- | Large Underground Xenon detector |
| LZ | LUX-ZEPLIN |
| MACHINA | Movable Accelerator for Cultural Heritage In-situ Non-destructive Analysis |
| MB | Megabyte |
| MCT | MOS-Controlled Thyristor |
| MDT | Monitored Drift Chambers |
| MeerKAT | Karoo Array Telescope in Meerkat National Park |
| MeV | Million electron Volt |
| MEXT | Ministry of Education, Culture, Sport, Science and Technology (Japan) |
| MKID | Microwave Kinetic Inductance Detector |
| MIT | Massachusett Institute of Technology |
| ML/DL | Machine Learning / Deep Learning |
| MLI | Multi Layer superInsulation |
| MOS | Metal-Oxide Semiconductor |

| | |
|---|---|
| MOSFET | Metal-Oxide Semiconductor Field-Effect Transistor |
| MOU | Memorandum of Understanding |
| MRI | Magnetic Resonance Imaging |
| MSc | Master of Science |
| MWA | Murchison Widefield Array |
| NASA | National Aeronautics and Space Administration |
| NCA | Neutron Capture Agents |
| NCES | National Center for Education Statistics |
| NCEPT | Neutron Capture Enhanced Particle Therapy |
| NCT | Neutron Capture Therapy |
| NDF | National Deuteration Facility |
| NGO | Non Governmental Organisation |
| NIKHEF | Nationaal Instituut voor Kernfysica en Hoge-Energiefysica |
| NMR | Nuclear Magnetic Resonance |
| NPV | Net Present Value |
| NRENs | National Research and Education Networks |
| NSF | National Science Foundation |
| n-TOF | neutrons Time of Flight facility |
| NTT | New Technology Telescope |
| OAIS | Open Archival Information System |
| OECD | Organisation for Economic Co-operation and Development |
| OFHC | Oxygen Free High thermal Conductivity |
| OpenQKD | Open Quantum Key Distribution |
| ORNL | Oak Ridge National Laboratory |
| OSOS | Open Schools for Open Society |
| OUP | Oxford University Press |
| PAFs | Phased Array Feeds |
| PandaX | Particle and Astrophysical Xenon detector |
| PB | Petabyte |
| PDBe-KB | Protein Data Bank Europe |
| PEP | Positron Electron Project |
| PET | Positron Emission Tomography |
| PETRA | Positron-Elektron-Tandem-Ringanlage (Positron Electron Tandem Ring Accelerator) |
| PIMMS | Proton-Ion Medical Machine Study |
| PISA | Program for International Student Assessment |
| PIXE | Particle-Induced X-rays Emission |
| POSTECH | Pohang University of Science and Technology |
| PS | Proton Synchrotron |
| PSB | Proton Synchrotron Booster |
| PTCOG | Proton Therapy Co-Operative Group |
| KEK | Japanese abbreviation of the High Energy Accelerator Research Organisation (Kō Enerugī Kasokuki Kenkyū Kikō) |
| QA | Quality Assurance |
| QC | Quality Control |
| QCD | Quantum Chromodynamics |
| RAL | Rutherford Appleton Laboratory |

| | |
|---|---|
| RBE | Relative Biological Effectiveness |
| R&D | Research and Development |
| REANA | Reusable and Reproducible data ANAlysis platform |
| REBCO | Rare-Earth Barium Copper Oxide |
| REID | Radiation Exposure Induced Death |
| RENO | Reactor Experiment for Neutrino Oscillation |
| RF | Radio Frequency |
| RHIC | Relativistic Heavy Ions Collider |
| RI | Research Infrastructure |
| RNA | Ribonucleic Acid |
| RPC | Resistive Plate Chambers |
| SACLA | SPring-8 Angstrom Compact free electron LAser |
| SARS-CO-2 | Severe Acute Respiratory Syndrome-Covid-2 |
| SBRT | Stereotactic Body Radiation Therapy |
| SCOAP | Sponsoring Consortium for Open Access Publishing in Particle Physics |
| S COOL | Student Cooperative Open Learning |
| SCP | Supernova Cosmology Project |
| SDG(s) | Sustainable Development Goals |
| SHARP | SolarHydrothermal Advanced Reactor Project |
| SKA | Square Kilometre Array |
| SLAC | Stanford Linear Accelerator Center |
| SLC | Social Learning Cycles |
| SM | Standard Model |
| SMPC | Secure Multi Party Computational protocols |
| SOC | Self-Organised Criticality |
| SppC | Super proton-proton Collider |
| SPEAR | Stanford Positron Electron Asymmetric Ring |
| SPS | Super Proton Synchrotron |
| SQL | Structured Query Language |
| SRIA | Strategic Research and Innovation Agenda |
| SSC | Superconducting Super Collider |
| SSS | Short Straight Section |
| STEM | Science, Technology, Engineering, and Math |
| STEMM | Science, Technology, Engineering, Mathematic, and Medicine |
| STFC | Science and Technology and Facilities Council |
| SuperCDMS | Super Cryogenic Dark Matter Search |
| SUPL | Stawell Underground Physics Laboratory |
| SUSY | SuperSymmetry |
| STAR | 2 MV Tandetron Accelerator (STAR) |
| SWAFS | Science With and For Society |
| TCP/IP | Transmission Control Protocol / Internet Protocol |
| TDR | Technical Design Report |
| TOTEM | TOTal Elastic and diffractive cross section Measurements |
| TP | Technical proposal |
| TSV | Through SiliconVia |
| TWAS | Third World Academy of Science |

| | |
|---|---|
| UA1 | Underground Area 1 |
| UA2 | Underground Area 2 |
| UBS | Union Bank of Switzerland |
| UN | United Nations |
| UNESCO | United Nations Educational, Scientific and Cultural Organization |
| UNRWA | United Nations Relief and Work Agency |
| UK | United Kingdom |
| US | United States |
| USA | United States of America |
| VISIR | VLT Imager and Spectrometer for mid-Infrared |
| VLA | Very Large Array |
| VLT | Very Large Telescope |
| W | watt |
| WBS | Work Breakdown Structure |
| WHO | World Health Organization |
| WIMP | Weakly Interacting Massive Particles |
| WMAP | Wilkinson Microwave Anisotropy Probe |
| WWW | World Wide Web |
| XFEL | X rays Free-Electron Laser |
| ZEPLIN | ZonEd Proportional scintillation in LIquid Noble gas detector |

# List of Contributors

**Dr Amalia Ballarino** is a Senior Staff Scientist at CERN and a section leader of the team responsible for the development, testing, and procurement of superconductors (strands and cables) for the CERN accelerators and the conception and design of superconducting devices such as HTS leads and electrical transmission lines.

**Professor Elisabetta Barberio** is a professor of High Energy Physics, at the University of Melbourne, the Director of the Centre for Dark Matter Particle Physics, and a member of the Experimental Particle Physics Group. Elisabetta spent much of her career as a researcher at CERN.

**Dr Michael Benedikt** is an accelerator physicist, who joined CERN in 1997 with a PhD in medical accelerator design expertise. He is a member of the CERN Proton-Ion Medical Machine Study group. Michael led the PS2 design study to design a new high-performance synchrotron and was the project leader for the design and construction of the accelerator complex for the Austrian hadron therapy centre.

**Dr Frédérick Bordry** was CERN's Director of Accelerators and Technology from 2014 to 2020 and the Chief Technical Officer of Gauss Fusion. He was responsible for the operation and exploitation of the whole CERN accelerator complex, with particular emphasis on the LHC and the development of new projects and technologies.

**Professor Tim Boyle** is Director, Innovation and Commercialisation at the Australian Nuclear Research Organisation (ANSTO) and is an innovation and economic development professional with 20 years of experience cultivating the interface of science and industry in technology transfer, research, and commercialisation roles. Tim is an adjunct professor at Swinburne University, Australia.

**Dr Beatrice Bressan** held the position of executive director of the UNRWA spin-off, GGateway IT Social Enterprise, Gaza Palestine, and she is now the Chief Executive Officer of the Global Challenges Forum Foundation and the Executive Advisor of Lynkeus. Educated as a physicist, she worked in the Technology Transfer group at CERN, where she is now the TOTEM and CMS/PPS Outreach Coordinator IPPOG resource advisor, and SHARP Project Leader. As a writer and European Union of Science Journalists' Associations (EUSJA) member, she has written on a wide range of subjects and authored, as well as edited scholarly books and publications.

**Professor Agustí Canals** works at the Universitat Oberta de Catalunya in Barcelona, where he is serving as Academic Director of the MSc in Strategic Management of Information and Knowledge in Organisations and leads the KIMO Research Group on Knowledge and Information Management in Organisations.

**Professor Mark Casali** is the Director of the Macquarie node of Australian Astronomical Optics (AAO-Macquarie) at Macquarie University, Sydney, Australia and has over 30 years of experience in the design and construction of world-class instrumentation and led the technology development and second-generation ELT instrument programmes for the European Southern Observatory (ESO).

**Dr Alberto Di Meglio** is the Head of Innovation in the CERN IT Department, Coordinator of the CERN Quantum Technology Initiative and former Head of CERN openlab. He has extensive experience in the design, development, and deployment of distributed computing and data infrastructures for high energy physics, medical, and space research.

**Professor Alan Duffy** is a Professor at the Swinburne University, Melbourne, Australia and the inaugural Director of the Space Technology and Institute. Alan is a professional astrophysicist who models universes on supercomputers to understand how galaxies form within vast halos of dark matter and has spun out technology from the search for dark matter into a company (mDetect Pty Ltd) he cofounded.

**Professor John Ellis** is a theoretical physicist and Clerk Maxwell Professor of Theoretical Physics at King's College London. He worked at CERN for many years in close connection with experiments and served as leader of the Theoretical Studies Division from 1988 to 1994. He participated in early studies of the physics capabilities of LEP and LHC, and currently studies the physics opportunities for the proposed Future Circular Collider.

**Dr Lyn Evans** is an experimental physicist who graduated from the University of Wales Swansea and obtained a PhD in Physics from the University of Wales for a combined theoretical and experimental study of the interaction of intense laser radiation with gases. Lyn was a research fellow in the Proton Synchrotron Division, Division Leader of the Super Proton Synchrotron and Project Leader of the LHC project. In 2012, he was appointed director of the Linear Collider Collaboration.

**Dr Pablo Gracia Tello** is the section head of new projects and initiatives at the CERN, EU office. He obtained his physics degree from the Complutense University of Madrid, Spain and his PhD in Material Science from the Basque Country University in San Sebastian, Spain.

**Dr James Gilbert** is a Visiting Fellow and former Lead Engineer at the Australian National University Research School of Astronomy and Astrophysics. His research covers infrared sensors, multi-object spectroscopy, opto-mechanics, and space instrumentation.

**Dr Steven Goldfarb** is a physicist working for the University of Melbourne on the ATLAS Experiment at CERN. He has served as the collaboration's Education and Outreach Coordinator and chaired the International Particle Physics Outreach Group from 2017 to 2022. He currently coordinates undergraduate research programmes for the University of Michigan at CERN.

**Professor Peter Jenni** obtained his Diploma in Physics at the University of Bern in 1973 and a Doctorate at the Swiss Federal Institute of Technology in Zürich (ETHZ) in 1976. He participated in CERN experiments at the Synchro-Cyclotron, at the Proton Synchrotron and as an ETHZ Research Associate at the Intersecting Storage Rings, the first high energy hadron collider. He is best known as one of the 'founding fathers' of the ATLAS experiment at CERN and was the architect at the ATLAS project of the LHC and the Spokesperson from 1995 to 2019.

**Professor Anita Kocsis** is the Director of Design Factory Melbourne, a transdisciplinary platform that creates the conditions for design-led innovation by brokering collaborative university industry-engaged research and leads DFM in a key role in developing innovation capability with partnerships across diverse university sectors.

**Professor Christine Kourkoumelis** is a physicist at the National and Kapodistrian University of Athens (NKUA). She holds a PhD in Physics from Yale University and has been working on various high energy physics experiments at CERN, Fermilab, and JeffersonLab (USA).

**Professor Shantha Liyanage** holds a biological science degree from the University of Colombo, Sri Lanka and a PhD from the University of Wollongong, Australia. He held professorial appointments in Business and Innovation with the University of Queensland, the University of Auckland, New Zealand, Macquarie University, the University of Sydney, and the University of Technology Sydney. As the Principal Researcher at the Australian Research Council's Centre for Research Policy at the University of Wollongong he contributed to innovation and organisational theory and practice. He was the Director of the Technology Management Centre, Chemical Engineering Department at the University of Queensland and developed e-learning programs for Master programme and was the Manager of UNESCO based Science and Policy Asian Network (STEPAN). As a professor of the Business School of the University of Auckland, New Zealand, he led the Technology, Innovation and Knowledge Management program (TEKIM). He contributed to leadership studies at the Graduate School of Management at the Macquarie University, Sydney. He held several visiting professorial appointments at the Nihon University in Japan, Copenhagen Business School, and Zeppelin University Germany. His research covers education, innovation and leadership including the management of Big Science experiments-ATLAS and CMS experiments at CERN. He is a senior researcher at the NSW Department of Education and the Editor-in-Chief of the *International Journal of Learning and Change.*

**Dr David Manset** is a French entrepreneur, and CEO of two software vendor companies (i.e. Be-ys Research and Cerebro). He is the CEO and co-founder of Be-ys Research, in collaboration with the ScimPulse research foundation and is affiliated to CERN openlab and is a CERN Alumni and a senior project coordinator, United Nations ITU (International Telecommunication Union).

**Professor Grace McCarthy** is a professor and dean of the Sydney Business School at the University of Wollongong, Australia. She has a PhD, MBA, MA, BA and is also a Graduate of the Australian Institute of Company Directors.

**Dr Markus Nordberg** heads the Development and Innovation Unit (IPT-DI) at CERN and has completed academic degrees in Physics and Business Administration. He coordinates multidisciplinary innovation projects at IdeaSquare at CERN, and the EU-funded sensor and imaging R&D&I initiative called ATTRACT (www.attract-eu.com) together with his colleague Pablo Gracia Tello with the goal of maximising the scientific and societal impact of disruptive co-innovation. Markus worked as the Resources Coordinator of the ATLAS at CERN for 12 years.

**Dr Panagiotis Charitos** is an astrophysicist from Imperial College London. He holds an MSc in Media and Communications from the London School of Economics. Following his studies in media and sociology, he went on to earn an MA and a PhD in Philosophy. He joined CERN in 2011 and since 2015 he has served as a scientific information officer for the EP department.

**Dr Ludovico Pontecorvo** is a physicist who joined the ATLAS experiment in 1992 and contributed to the construction of the tracking chambers for the ATLAS muon spectrometer and to the installation and commissioning of the Barrel muon spectrometer. He is currently involved in the upgrading of LHC projects—both for phase 1 and phase 2 (HL-LHC).

**Professor David Reitze** holds joint positions as the Executive Director of the LIGO Laboratory at the California Institute of Technology and as a Professor of Physics at the University of Florida. His research focuses on the development of gravitational-wave detectors.

**Dr Mitra Safavi-Naeini** has been a researcher in the field of particle physics and medical radiation physics for 10 years and is currently a research project leader at the Australian Nuclear Science and Technology Organisation (ANSTO).

**Dr Faiz Shah** is the executive director of the Yunus Centre, Asian Institute of Technology (AIT), with a faculty affiliation at the School of Management. He steers the Yunus Master's in Social Business & Entrepreneurship and is the head of Development Management. He coordinates the AIT Healthcare Resources Group, Energy Development, and Services Management & Technology programmes.

**Dr Suzie Sheehy** is an accelerator physicist and Associate Professor at the University of Melbourne where she leads the Medical Accelerator Physics group, and a Visiting Lecturer at the University of Oxford. She is an award-winning science communicator, TV presenter, and author of the highly acclaimed popular science book with Bloomsbury Publishing in 2022 *The Matter of Everything: Twelve Experiments that Changed Our World*.

**Professor Viktorija Skvarciany** holds a PhD in Economics and works as a Vice-Dean for Research and Innovation and a Professor at the Faculty of Business Management at Vilnius Gediminas Technical University. Her research interests include the digital and circular economy as well as decision-making processes.

**Dr Marilena Streit-Bianchi** obtained her doctorate in the Biological Sciences at the University La Sapienza in Rome. She commenced her career at CERN in 1969 and studied the biological properties of accelerated particle beams for the treatment of tumours. She is the Vice President of the international association ARSCIENCIA and a member of the Italian Physics Society (SIF) and she has published recent books: "Mare Plasticum-The Plastic Sea-Combatting Plastic Pollution Through Science and Art" Springer 2020; "Advances in Cosmology" Springer nature 2022; and "New Challenges and Opportunities in Physics Education" for Springer book series: Challenges in Physics Education, 2023.

**Professor Geoffrey Taylor** is a Redmond Barry Distinguished Professor, Faculty of Science, University of Melbourne, and the Director, ARC Centre of Excellence for Particle Physics at the Terascale. Australian physicists, led by Professor Taylor, made important contributions to the discovery of the Higgs Boson. He played a major role in the design and construction of the advanced detectors for the Large Hadron Collider at CERN right from the initial 1989 ideas. Professor Taylor's work on ATLAS and Belle/Belle II at KEK in Japan, is just a part of his distinguished career in Experimental Particle Physics going back several decades. He is currently (2018–20) Chair of ICFA (International Committee for Future Accelerators, and Asia-Pacific representative on the International Linear Collider IDT (International Development Team).

**Professor Christine Thong** leads the academic direction of the Design Factory Melbourne at Swinburne University and is involved in a range of teaching, research, and strategic initiatives. She is passionate about coaching the next generation of globally responsible design innovators.

**Professor Tejinder Virdee** is an experimental particle physicist and a Professor of Physics at Imperial College London. He is best known for originating the concept of the Compact Muon Solenoid (CMS) at the LHC with a few other colleagues and has been referred to as one of the 'founding fathers' of the CMS. He was elected a Fellow of the Royal Society and the UK Institute of Physics in 2012 and was knighted in the Queen's Birthday Honours in 2014.

**Professor Ruediger Wink** is an economist at Leipzig University of Applied Sciences and he has led several European and transnational research projects at the interfaces between basic science, innovation processes, and economic impact based on interdisciplinary research from engineering, nature science, and diverse social sciences.

# Introduction

## Big Science, Innovation, and Societal Contributions

*Shantha Liyanage, Markus Nordberg, and Marilena Streit-Bianchi*

This book, *Big Science, Innovation and Societal Contributions*, is a sequel to our previous book, "*Collisions and Collaboration—The organisation of learning in the ATLAS experiment at the LHC*" (Boisot et al., 2011). Looking beyond CERN and the ATLAS experiment, the current book explores the increasing roles of Big Science and its contribution to fundamental science, experimental research, technological innovation, and organisational development for societal contributions.

'Big Science' was a term used by the physicist Alvin Weinberg (Weinberg, 1961) and it epitomises the transformation of the scientific enterprise from largely individual research pursuits to well-organised collaborative, interdisciplinary endeavours involving large-scale common experimental faculties developed over time.

The book addresses a multi-disciplinary audience, who is interested in some of the most intriguing questions that science and society encounter. Answering complex questions attempted in Big Science experiments helps to progress fundamental scientific knowledge using cutting-edge technology, power of knowledge of thousands of researchers, large public sector funding, and a long-time scale to complete. Such fundamental questions are simply too complex and large for any single organisation or individual to solve (Cramer and Hallonsten, 2020). This book is mostly written for scientists, policymakers, institutional managers, university students, graduates, and academics who are interested in understanding design, development, implementation of Big Science experiments, and their contribution to society. It explores the rationale behind such large-scale collaborations and experimental facilities that require complex organisation, substantial funding, and human resources.

Big Science often deals with fundamental research questions. Fundamental research (sometimes referred to as basic research) is defined as: 'Experimental or theoretical work undertaken primarily to acquire new knowledge of the underlying foundations of phenomena and observable facts, without any particular application or use in view' (OECD, 2015, *Frascati Manual*, p. 45).

One of the most well-known examples of Big Science experiments is the Large Hadron Collider (LHC) at CERN in particle physics field, which discovered the Higgs boson in 2012. Another example is the Laser Interferometer Gravitational-Wave Observatory (LIGO), which detected gravitational waves in 2015. These experiments and related accelerator and detector experiments such as ATLAS and CMS

Shantha Liyanage et al., *Introduction: Big Science, Innovation, and Societal Contributions*. In: *Big Science, Innovation, and Societal Contributions*. Edited by: Shantha Liyanage, Markus Nordberg, and Marilena Streit-Bianchi, Oxford University Press. © Shantha Liyanage et al., (2024). DOI: 10.1093/oso/9780198881193.003.0001 detectors in particle physics and several telescopes in astronomy provide the most valuable insights into the behaviours of complex systems such as origins of universe, natural forces, and nature and behaviour of dark matter and dark energy.

The increasing complexity of certain research fields in high energy physics, astrophysics, biomedical research, and climate change warrants large-scale international collaborative research efforts. Since the Cold War era, such extensive collaborations have not been observed (Aronova, 2014). Big Science requires a collective group of expert scientists from different disciplines. Often these efforts require large-scale multi-national investments, complex technological systems, and experimental processes that are capable of expanding the boundaries of science.

For this book, Big Science is defined as large-scale fundamental scientific research aiming at major scientific challenges that require advanced and complex technology, significant interdisciplinary and international collaborations, sizeable funding, extensive human resources and public support. This definition has a similarity to Galison and Hevly's (1992) definition of Big Science as large-scale, collaborative scientific research projects requiring significant financial and human resources. Invariably, Big Science organisations and experiments lead to a large volume of data, information, and knowledge. Such knowledge is ideally consumed in public domains and should contribute to human progress (Yin et al., 2022).

Selecting Big Science themes is indeed a complex and difficult process. Such themes are selected after intense scientific scrutiny and with a thorough understanding of the status of knowledge in selected fields. It certainly is not a haphazard process, and it usually goes through rigorous scientific scrutiny before agreeing on the most pressing scientific questions, taking into account scientific and technological feasibility as well as political and social support. Once agreed upon a particular theme, scientists have to work out ways to develop advanced analytical tools, sophisticated scientific and technological equipment, and innovative problem-solving processes to solve research questions.

Building consensus among the scientific community to tackle fundamental questions is also a dynamic process and an arduous task. Complex Big Science questions often include fundamental aspects of life on earth, the universe, and the cosmos and those include: What is the origin of matter? What are the fundamental components of matter? Why is there matter-antimatter asymmetry in our universe? Why is the gravitational force so weak compared to the other forces? What is the nature of Dark Matter? What is the nature of black holes and how do they form? What is the nature of genetic sequence and genetic code? Choosing Big Science experiments to deal with any one of these questions requires negotiations with numerous stakeholders, who may have different priorities, values, and perspectives that can lead to disagreements and conflicts.

Big Science questions continue to fascinate scientists and intellectually challenge scientists across multiple fields of science, including high energy physics, astrophysics, space science, biological science, and information science. Asking the right questions leads to the expansion of knowledge and provides enormous challenges and opportunities for human progress.

Big Science experiments are expensive and involve heavy capital, research, and technology investments. For example, the LHC machine, which is necessary to carry out the experiments, and the associated huge experimental set-up took a long time to complete and required the commitment of highly competent scientists and engineers. These motivated individuals have to dedicate most of their working lives to solving these questions. The legacy of Big Science remains with the construction of large-scale research infrastructure, including the Large Hadron Collider, the International Space Station, the Human Genome Project, and the Square Kilometre Array radio telescope.

Big Science projects are long-term. The Human Genome Project (HGP), which involved numerous scientists and research facilities all over the world, took almost 13 years to complete. HGP was made possible by the collaboration of scientists from around the world to investigate and develop technologies for preparing, mapping, and sequencing DNA, which ultimately contributed to advances in human health and medicine.

Several Big Science facilities including the Brookhaven National Laboratory (BNL), Lawrence Berkeley National Laboratory (LBNL), the Stanford Linear Accelerator Centre (SLAC), and Fermi National Accelerator Laboratory (Fermilab) in the US are testimonies to their commitment to Big Science challenges. In Europe, the European Organisation for Nuclear Research (CERN) in Switzerland and the European Southern Observatory (ESO), European Molecular Biology (EMBL) in Germany, European Space Agency (ESA) in France are some shining examples of advanced research faculties dedicated to Big Science. There are several other facilities in Asia, Latin America and Africa, which can be considered as Big Science investments aimed at advancing fundamental research.

Although this book heavily draws on examples from CERN, ESO, and gravitational wave observatories LIGO, it is not intended to highlight these institutions, but rather to outline the research ecosystems that underpin Big Science organisations as well as their relationship to smaller laboratories.

Serendipitously, Big Science contributes to industrial, social, and business innovations. For example, the invention of the World Wide Web (WWW) in 1989 was originally meant to share information among scientists using the WWW as an information management system. In 1993, CERN decided to make available WWW software available in public domains for rapid dissemination of information and knowledge.

Besides large instruments, machines, telescopes, science and technology organisations, Big Science operations are driven by human and social interactions. Without dedicated scientists, none of the Big Science experiments would have been possible. Big Science highlights the importance of the nexus between science for diplomacy and what humans can achieve by collaborating and converging useful ideas and aspirations to solve common problems.

Big Science is not without controversy. Some outcomes in nuclear and human genome research have raised ethical concerns for the future of humanity. Some argue Big Science takes more resources and starves traditional research fields and organisations. Big Science, unlike other forms of science, generates and tries to provide an answer to incommensurable forms of knowledge quests with epistemic and non-epistemic factors that are inextricably entangled (Pickering, 1988). Research, by its nature, may lead to dead ends and departures from the assumed. Self-correcting paths to look for the next breakthrough are a stark reality in science, irrespective of the scale. Big Science aims to look for reliable evidence and truth. As Bronowski (1988) suggested, knowledge is uncertain and there is no absolute truth, even in science.

The least well-understood subject related to Big Science is the nexus between Big Science and social contribution. This strikes us as odd, given that Big Science organisations make significant contributions to society in ways other than the advancement of knowledge. Big Science projects produce continuous innovations in a wide range of fields, from digital to electronic, applied sciences and medicine, and, more recently, climate change research, sustainable energy initiatives, and epidemics.

After the authors undertook to write this book back in 2019, the world confronted an enormous health challenge due to the Covid-19 pandemic. Covid-19 caused many deaths and disruptions to normal life. Its widespread impact on human health and well-being and on global trade, economies, and social structures resulted in many tragedies. National borders were closed, resulting in economic crises in some countries. Schools, universities, and workplaces were shut down and many people had to work and interact remotely.

Research was confined to virtual meetings using digital communication tools. Airlines were grounded, making it almost impossible to travel internationally. The world would have come to a standstill if it had not been nations, for the online communication tools based on the World Wide Web, which had been invented at CERN some 30 years earlier to enable global collaboration in Big Science. The global science community also enabled the pandemic to be brought under some level of control with the rapid development of new vaccines, which was facilitated by online exchange of new research results.

In early 2022, the world was struck yet again by another disaster, this time a human-made disaster, when a Russian invasion of Ukraine exacerbated already-existing tensions between Russia on the one hand and Ukraine and neighbouring countries on the other, flaring up into a full-blown war. Besides its devastating effect on both Ukrainian and Russian science, this conflict triggered sanctions on Russia by many countries that severely disrupted scientific collaboration.

It is worth noting that CERN was established in the aftermath of World War II to bring harmony to people and nations and the peaceful pursuit of knowledge is enshrined in the core values of CERN. Consequently, the CERN Council, in a recent meeting in May 2022, declared its intention to terminate cooperation agreements with the Russian Federation and the Republic of Belarus at their expiration dates in 2024, as a response to the invasion of Ukraine (CERN, 2022).

These global events and conflicts, beyond the control of science, in essence, should not hinder research efforts on an international scale. Scientists should not be obstructed from working freely across international boundaries. However, the stark reality of politics and social distrust prevent doing business as usual. It is regrettable that, at present, international collaborations on Big Science projects between Western and Eastern Bloc countries are hampered in ways that did not exist even during the Cold War.

Despite these negative developments, the hope remains that Big Science projects will nevertheless be able to continue bridging cultures, synthesising complementary ideas and perspectives in a peaceful way, for the good of humanity. For example, the Covid-19 pandemic has inspired researchers in both the public and private sectors to share knowledge among various groups of researchers in a broad collaboration, demonstrating that this enables the development of many new Covid-19 drugs and patient treatment methods that are beneficial to humanity.

Organisations and individuals face some constraints when it comes to Big Science operations. Limited incentives for young and upcoming scientists, the long-term nature of Big Science, rigid publication policies and arrangements, and limited technology transfer opportunities are some challenges. Furthermore, issues such as varying project management quality, complex administrative and leadership issues, budget constraints, cost overruns, and recruiting experienced scientific staff to maintain core facilities continue to be challenges.

Due to the breadth of content, this book was structured into three parts: the first part (Ideation) addresses the nature of and the relationship between scientific and technological knowledge, based on available literature and dedicated accounts from Big Science challenges (Chapters 1–5); the next four chapters in the second part (Science at Work) discuss the different aspects related to technology artefacts and innovative applications that can favour dissemination in society (Chapters 6–9); the third part (Societal implications) outlines the connections between science and society and the connectivity between physics research and society (Chapters 10–14). In doing so, the role of emerging scientific systems and future physics research are examined. Chapter 15 builds on the discussions and findings of the preceding chapters.

Collectively, these chapters explore several interconnected aspects of theoretical and experimental discoveries as they relate to complexity theories (Merz and Sorgner, 2022), serendipity (Wareham et al., 2022), collaboration and networking (Leogrande and Nicassio, 2021). These concepts provide valuable insights into the operations and developments of Big Science ventures, collaboration processes, and ways in which organisations and society can benefit from Big Science initiatives.

We are left with many questions for the reader to ponder: Why do we need Big Science?; Why do we need to support it?; What are the benefits of Big Science to society?; What types of leadership are applicable to Big Science organisations and collaborations?; What organisational structures are effective for Big Science operations?; What is the nature of epistemic culture in Big Science organisations?; What types of educational initiatives are necessary to transfer Big Science knowledge?

Not all questions have simple answers and the authors aim to address some of these questions in the following chapters. We hope you find the book worthwhile and a pleasure to explore.

# PART 1

# BIG SCIENCE OPPORTUNITIES AND CHALLENGES

To improve our understanding of a vast and complex universe, high energy physicists have created increasingly ambitious new machines and experiments. Such scientific work is never easy. Part I explains how Big Science experiments are put together and what it takes to create such experiments.

# 1
# Big Science and Society as Seen through Research Lenses

*Markus Nordberg, Shantha Liyanage, and Marilena Streit-Bianchi*

## 1.1 Introduction

For decades, debates raged over Big Science operations and the support given to fundamental research. In his 1945 report to President Truman, 'Science, the Endless Frontier', the first presidential science adviser, Dr Vannevar Bush advocated for an expansion of government support for science and recommended the creation of the National Science Foundation.

The report was highly influential in underpinning public support for fundamental research. Several scholars have also examined the role of fundamental research (also known as basic research or pure basic research), its social and economic impacts, tensions, and relevance for security and global peace. Such research enhances our understanding of nature and its laws. How science functions as a social institution was explained by several scholars (Polyani, 1962; Kitcher, 1993; Gibbons, 1994; Pavitt, 1998; Aronova, 2014).

Big Science refers to large-scale instruments and facilities, funded by national and international governments and agencies where research is conducted by specialised teams or groups of scientists and technicians on a common and significant problem. Large-scale public investments enable Big Science to produce public goods (Weinberg, 1961). Particle physics is a good example with significant social implications, but it is not the only one. Developing nations such as India provide evidence for this. Homi Bhabha, for instance, who was a theoretical physicist, initiated the revival of Big Science programmes (in relative terms) at the Tata Institute of Fundamental Research in 1945 to support physics research (Wadia, 2009).

Big Science investments grew further over time. It was a growth driven by the need to enable large numbers of scientists to assemble diverse expertise to collectively resolve major research questions. The establishment of international science organisations such as CERN in 1954, for instance, opened space for an extensive research community. The acronym CERN is also used to refer to the laboratory; in 2021, it had 2,676 scientific, technical, and administrative staff members and 783 fellows. In addition, CERN hosted about 12,731 associated members and users from institutions in more than 110 countries with a total number of personnel recordings

Markus Nordberg et al., *Big Science and Society as Seen through Research Lenses*. In: *Big Science, Innovation, and Societal Contributions*. Edited by: Shantha Liyanage, Markus Nordberg, and Marilena Streit-Bianchi, Oxford University Press.
© Markus Nordberg et al., (2024). DOI: 10.1093/oso/9780198881193.003.0002

16,190 by 2021[1]. The CERN experiments are supported by a larger group of scientists and engineers from various countries, the majority of whom are connected to national laboratories and institutions in their home countries.

The demonstrated success of laboratories like CERN suggests that reliable collaboration is possible with advanced communication tools, structured workshops, and effective interactions among the Big Science research community and technical staff. How Big Science operates and continues to function efficiently, nonetheless, remains yet another puzzle for many.

## 1.2  Big Science as Described in the Literature

Big Science, breakthrough innovation, and societal benefits are often tightly linked with each other (Bach and Lambert, 1992; Autio et al., 2004; Vuola and Hameri, 2006; Liyanage et al., 2007). In fact, the term 'Big Science' was coined into the vocabulary of scientific enterprise in the last century. Lawrence's Cyclotrons and the University of California Radiation Lab established in 1930 are classic examples of the emergence of Big Science concepts. The advent of Big Science is a major step forward in human inquiry into nature and it extends beyond what individuals can do with structured and organised exploration of nature and nature's phenomena including the existence of life and biomedical and astronomical events.

The term Big Science specifically originated in the US during World War II. However, it was subsequently used in more general expressions to refer to significant scientific advances, which, when considered by their order of magnitude, achieved complex goals that otherwise would have remained unattainable (Bush, 1945; Price, 1963; Weinberg, 1968; Etzkowitz and Kemelgor, 1998). Naturally, Big Science demands big investments, intense international collaboration, and the complex organisation of leading scientists, which entails some risk-taking that can be overcome by carefully crafted collective decisions. Collaborative organisation of science thus has its inherent advantages (Hicks and Katz, 1997; Etzkowitz and Leydesdorff, 2000; Giudice, 2012). Typically, Big Science projects require dedicated and technologically advanced infrastructure and a set of project management skills which were new at the time to the contributing scientists and engineers. Although their research goals could be described as 'high risk, high gain', these laboratories and collaborations were assigned oversight structures to ensure that adequate risk mitigation practices were in place. Collaborations in Big Science require building connections with leading scientists (see Figure 1.1).

---

[1] CERN Personnel Statistics 2021, Human Resource Department, March 2022—https://cds.cern.ch/record/2809746/files/CERN-HR-STAFF-STAT-2021-RESTR.pdf.

**Figure 1.1** Master Builders of Big Science—Steven Weinberg visiting CERN and ATLAS (with Peter Jenni, former ATLAS spokesperson)

*Source:* © CERN

Big Science laboratories are clustered around nuclear and later particle physics research, astronomy and, more recently, in areas of the life sciences (Galison and Hevly, 1992). These dedicated, large-scale technological infrastructures also offer potentially interesting opportunities for industries interested in advancing R&D (Hameri, 1997). Most of these laboratories are geared towards solving some of the most challenging scientific puzzles of today. They probe into the origin, density, structure, and distribution of mass (energy) in the universe and explore the early stages and structure of space–time, the origins and evolution of massive stars, and the origins of life on Earth. These research questions, among many others, bring together hundreds of research institutions, creating complex, interacting networks across diverse disciplines (Nature Index, 2019).

The quest to understand the birth of the universe builds upon and complements research data created by big accelerators like the LHC at CERN, arrays of telescopes operated by ESO and research operated by ESA and NASA such as the Hubble Telescope. Recent Planck results (Planck, 2019) of the cosmic microwave background (CMB) fluctuations support, among other things, the Standard Model of neutrinos consolidated by recent results found by the particle physicists at the LHC and elsewhere (DUNE, 2020; IceCube, 2020), the concept of an accelerating expansion of the universe measured by the SCP astronomer community in the early 1990s, and the absence of spatial curvature suggested by the earlier CMB measurements. The results, in fact, reveal deeper connections between the Higgs particle and the accelerating universe (Steinwachs, 2019).

Several communities are working on the connections between black holes, space–time curvature and gravitational waves, most notably the LIGO experiment (Gregoris et al., 2019). The thresholds of the formation of massive early stars which end up in Black Holes once they have run out of the nuclear fusion processes combine a wide range of communities across astronomers and nuclear physicists and they are addressing fundamental questions in the emergence of visible matter, the organisation of subatomic matter, and their interactions within (NRC, 2013). Such studies are possible thanks to a few accelerators and instrumentation available in nuclear physics facilities, such as HIAF in Australia (HIAF, 2020).

Large-scale facilities like ESRF in France are running experiments to reveal the fundamental nature of space–time symmetries and are working together with life-science laboratories like EMBL in Europe (EMBL, 2020) on the exploration of living matter in diverse disciplines such as chemistry, structural biology and medical applications, environmental sciences, information science, and nanotechnologies (ESRF, 2020). Powerful X-ray lasers at large facilities such as the European XFEL in Germany (XFEL, 2020), LCLS (SLAC, 2020), and SACLA in Japan (Riken, 2020), unveil the composition and structure of complex biomolecules and materials on the atomic scale.

Big Science laboratories thus act as catalysts for the many different scientific communities using them. These laboratories offer shared technical and scientific facilities providing necessary technological infrastructure and knowledgical knowhow.

These laboratories provide instrumental support for scientific and technological investigations (Beck and Charitos, 2021).

Yet, in the light of history, the processes and interconnections between science, innovation, and society are not very easy to untangle, despite several compelling examples provided. Electricity and radio waves were harnessed in the early part of the twentieth century based on experimental work carried out some half a century earlier; the transistor and the laser were developed after World War II based on observations and theories made decades earlier about the behaviour of atoms and molecules (Johnson, 2010). These disruptive innovations resulted mainly from sequential contributions made by individuals and small teams. Big Science laboratories provided advanced technical facilities and dedicated teams to leapfrog and scale up discoveries and technological advances of much grandeur. In that sense, Big Science facilities herald as next-generation investments in technological innovation.

A rapid and steady growth of more complex scientific collaborations, therefore, took place forming new and expensive laboratories and partnerships, involving industrial companies (Krige, 1993; Kronegger et al., 2011; Qi Dong et al., 2017). At the forefront were the versatile domains of physics, biomedicine astrophysics, and data science. In time, Big Sciences model rapidly expanded to include climate sciences, ecology, oceanography, astronomy, gravitation, neutrons, synchrotron light and laser physics, fusion research, artificial intelligence, and other disciplines.

Noteworthy among the more recent examples is the unravelling of the composition of the human genome in 2000, paving the way for new drug discoveries. This discovery relied on the foundations of genetics dating back half a century and the use of massive computing power made possible by advances in computer chip development (Davies, 2002). A second example is the discovery of the Higgs-particle at CERN (Figure 1.2) in 4th July 2012 (jointly announced by both the CMS Collaboration, 2012 and the ATLAS Collaborations, 2012), than half a century earlier and using a massive amount of computing power to analyse and find the particle which stimulated the development of cloud computing (Chandrasekaran, 2015). The discovery of the Higgs boson was a major achievement in the field of particle physics because the Higgs boson is extremely unstable and rapidly decays into other particles (see Figure 1.2).

Big Science stretches across borders with laboratories and collaborations having a global reach because of the nature of the scientific work they foster (Holden, 1985). In most cases, the host labs act as the host organisation for their research community and connect with several other research laboratories and universities. At least in the domain of physics, it is not unusual that different Big Science labs host overlapping scientific visitors and users. For example, the ATLAS and CMS collaborations have more than 180 institutions each, including all major Big Science labs in particle and nuclear physics from over 40 countries from all continents (ATLAS Collaboration, 2020; CMS Collaboration, 2020).

Another example is Australia's Nuclear Science and Technology Organisation (ANSTO). ANSTO occupies much of Australia's landmark infrastructure including

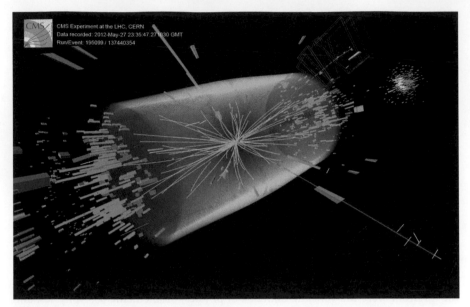

**Figure 1.2**  Tell-tale sign of the Higgs boson
*Source:* © CERN- CC BY 4.0

modern nuclear research reactors, a comprehensive suite of neutron beam instruments, the Australian Synchrotron, the National Research Cyclotron, and the Centre for Accelerator Science. It has over 60 research, technology, and regulatory partners all over the world, including CERN and ATLAS (ANSTO, 2020).

A third example that depicts the nature of interconnectivity and embedded network structure linked to Big Science is ATTRACT,[2] an EU-funded framework for promoting early-stage detection of the pathology of disease and the associated imaging technologies in Europe (ATTRACT, 2020). It is coordinated by six leading European Big Science facilities with the intent to seed-fund and cross-link the different stakeholders across detection and imaging, with the objective of creating an innovation platform for Europe. These types of activities are expected to breed innovation through collaborative research networks (Liyanage, 1995; Liyanage et al., 1999).

Understanding how Big Science collaborations are structured and managed is also becoming increasingly important in gauging their effectiveness (Bammer, 2008; Hsu and Huang, 2011; Canals et al., 2017). Obviously, that needs to take into account cultural, geographical and historical factors (Gazni et al., 2012; Ortoll et al., 2014), nonetheless that alone is not enough. It should be equally taken into consideration how they manage to scale up, how individual researchers can act and respond within the project structures and finally how to arbitrate possible internal disputes or conflicting requirements.

---

[2] The ATTRACT project has received funding from the European Union's Horizon 2020 research and innovation programme under grant agreements No. 101004462 and No. 777222.

Some insights have been offered in the context of CERN (Knorr-Cetina, 1999; Tuertscher et al., 2014; ATLAS, 2020) and LIGO (Collins, 2003) and the emphasis is on the well-articulated and strong shared goals, well-crafted procedures of conduct and collegial, and lastly the rotating management structures. However, having a better understanding of how processes work—or do not work—within collaborations, as well as how they deal with unforeseen problems or unpleasant surprises, is more important. For example, the LHC uses dipole magnets (see Figure 1.3) to bend the paths of circulating high energy proton beams which generate enormous energy, hence these superconducting magnets need to be cooled to extremely low temperatures (about −271.3 degrees C).

It should be noted that the management structures of—and leadership issues related to—Big Science labs differ from those of the scientific collaborations they foster. Running large-scale laboratories is a more top-down approaches where the governances is determined by the funding agencies and governments, and where adequate resources are allocated for supporting infrastructure and research projects (Mark and Levine, 1984; Kinsella, 1999; Geles et al., 2000; Anadon et al., 2016; Fabjan et al., 2017). In contrast, management of individual projects—even if large in scale—is more bottom-up, and often governed by the network of contributing universities and funding agencies; in the capacity of users of these large facilities (Robinson, 2021).

In addition to the observed time-lag of decades between scientific theory formulation, discovery and ultimate recognised value for society (Goddard, 2010), the process from discovery to practical use is often non-deterministic or could be

**Figure 1.3** LHC dipole magnets in the underground LHC tunnel
*Source:* © CERN- CC BY 4.0

serendipitous by nature. Well known and often cited examples are the discovery of penicillin (ACS, 1999) and the invention of the World Wide Web (Hameri and Nordberg, 1998; Gillies and Cailliau, 2000), where the final, revolutionary product resulted from addressing initial needs or challenges elsewhere. The importance of understanding serendipity in scientific discovery or innovation processes is well known (see e.g. Merton and Barber, 2004; Garud et al., 2018; Yaqub, 2018) but it has not been given due consideration with regard to the societal impact of Big Science.

Provided that the Oxford English Dictionary definition of serendipity is 'the occurrence and development of events by chance in a happy or beneficial way' (Oxford English Dictionary, 2020), this approach has so far not been essentially included in studies on the economic or societal value of Big Science, although its presence has been acknowledged (OECD, 2008, 2014).

A more classic, cost-benefit analysis approach has been applied to estimate the economic returns of investment in Big Science laboratories and collaborations (Science Business, 2015; Florio, 2019). Although these methods consider variations in the Net Present Value (NPV) and rate of return criteria, the actual benefits of investing in Big Science research are difficult to quantify. For example, somewhat unexpectedly, the single most significant generator of socioeconomic impact from such endeavours is training. This finding emerged from Cost-Benefit Analysis studies for the LHC and the High-Luminosity LHC (HL-LHC) upgrade as well as from lessons concerning the socioeconomic impact that these facilities have beyond the core scientific mission. (Gutleber, 2021). Other studies have shown that the applied discount rates have positive implications for big projects like the LHC at CERN (Florio and Sirtori, 2016). However, these studies do not describe the process of knowledge creation or variations within, of Big Science designing and building, and of making use of their instruments in interaction with the different stakeholders in society. Within Big Science experiments, there are sophisticated instruments and technologies. For example, the Pixel Detector of the CMS experiment at LHC consists of advanced electronics and silicon sensors as shown in Figure 1.4.

Alternative attempts have been made, for example by Boisot et al. (2011), using options thinking (McGrath and MacMillan, 1999; van Putten and MacMillan, 2004; MacMillan et al., 2015) to capture the potential future value of Big Science undertakings. This approach is based on knowledge or the information economy (Nelson and Romer, 1996, Romer, 1990; David and Foray, 2001) but assets can be described as a dynamic cycle from creation to their oblivion using the so-called Information Space (I-Space) framework (Boisot, 1998; Child et al., 2014). In this approach, Big Science projects, while pushing the envelopes of science and technology to leap forward, create options that may or may not be realised ('executed') by the different stakeholders, acknowledging at the same time the act of serendipity. Despite the promise of this approach, not much progress has been seen during recent years on this front, even if there are documented case studies about dynamics and structures within Big Science collaborations (Knorr-Cetina, 1999; Glänzel and Schubert, 2004; Tuertscher et al., 2008; Canals et al., 2017) and on supplier relations (Nordberg and Verbeke, 1999;

**Figure 1.4** The CMS experiment at the LHC
*Source:* © CERN

Autio et al., 2003; Vuola, 2006). More recently, Big Science and economy were closely scrutinised from various knowledge and intellectual property angles (Beck and Charitos, 2021).

While Big Science laboratories and collaborations have been focusing on their well-defined research missions, policymakers and governments have been increasingly calling upon the scientific communities to also address pressing societal issues (EU, 2015). This is not a new call—it was noted already in the 1960s that while we can reach for the moon, we still have ghettos (Nelson, 2011). But more recently, impelled particularly by the Covid-19 pandemic, governments are increasingly turning towards scientists to know how the advancements in their respective fields help resolve complex, 'wicked' societal problems (Skaburskis, 2008), thus introducing a conditional element to their research funding. This top-down versus bottom-up projection of objectives can be hard to align because of the diversity of the dynamics of social and natural phenomena.

A leading sociologist of collaborative networks said: 'Particles do not yell back at you' (Grey, 2003). Although concepts like 'social physics' (Pentland, 2014) can be helpful in guiding how scientific methods can be used to influence human behaviour, a fundamental layer still appears to be missing, despite good efforts, that is able to capture the process of doing science itself to the dynamics of innovation and eventual societal impact (Cardinal et al., 2001; Caraca et al., 2009). Yet, the impact of public funded Big Science research has been a central concern for many scholars, policy makers, and research managers (Cohen and Noll, 2002; Mazzucato, 2013; Kokko et al., 2015; Maroto et al., 2016; Gutelber, 2021). Some of the advances in medical technology are obvious. The Linac booster (LIBO) is used to produce particle beam for cancer therapy (Figure 1.5).

There are two ingredients to consider here: the open nature of science and the design or 'fabric' of the scientific process itself. The methodology used in modern science dates back to the Greeks and was consolidated by Francis Bacon in the seventeenth century, using inductive reasoning based on data and the subsequent verification or rejection of a set hypothesis, to be openly shared with the scientific community for debate and reflection (Kuhn, 1962; David, 1998; Gribbin, 2002). Publishing in scientific journals, which offers a system of trust and earned scientific reputation, serves as the primary channel for communicating results (Merton, 1957). The impact of scientific work is also increasingly visible through this channel (Benavent-Pérez et al., 2012). This principle of 'Open Science' (David, 1998) is deeply rooted not only in the way Big Science labs and collaborations operate but also in the way they innovate their scientific instruments. The latter is captured by the principle of 'Open Innovation' (see e.g. Chesbrough, 2003; Enkel et al., 2009; Baldwin and von Hippel, 2011) which took inspiration from the practices of software communities openly sharing their code for enhanced development and applying gentle, collegial coordination—for example, the Linux operating system (Henkel, 2006). The key idea is that (external) communities are stronger than (internal) organisations in innovating new, breakthrough concepts, products, and services. This has been further enhanced by the use of online collaborative platforms that permit citizen participation to solve specific technical challenges (Seltzer and Mahmoudi, 2013;

**Figure 1.5** The Linac booster (LIBO) for producing particle beams for cancer therapy
*Source:* © CERN

Sloane, 2011). The principles of 'Open Innovation' can also be applied to support actual scientific processes as well (Beck and Charitos, 2021).

Although Big Science thrives on the above dynamics of 'Open Science and Open Innovation', which is also echoed by the science and technology policies of many countries (EU, 2016, 2020; Science Business, 2019), it can also inadvertently result in a kind of an 'innovation paradox': by openly sharing the technology to invite the research and other communities to substantially enhance its performance, making it harder for others, later on, to commercialise it due to unclaimed or diluted intellectual property rights (IPR). Putting aside here the relationship between—and implications of—open innovation and IPR policies (Bogers and Santos, 2021; Bogers et al., 2012; Granstrand and Holgersson, 2014), it is noted that in general, being publicly funded, Big Science labs and collaborations tend to follow rather loose IPR policies. They make good use of open software and hardware repositories for sharing their work in addition to their usual channels of scientific publishing (Murillo and Kauttu, 2017; Pujol, 2020). This would imply that classical measurement tools like patent-counting may not be that applicable and that the emphasis is more on the transfer of knowledge than on the transfer of concrete, identifiable products.

As noted above, the nature of the scientific process and its relevance to the design of the innovation process have not been extensively studied. The issue of design in science has been raised from an engineering perspective (Cross, 1993). The question of the architecture of complex organisational structures—which could be relevant in some Big Science endeavours—has also been addressed (Simon, 1962). Yet the role of the potential end-users has not been thoroughly examined, apart from recognising the importance of lead users in expanding the use of scientific equipment (von Hippel, 1988).

Starting with the societal challenges facing citizens, it has not been systematically examined whether the diverse cumulative knowledge and technology available in Big Science organisations and experiments can be well used in solving complex social problems. Recognising that making such a direct link between Big Science and social benefit would be difficult, user-centric techniques are available to transfer knowledge. Technology enters only at the end of the process, and not at the start, which is usually the case in the more classical thinking of technology transfer (Harmon et al., 1997).

The approach used in this book is also inspired by Design Thinking (Brown, 2008 and 2009) where cross-disciplinary MSc-level university student teams are assigned sustainable development goals (SDG) -related projects (UN, 2020) and are then exposed to Big Science surroundings to look for potential solutions (CERN, 2019).

The students come from different backgrounds, ranging from product design to business management and engineering, and are mostly from a global network of Design Factories (DFGN, 2020). Although the primary motivation for this type of approach is educational, the project results do suggest that tools and technologies developed by Big Science labs and collaborations can contribute to pressing challenges related to topics such as climate change, pollution, and health care (CBI, 2020).

As governments are launching more Big Science-type 'moonshot' initiatives to solve societal problems such as climate change or to conquer cancer (EU, 2018), the question should be asked how current Big Science laboratories will be able to adapt, without compromising their defined scientific mission and focus. The current collaborative, bottom-up project-like structures around Big Science facilities suggest by themselves an agile approach: if participating countries are willing to fund their scientists in global projects hosted or coordinated by Big Science laboratories. But that might come with more strings attached, notably demanding that collaborations to involve societal stakeholders outside their primary scientific fields. If so, Big Science laboratories and collaborations will need to think about how these new actors could be best integrated and what the rules of engagement will be. For example, some indication of this line of thinking can already be observed in the planning of CERN's Future Circular Collider project (FCC, 2020) where different kinds of societal benefits are envisaged stemming from the technology development work, including medical applications, energy transfer, and storage and engineering software. Also, engaging a wide range of students from different fields is foreseen. In that respect, the current Sustainable Development Goals (SDG) driven student projects at CERN's IdeaSquare (CERN, 2019) might provide some insights into how this could be scaled up, if needed.

Finally, the capabilities and role of Big Science labs in responding to acute and unforeseen disruptions in society in the future need to be considered. The most recent and most vivid example is obviously the Covid-19 pandemic, which in 2020 shut down major parts of world economies, with ripple effects lasting for a long time. Although the ultimate research missions of Big Science labs will remain unchanged, the infrastructure available at Big Science laboratories could be used for rapid response to crisis, such as using scientific instrumentation and computational facilities, as was the case for Covid-19 (CERN, 2020; EMBL, 2020; ESFR, 2020; ESS, 2020; ILL, 2020; XFEL, 2020). In the future, Big Science will be able to accelerate cross-connecting of new and complementary parts of their user communities to speed up development work, i.e. contributing scientific networks in the spirit of open science and open innovation (Berkley, 2020; Chesbrough, 2003; 2017; and 2020).

## 1.3 Conclusions

Big Science, often refers to large-scale scientific projects, covering a broad spectrum of scientific, technical, economic, knowledge transfer, and science and society issues. Since the publication of Vannevar Bush's thesis, 'Science—the Endless Frontier' in 1945, a plethora of research publications about Big Science have covered fundamental research, the role of government and industry, the impact of science on society, and the ethics and morality of science.

The main purpose of this review is to outline some core practices, underlying theories, and concepts related to Big Science. What has been covered within a narrower scope, are accounts of Big Science undertakings from a practitioner perspective, i.e. shared experiences about the challenges faced—scientific, technological, administrative, or political—and how these are being addressed and resolved.

The literature review suggests a rather wide range of empirical evidence on the importance and role of Big Science and its impact on society. Several studies confirm Big Science projects are highly efficient, capital intensive and complex research processes. Coordinated multidisciplinary groups using the latest technology and experimental systems are necessary to solve fundamental questions attempted in Big Science organisations and experiments.

Several themes related to the future of disciplines, economics, and ethics are emerging from the literature review. Open Science and Open Innovation play a central role, and various aspects of big data and digital information systems are often highlighted. In addition, several studies outline technology transfer, design, and innovation in transferring fundamental knowledge to useful social benefits, including significant advances in medical science.

Big Science is a dominant mode of conducting fundamental research with growing international collaborations of increasing size. Indeed, there are concerns covering equity, ethics, and the role of collaboration and competition in Big Science.

The authors conclude that there is scarce literature offering examples of how Big Science can connect with society. Although there are anecdotal examples, there is scarce research literature on innovation in Big Science, future development of scientific methodology, strategic development of technological tools, recognising the role of industry, identifying educational models for the diffusion of knowledge opportunities, and impact on society.

In the light of the above review, the authors of this book saw an opportunity—that is to adopt a more holistic, process-driven practitioner-approach. Based upon the literature reviewed, one can identify three phases of Big Science processes: ideation, science in progress, and as a process, connecting with society.

Our hope is that our selection of this path will inspire further research on this intriguing topic.

# 2
# Chasing Success

## The ATLAS and CMS Collaborations

*Peter Jenni, Tejinder S. Virdee, Ludovico Pontecorvo,
and Shantha Liyanage*

## 2.1 Introduction

A Toroidal LHC ApparatuS (ATLAS) and Compact Muon Solenoid (CMS) are
two general-purpose experiments in the Large Hadron Collider (LHC) at CERN,
Geneva, Switzerland. These were the two important Big Science collaborations primarily aimed at exploring the highest energy collisions and detecting the elusive
Higgs boson. The years immediately following the ATLAS and CMS Letters of Intent
in October 1992 were a time when the two collaborations grew most rapidly in terms
of people and institutes (CERN Courier, June 2013, p. 22).

This chapter aims to cover the long journeys of conception and construction of
the two general-purpose LHC detectors, the ATLAS and CMS. A brief historical setting of the physics landscape leading to a global consensus for a hadron collider at the
energy frontier, extend for some 35 years. The considerations leading to the two complementary detectors, the conception, evolution, and construction of the ATLAS and
CMS detectors were a long tale. Their specific technologies and their internal organisations for managing the world-wide collaborative effort over 15–20 years, from the
first R&D to the first operation, were an arduous and interesting journey for particle
psychists. A few examples of success stories or failures, and the lessons learnt, are nar-
rated; more importantly, human factors are explained that contributed to the success
of the ATLAS and CMS, bringing together an extensive and world-wide collaborative
effort.

Much has been written about the technical details of these two detectors and readers can find those in a number of books and articles (Della Negra et al., 2018; ATLAS,
2019). Designing of detectors is complex and they are designed to capture the collision events (produced particles and their energies) of all proton (p–p) interactions
occurring during encounters ('crossings') between circulating bunches of protons,
rejecting those which are not interesting and disentangling individual, interesting
p–p collisions ('events') from uninteresting ones, reconstructing the outcome of

Peter Jenni et al., *Chasing Success*. In: *Big Science, Innovation, and Societal Contributions*. Edited by: Shantha Liyanage,
Markus Nordberg, and Marilena Streit-Bianchi, Oxford University Press. © Peter Jenni et al., (2024).
DOI: 10.1093/oso/9780198881193.003.0003

each, and identifying distinctive signatures (of known and new phenomena) in the set of interesting events.

This chapter outlines the personal experiences and insights of the two principal architects of these experiments. One cannot emphasise enough the importance of the many years of research and development (R&D), design and prototyping that preceded the construction of the detectors. Technologies had to be developed far beyond the state of the art in the early 1990s in terms of granularity, speed of readout, radiation resistance, reliability, and very importantly the cost.

For many of the detector subsystems, several technologies were initially considered, as it was far from certain which technologies would be able to attain the required physics performance. The construction of the LHC, comprising the accelerator, the detectors, and the computing systems, pushed many technologies to their limits. More information can be found in Della Negra, Jenni, and Virdee (2012, 2018) and Evans (2018). How these detectors were successfully accomplished, despite encountering many challenges, is a major thrust of this chapter.

## 2.2 Physics Case for ATLAS and CMS

In the early 1980s, the Standard Model (SM) of particle physics was given a significant boost by the discovery of the $W$ and $Z$ vector bosons by the UA1 and UA2 experiments (denotes Under Ground Area 1 and 2) at the CERN proton–antiproton collider (see Figure 2.1).

The predictions of the SM have been experimentally verified with high accuracy in many generations of experiments, at both low and high energies. However, two particles were needed to complete the particle content of the Standard Model (SM):

(a) the heaviest of the six quarks, the top quark; and
(b) the particle that could confirm the proposed spontaneous electroweak symmetry breaking mechanism via a scalar field that permeates the entire Universe.

This mechanism, termed the Brout–Englert–Higgs mechanism, gives the $W$ and $Z$ bosons their large masses and leaves the photon massless. The interaction of quarks and leptons with this scalar field imparts them with masses that are proportional to the strength of their couplings to this field.

In the early 1990s, the top quark was expected to be found by existing or just-starting colliders because it could be relatively abundantly produced in what were at the time the available high-energy proton–proton collisions. Also, its mass was progressively better constrained by the electroweak theory and the experiments. In 1995, the top quark was discovered at a mass of 175 GeV (Giga-electron volts) by the CDF (Collider Detector at Fermilab) and D0 experiments at the Tevatron collider at Fermilab near Chicago, USA.

**Figure 2.1**  UA2 particle detector
*Source:* © CERN

The scalar Higgs boson is not as abundantly produced and depending on its mass would have differing signatures, most of which were quite difficult to discern from the backgrounds of other SM processes. Furthermore, there is no precise prediction for its mass; its value ranged from some tens of GeV (from experiment) up to 1 TeV Tera-electron volts (from theoretical considerations), where it would cease to be visible as a peak in any mass distribution. It turned out that the Higgs boson was well beyond the reach of any experiment at the time.

The hunt for the Higgs boson became a central theme in discussions concerning the future of particle physics, as well as a primary motivation for the Large Hadron Collider (LHC) and its experiments (Evans, 2018). By the late 1970s, the idea that the Large Electron–Positron Collider (LEP) tunnel should be able to house a future Large Hadron Collider (LHC) had already been put forward.

The scientists leading CERN at the time planned for a tunnel with a large enough circumference to accelerate particles to very high energies and with a large enough diameter to eventually accommodate high-field cryomagnets for a high energy hadron collider that was to become the LHC (for details see Chapter 3).

In the broader physics community, enthusiasm for the LHC clearly emerged in 1984, promoted in part by members of the successful UA1 and UA2 experiments at a workshop in Lausanne, Switzerland on the 'Large Hadron Collider in the LEP Tunnel' (CERN, 1984). Researchers formed working groups comprising accelerator experts, theorists, and experimentalists. With the realisation of the large physics potential of an LHC, a series of workshops and conferences followed. They included a 1987 workshop in La Thuile, Italy, under the auspices of the Rubbia Long-Range Planning Committee (CERN, 1987) that recommended the LHC as the next accelerator for CERN; the 1989 study week in Barcelona, Spain, on 'Instrumentation Technology for High-Luminosity Hadron Colliders' (Jarlskog and Fernández, 1989) and the 1990 'LHC Workshop' in Aachen, Germany (Jarlskog and Rein, 1990).

During these years, the design and construction of the Superconducting Super Collider (SSC) were under way in the United States with proton–proton collisions foreseen at 40 TeV. The LHC, with its initially planned energy of 16 TeV, could be competitive with the SSC only if its instantaneous luminosity were an order of magnitude higher (i.e. $10^{34}$ cm$^{-2}$ s$^{-1}$). The formidable experimental challenges for detectors at such a high-luminosity hadron collider started to appear tractable at the workshop in Aachen, provided that sufficient R&D and prototyping could be carried out on the necessary instrumentation. Soon afterwards, R&D and prototyping for the LHC experiments were undertaken under the auspices of CERN's Detector R&D Committee, which was formed in 1990. The SSC was ultimately cancelled in October 1993. Table 2.1 presents the timeline of the LHC project.

**Table 2.1** LHC Timeline

| | |
|---|---|
| 1984 | Workshop on a Large Hadron Collider in the LEP tunnel, Lausanne, Switzerland |
| 1987 | Workshop on the Physics at Future Accelerators, La Thuile, Italy. The Rubbia 'Long-Range Planning Committee' recommends the Large Hadron Collider as the right choice for CERN's future |
| 1990 | LHC Workshop, Aachen, Germany (discussion of physics, technologies, and detector design concepts). |
| 1992 | General Meeting on LHC Physics and Detectors, Evian-les-Bains, France (with four general-purpose experiment designs presented) |
| 1993 | Three Letters of Intent were evaluated by the CERN peer review committee LHCC. ATLAS and CMS selected to proceed to a detailed Technical Proposal (TP) |
| 1994 | The LHC accelerator was approved for construction, initially in two stages |
| 1996 | ATLAS and CMS technical proposals approved |
| 1997 | Formal approval for ATLAS and CMS to move to construction (materials cost ceiling of 475 MCHF) |
| 1997 | Construction commences (after approval of detailed Technical Design Reports of detector subsystems) |
| 2000 | Assembly of experiments commences, LEP accelerator is closed down to make way for the LHC |
| 2008 | LHC experiments ready for pp collisions. LHC starts operation. An incident stops LHC operation |
| 2009 | LHC restarts operation, pp collisions recorded by LHC detectors |
| 2010 | LHC collides protons at high-energy (centre of mass energy of 7 TeV) |
| 2012 | LHC operates at $\sqrt{s}=8$ TeV: discovery of the Higgs boson |
| 2015 | LHC operates at $\sqrt{s}=13$ TeV for Run 2 (2015–2018) |
| 2022 | LHC operates at $\sqrt{s} = 13.6$ TeV for Run 3 (scheduled until 2025) |

Around 1990, the future of ATLAS and CMS collaborations was conceived. Initially, there were four prototype collaborations, all with the goal of exploring the TeV energy scale with general-purpose designs. It was clear that no more than two large general-purpose experiments could be affordable. Four proto-collaborations presented these designs in expressions of interest (EoIs) at a meeting in Evian-les-Bains, France (CERN, 1992) entitled 'Towards the LHC Experimental Programme'. The four were eventually whittled down to two large experiments, the ATLAS and CMS.

As always, the physics to be probed drives the design of an experiment. The designing and building an experiment that was able to perform well at an instantaneous luminosity of $10^{34}$ cm$^{-2}$ s$^{-1}$ indeed represented a challenging task (Ellis and Virdee, 1994). In the case of the LHC, probing physics at the 7 TeV scale inevitably meant that the search for the Higgs boson set the primary benchmarks. Discovery of this particle would involve precise measurement of the energy and momentum, over a wide range, of electrons, photons, muons, jets, and missing transverse energy.

In the late 1980s, there was clear theoretical motivation for physics beyond the Standard Model (BSM). The most popular candidate being the supersymmetry (SUSY) with its characteristic missing transverse energy signatures due to the escaping lightest neutral SUSY particle. Other candidate models predicted new heavy resonances, leptoquarks, substructure to quarks, and so forth. The design of the experiments had to allow the discovery of such states, which also served as benchmarks (see also Chapter 5).

SUSY predicted copious production of bottom quarks requiring pixel detectors to be placed close to the interaction point; heavy resonances (e.g. heavy gauge bosons) necessitated the detection of massive (multi-TeV) $Z$ bosons requiring the measurement of the momenta of muons or electrons at the TeV scale. Measurement of the charge of TeV leptons sets a challenging benchmark for momentum resolution of ~10% at a transverse momentum ($p_T$) of ~1 TeV, with the commensurate requirement of a high enough magnetic field.

The experimental search for the Higgs boson across the entire possible range of mass was fully explored for the first time. Among the proponents of SUSY, there was a prevalent belief that the mass of the Higgs boson ($m_H$) was less than 135 GeV. This low end of the mass range was considered to be especially difficult to probe at hadron colliders unless very good photon, electron, and muon energy and momentum resolution could be attained. For this reason, the LHC experiments had to pay particularly close attention to the performance requirements imposed by the search for the Higgs boson in this low-mass range.

Given that the decay width of the Higgs boson is very small (<10 MeV for $m_H$ < 150 GeV), the width of the reconstructed mass distribution, and hence the signal-to-background ratio, would be limited by the electron–photon energy resolution of the electromagnetic calorimeter, as well as by the charged-particle momentum resolution of the inner tracker and the muon spectrometer. As a consequence, these subsystems, as well as the magnetic field strength, were considered highly important. The search for the high-mass Higgs boson, particles predicted by SUSY, and other exotic states required excellent resolution for jets and missing transverse energy, implying full $4\pi$ calorimeter coverage.

## 2.3 Genesis of Two Complementary General-Purpose Experiments

In 1964 a conjecture was put forward that could explain why the electromagnetic and weak forces were so different in strength in experiments conducted previously. The conjecture predicted the existence of a new and unusual particle that eventually became known as the Higgs boson. The scientific method, at the heart of which is empiricism, required this hypothesis to be verified by experiment before being accepted as a true description of nature. The essence of scientific observation and establishing acceptability is reproducibility, i.e. the principle that if the experiment were to be conducted under the same conditions again, the result would be the same.

The 1964 conjecture said nothing about what should be the mass of the Higgs boson. So, when the search was eventually to be made, the full range of mass required exploration. The signature of the presence of the Higgs boson changes with mass, making the experimental design a particular challenge. Hence the LHC experiments had to pay particular attention to the performance requirements imposed by the search for the Higgs boson, particularly in the low mass range ($M_Z < M_H < 2M_W$) where there was a prevalent prejudice of the protagonists of supersymmetry that $m_H$ should be smaller than 135 GeV. As a consequence, much importance was placed on the tracking (inner and muon), as well as the magnetic field strength, and the electromagnetic calorimeters.

In general, as mentioned in Section 2.1, the search for the high-mass Higgs boson, particles predicted by SUSY, and other exotic states mentioned above, required excellent resolution for jets and missing transverse momentum (pTmiss), requiring full solid angle calorimeter coverage.

A saying prevalent in the late 1980s and early 1990s captured the challenge: 'We think we know how to build a high energy, high luminosity hadron collider—but we don't have the technology to build a detector for it.' Making discoveries in the unprecedented high collision rate environment generated by around one billion proton–proton interactions per second, with several tens of simultaneous collisions per bunch crossing, would require extraordinary detectors. Many technical, financial, industrial, and human challenges lay ahead, which eventually were all overcome, to yield experiments of unprecedented complexity and power (Evans, 2018).

Given the cost of the accelerator and the long construction time for both the accelerator and the experiments, it was clear that a minimum of two experiments would be needed. Given the complexity of the experiments, that each would require the talents of thousands of physicists and engineers to design, carry out the R&D and prototyping, and then build them, and the costs incurred (material cost of around 0.5 billion CHF (Swiss Francs) each not counting the staff costs), it was clear that at most only two general-purpose experiments would be approved and built. Several state-of-the-art technologies, or those pushed to their very limits, would be necessary.

Therefore, it was not surprising that complementary designs and techniques would have to be favoured to maximise the chance of success in at least one of the two experiments. Furthermore, were a discovery to be made it would be important for there to be confirmation, in parallel or soon thereafter, by a completely independent experiment by an independent team of scientists and engineers using independent analysis methods and computer code.

Another reason for complementary design and choices was that were a discovery to be made, the systematic uncertainties due to the differing techniques used would by construction be different, minimising the possibility of wrongly claiming a discovery through a poorly understood common systematic or instrumental effect or a bias in the way the analysis was performed.

At the Evian-les-Bains meeting (CERN, 1992), four experiment designs were presented: two deploying toroids (one with a superconducting magnet in the barrel) and two deploying superconducting high-field solenoids. The choice of the magnetic field configuration determined the overall design of the experiments.

The collaborations deploying toroids merged to form the ATLAS Collaboration. The ATLAS design (CERN, 1994d) was based on a very large superconducting air-core toroid for the measurement of muons, and supplemented by a superconducting two Tesla solenoid to provide the magnetic field for inner tracking, comprising two technologies—silicon microstrip sensors for the innermost parts and gaseous detectors using straw-like tubes augmented with the capability of detecting transition radiation, followed by a cryogenic liquid-argon/lead electromagnetic calorimeter with a novel 'accordion' geometry.

The CMS design (CERN, 1992b) was based on a single large-bore, long, high-field superconducting solenoid for analysing muons, together with an entirely silicon microstrip-based inner tracking device and an electromagnetic calorimeter comprising scintillating crystals. CMS has a single, very high field (4 Tesla); long, large bore solenoid surrounds all the tracking and calorimetry and an instrumented iron flux return yoke constitutes the muon measurement system and this solution is relatively compact, hence it was called the 'Compact Muon Solenoid' (CMS).

In both the experiments precise measurement of muon trajectories was performed by large area, but using different technology, gas detectors: primarily drift tubes and thin gap chambers in ATLAS and drift chambers and cathode strip chambers in the case of CMS. Even the selection of interesting collision events was complementary with only one hardware level in CMS but two in ATLAS. Clearly, the analysis teams also were independent, by construction, and when physics analysis started, much care was taken to minimise any flow of discovery-sensitive information from one experiment to the other, to respect independence as one of the fundamentals of the 'scientific method'.

When the announcement of the discovery of the Higgs boson was made in July 2012 at CERN in a common session, neither experiment had prior knowledge of what the other experiment had observed. Once the discovery of the Higgs boson had been made, the results of the two experiments were eventually combined, taking proper account of the statistical and systematic (many differing) uncertainties. The combined measurements are shown in Table 2.2 (Khachatryan et al., 2015)

**Table 2.2**  Measured global signal strength

| Best fit μ | | Uncertainty | | | | |
|---|---|---|---|---|---|---|
| | | Total | Stat | Expt | Thbkg | Thsig |
| ATLAS + CMS (measured) | 1.09 | +0.11 | +0.07 | +0.04 | +0.03 | +0.07 |
| | | −0.10 | −0.07 | −0.04 | −0.03 | −0.06 |
| ATLAS + CMS (expected) | | +0.11 | +0.07 | +0.04 | +0.03 | +0.07 |
| | | −0.10 | −0.07 | −0.04 | −0.03 | 0.06 |
| ATLAS (measured) | 1.20 | +0.15 | +0.10 | +0.06 | +0.04 | +0.08 |
| | | −0.14 | −0.10 | −0.06 | −0.04 | −0.07 |
| ATLAS (expected) | | +0.14 | +0.10 | +0.06 | +0.04 | +0.07 |
| | | −0.13 | −0.10 | −0.05 | −0.04 | −0.06 |
| CMS (measured) | 0.97 | +0.14 | +0.09 | +0.05 | +0.04 | +0.07 |
| | | −0.13 | −0.09 | −0.05 | −0.03 | −0.06 |
| CMS (expected) | | +0.14 | +0.09 | +0.05 | +0.04 | +0.08 |
| | | −0.13 | −0.09 | −0.05 | −0.03 | 0.06 |

The signal strength, μ, is defined as the ratio of the measured Higgs boson rate to its Standard Model (SM) prediction. The uncertainties are labelled as: Stat for statistical, Expt for experiment-specific, Thbkg and Thsig for the estimates of background and signal from theoretical uncertainties. The obvious advantage of conducting two independent experiments can be discerned from the Table 2.2 as the measured value μ is not only compatible between the two experiments but the error is lower by a factor of $\sqrt{2}$, as each measurement individually, at the time, was dominated by statistical uncertainty. Table 2.2 illustrates the measured global signal strength μ and its total uncertainty, together with the breakdown of the uncertainty into its four components.

The results are from Run 1 of the LHC and shown for the combination of ATLAS and CMS, and separately for each experiment. The expected uncertainty, with its breakdown, is also shown.

## 2.4  The Role of ALICE and LHCb Experiments

In addition to the two large general-purpose experiments, ATLAS and CMS, there are two relatively smaller experiments, ALICE (ALICE Collaboration et al., 2008), and LHCb (LHCb Collaboration et al., 2008). These two experiments are specialised in the sense that they were designed to study two topics: Alice for properties of subatomic particles like gluons and quarks; and LHCb for the difference between matter and antimatter, also of profound interest.

In our early universe, some $10^{-10}$s after the Big Bang, physical processes occurred that led to the dominance of matter over antimatter that we observe today.

At the moment of the conjectured creation of our universe, matter and antimatter were equally abundantly produced. A fundamental question arises—how did the antimatter disappear?

The necessary conditions for this to have occurred were laid down by the Russian physicist, Andrey Sakharov.[1] One of the three necessary conditions is that there be a violation of the symmetry represented by the product of charge conjugation, C, i.e. changing all particles to their antiparticles, and parity, P, i.e. effectively observing the physical processes in a mirror. LHCb was primarily designed to study Charge/Parity (CP) violation in the b-quark system.

Around one microsecond after the Big Bang, as the universe was expanding and cooling down, a phase consisting of a fluid of quarks and gluons, interacting freely, underwent a transition to one containing hadrons (e.g. pions, protons, neutrons, etc.). The quark-gluon fluid can be created at the LHC by colliding lead ions at high energies in the same LHC accelerator. These high energy collisions are akin to heating up these lead nuclei and melting their constituent structures (neutrons and protons). This leads to a sort of time reversal of what happened in our universe around one microsecond after the Big Bang. ALICE was specially designed to enable the study of the quark-gluon fluid and its transition into particles such as pions, protons, neutrons, etc. ATLAS and CMS also have targeted and specific capabilities in the study of these two areas and subatomic particles, in particular Higgs boson.

The four experiments neatly cover essentially all the physics that can be extracted from the LHC.

## 2.5 The Conception of the Experiments

### 2.5.1 ATLAS

It was natural that the two teams, named as ASCOT and EAGLE, with toroidal magnets seek a common approach. Merging of the two toroid-based detector concepts and the two proto-collaborations was not straightforward and needed a lot of good will and careful planning. The efficient progress and success of the process were underpinned by the common motivations for LHC physics, and by the fact that CERN's newly formed peer-review committee, the LHC Experiments Committee (LHCC), set an early deadline of 1 October 1992 for submitting Letters of Intent (LoI) only six months after the presentation of the Expression of Interests (EoIs). The ASCOT detector, with some 20 interested institutes, and the EAGLE, with some 70 interested institutes, joined forces over the summer of 1992 and submitted LoI under a new, democratically chosen name, the ATLAS, signed by 88 institutes comprising about 850 scientists.

---

[1] Andrey Sakharov is regarded as the father of the Soviet hydrogen bomb. He was awarded the Peace Prize in 1975 for his opposition to the abuse of power and his work for human rights.

Even though a toroid magnet configuration was common to both concepts, they differed considerably in their approach to the challenges posed at the LHC. ASCOT's design featured a novel superconducting air-core toroid magnet system in the barrel with excellent resolution for the measurement of the momenta of muons, even in stand-alone operation, at the highest instantaneous luminosities. EAGLE's approach was based on a cheaper, less performant, magnetised iron toroid, reserving more resources for powerful electromagnetic and hadronic calorimetry. In both cases, a superconducting solenoid was foreseen around the inner tracking system.

Combining the best aspects of both approaches was of course attractive and desirable as it would make a general-purpose LHC detector better able to satisfy the ambitious LHC physics goals. Pooling the resources within the newly formed ATLAS collaboration was expected to make such a complete instrument more probable and enhance its attractiveness to new collaborators and contributors. Indeed, seen from today's view point, this strategy was successful, though it was not just a simple and smooth journey to reach all physics goals and converge on today's ATLAS detector. The ATLAS detector layout is sketched in Figure 2.2 in a longitudinal cutaway view.

During the few months available to produce the LoI-letter of intent (ATLAS Collaboration, 1992; 1994), not all technological choices could be made, and many more studies and R&D were required. Therefore, options were presented in several cases, together with an indication of how these would be resolved. It took several rounds of intense negotiations among the collaborators and the LHCC referees before the Committee decided at its seventh meeting held from 8 to 9 June 1993 to recommend the ATLAS (as well as CMS) to proceed to a technical proposal (TP).

Eventually the ATLAS was constructed with many complexities of design, development, and installation. Figure 2.3 was a photograph (February 2007, during the

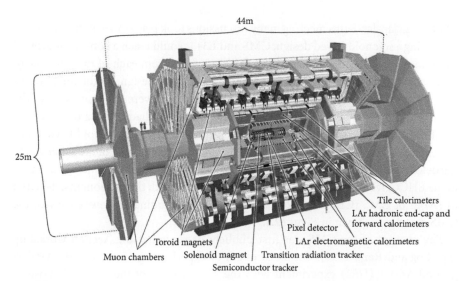

**Figure 2.2** Structure of the ATLAS detector
*Source:* © CERN

**Figure 2.3**  The ATLAS detector in the LHC cavern
*Source:* © CERN

installation phase) of one end of the barrel part with the calorimeter endcap still retracted before its insertion into the barrel toroid magnet structure and subsequent lowering of the endcap toroid into place.

## 2.5.2  CMS

After the Evian-les-Bains meeting many discussions took place to see if the two teams proposing a solenoid-based design, CMS and L3+1, could reach a consensus around a common design. A team comprising three scientists from each of the two collaborations was tasked with finding an acceptable common design. Unfortunately, this proved not to be possible. The main areas of divergence were the placement of the high field solenoid, in between the electromagnetic and hadronic calorimeters or behind all of the calorimetry; the inner radius and the extent of the rapidity coverage of the precision electromagnetic calorimeter; and the exact layout of the inner tracking detectors. In October 1992, the two collaborations separately submitted their LoIs to the LHCC, who, after several rounds of interactions with the proponents, declared in favour of CMS (CERN, 1992b). A Schematic longitudinal cut-away view of the CMS detector is shown in Figure 2.4.

The concept of the CMS was first publicly presented at the Aachen workshop (Jarlskog and Rein, 1990). On the strength of experience gained from the Underground Area 1 (UA1) experiment with the co-discovery of the W and Z bosons, several of its physicists turned their attention to the design of a detector for the LHC.

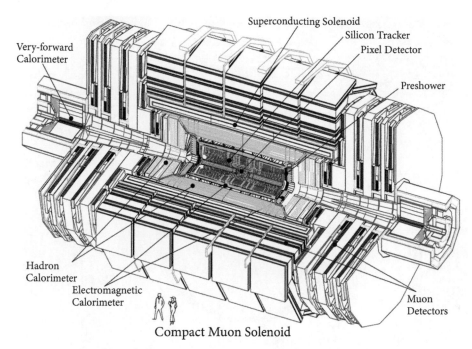

Very-forward
Calorimeter

Superconducting Solenoid

Silicon Tracker

Pixel Detector

Preshower

Hadron
Calorimeter

Electromagnetic
Calorimeter

Muon
Detectors

Compact Muon Solenoid

**Figure 2.4** Schematic longitudinal cut-away view of the CMS detector
*Source:* © CERN

In UA1, with inner tracking bathed in a magnetic field, discovering the W boson through its W→ e$\nu$ mode turned out to be remarkably easy. This was due to the presence of 4$\pi$-coverage 'hermetic' calorimetry, and a good electromagnetic calorimeter in addition to inner tracking in a magnetic field. However, the discovery in the W→$\mu\nu$ mode proved to be a lot more difficult. High $p_T$ muons suffer from poor momentum resolution in a low magnetic field (UA1 had a 0.7 T dipole field). Furthermore, charged pion decays ($\pi$→$\mu\nu$) can fake high $p_T$ muons and induce fake missing transverse energy. Low $p_T$ muons on the other hand have an advantage over electrons as they can be detected inside jets and B-physics at hadron colliders with such a characteristic signature was pioneered by UA1. These observations led to the first ideas for an LHC detector.

The design of a powerful muon triggering and reconstruction system was considered to be a key element, requiring a high magnetic field. A high value was sought for the parameter $BL^2$ driving the design, where B is the magnetic field strength and L is the muon's path length in the field. The highest field engineering-wise possible was desirable. Several field configurations were examined—solenoid, toroid, magnetised iron box.

After discussions with the magnet group in Saclay Laboratory, it was understood that only a solenoid could resist the very large magnetic forces exerted on the conductor in a very high magnetic field. CMS was thus designed around a high-field (4T) solenoid with large dimensions: a cylinder with a 6m diameter and 15m length.

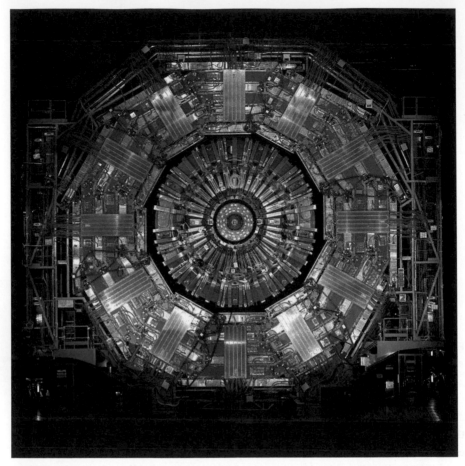

**Figure 2.5** Transverse section of the barrel part of CMS illustrating the successive layers of detectors
*Source:* © CERN

The large bore and long solenoid configuration have the additional advantages of not requiring an additional magnet for the inner tracking volume or in the forward region. For a high $BL^2$ the solenoid field configuration also leads to the most 'compact' design possible, moderating the size, and thus the cost, of the rest of the detector. The bore was chosen to be large enough to accommodate almost the full calorimetry inside the coil of the solenoid (Figure 2.5).

## 2.6 The Evolution and Construction of the Experiments

The ATLAS and CMS experiments underwent a long period of R&D, prototyping, and construction, though the basic designs were settled soon after the submission of the Technical Proposals in 1994 (ATLAS collaboration, 1994;

CMS collaboration, 1994). The two Collaborations opted not only for the magnet systems' different configurations but also complementary approaches for the active detector components. This complementarity turned out to be a wise approach for CERN, strengthening the scientific case for the Higgs boson discovery with two independent measuring methods, see Table 2.2.

In the early 1990s, there were only two complementary possibilities for the electromagnetic calorimeters that could perform in a high radiation environment and were sensitive enough to detect the two-photon decay of the SM Higgs boson at low mass: the lead-liquid argon sampling calorimeter, chosen by ATLAS, and fully sensitive dense scintillating crystals, chosen by CMS. Both were novel techniques, and each was tested and developed over many years before mass production could commence.

The hadron calorimeters in each detector are similar and based on known technologies, albeit in novel geometries; alternate layers of iron or brass absorbers in which the particles interact, producing showers of secondary particles and scintillator plates that sample the energy of the shower. The total amount of scintillation light detected by photodetectors is proportional to the incident energy.

The muon detectors used complementary technologies based on gaseous drift chambers, which provide precise position measurement (and also provide the trigger signal in the case of CMS), and thin-gap chambers and/or resistive plate chambers that provide precise timing information as well as a fast trigger signal.

The electronics on the detectors, much of which were manufactured using radiation hard technology, represented a substantial part of the material cost of the LHC experiments. The requirement of radiation hardness was previously found only in military and space applications.

The construction of the various components of the detectors took place over about ten years in universities, laboratories, and industries, and was then sent to CERN in Geneva. This chapter can only do partial justice to the technological challenges that had to be overcome in developing, constructing, and installing all the components in the large underground caverns. All the detector elements were connected to the off-detector electronics and fed data to computers housed in a neighbouring service cavern. Each experiment has more than 50,000 cables with a total length exceeding 3,000 km, and more than 10,000 pipes and tubes for services (cooling, ventilation, power, signal transmission, etc.). Access to repair any significant fault or faulty connection buried inside the experiment would require months to just open the experiments. Hence, a high degree of long-term operational reliability, which is usually associated with space-bound systems, had to be attained.

## 2.6.1 Some Comments Specific to ATLAS

Making technology choices is not a straightforward process. It was therefore important to establish in advance a clear and transparent process for arriving at decisions. Detector Review Panels were established, chaired by, and included senior collaborators not directly involved in developing the technologies in question, together with expert proponents. The evaluations of differing technologies included the ability to satisfy the requirements imposed by physics, as well as aspects concerning schedule, cost, available resources and expertise, the capacity for carrying out the construction, and the suitability of integration into the ATLAS detector. The progress of these evaluations was regularly reported, to all levels of ATLAS, including the project management as well as to the plenary meetings of the Collaboration. They concluded with a recommendation to the Collaboration Board where each ATLAS institution has one voting representative for the final ratification.

A very important decision had to be taken very early on concerning the choice of the technology for the toroid magnet system. The choice was between two configurations: whether to deploy conventional iron-core or superconducting air-core coils. In April 1993, only six months after the submission of the LoI, the Collaboration settled on the air-core design that yields a better physics performance but at a higher cost, hence needing a round of optimisation of the costs of the rest of the detector. Indeed, in July 1993, the initial design with 12 coils around the beam axis was reduced to a cheaper 8-coil design. The slightly increased magnetic field inhomogeneity was considered acceptable for physics. The choice also turned out to be advantageous in terms of increasing the solid angle in which muons do not pass through the material of the coils.

Also, within the first year after the submission of the LoI, following intensive evaluation work by the Calorimeter Review Panel, the calorimeter technologies and the layout for the barrel and endcap regions were chosen from the various options presented in the LoI. At the end of 1993, ATLAS settled on a configuration with a central and two extended barrel cylinders, one on either side of the central one, with fine-grained iron-scintillator tile modules for hadronic calorimetry. This combination formed an outer calorimeter cylinder around the three LAr cryostats housing, in the barrel a novel, highly granular electromagnetic lead sampling calorimeter with 'accordion' geometry including a pre-sampler, and in the endcaps, both the electromagnetic and the hadronic LAr calorimeters with lead and copper absorbers respectively. Somewhat more time was required for converging on the choice of a LAr forward calorimeter integrated within the endcap cryostats.

Another major choice concerned the technology and the configuration for instrumenting the muon spectrometer which separately employs precision and (fast) trigger chambers. The Muon Chamber Panel formed in 1994 helped to focus the muon community to converge on a baseline, well before the submission of the technical proposal (TP) which was implemented in the ATLAS detector. However, in this case the decision path was not straightforward. An intense R&D programme was launched in the early 1990s on three competing technologies: High Precision Drift Tubes (HPDT), Honeycomb Strip Chambers (HSC), and Jet Cell Drift Chambers (JCDC). Each technology aimed at very high precision in the single point measurements but used very different approaches to maintain the precision over the huge system against environmental variations. The JCDC used for example a very stiff carbon fibre support structure, whereas the HSC were not rigid objects but included an alignment system that was measuring online the chamber deformations such that it would be possible to correct offline any deformations. The HPDT had a very good intrinsic special resolution, but the operation mode ('streamer mode') could become a problem over the lifetime of ATLAS. The Muon Chamber Panel, evaluating all R&D results and design aspects, came up with a recommended choice (HSC) that was a surprise for the muon community and other experts, who considered this as very problematic for the long-term system performance of the experiment.

The ATLAS management, under the leadership of the Technical Coordinator, gathered the muon community and some experts in a workshop spanning several days at a somewhat remote location, La Mainaz, away from CERN, to review all aspects of the panel recommendations and comments. During these days, a consensus emerged for a concept called 'Monitored Drift Tube chambers' (MDT) that combined many of the best characteristics of the competing developments. At the end of the workshop the whole muon community was quite surprised by the new solution but accepted it and embarked on the development and construction of today's very successful MDT technology, realised in 1200 chambers that were constructed in an excellent collaborative spirit in many countries all over the globe.

Particular attention had to be paid to the technologies and layout of the inner detector (ID) tracking system, given the harsh high-rate environment of the LHC. The submission date of the TP, set at the end of 1994, was too soon to arrive at a definite choice for the layout that could be fully substantiated as the R&D was still going on. The Inner Detector Panel functioned up to September 1995 when ATLAS adopted a layout with pixel detectors in the innermost part, followed by silicon microstrip detectors, both providing discrete precise points, and at larger radii by a straw tube gaseous detector, essentially enabling continuous tracking with built-in transition radiation detection to enhance electron identification.

By this point, the baseline concept was settled, the technologies were chosen, and the project was ready to enter the construction phase. In passing we note that only much later, for the second LHC run starting in 2015, an additional layer of advanced pixel technology was added to further improve secondary vertexing capabilities at high luminosities. However, there were still many developments that took place and decisions made during the decade of construction. A detailed account of the ATLAS history is documented (ATLAS Collaboration, 2019).

## 2.6.2 Some Comments Specific to CMS

The main design goals of CMS, set during the design stage were:

  i) a robust and redundant muon system;
 ii) the best possible electromagnetic calorimeter consistent with (i);
iii) a high-quality central tracking system to achieve (i) and (ii);
 iv) a detector costing less than 475 million CHF.

The basic CMS detector design did not alter much from that presented in the LoI, though later, specific technologies were selected from the options presented.

CMS' design is defined by its magnet system, which is the main component in terms of its size, weight, and structural rigidity. The magnet system therefore plays a natural role as the structural element supporting all the other components, which are either mounted inside the vacuum tank of the solenoid coil or attached to the iron yoke return.

An important feature of CMS, which bore strongly on the magnet design, is the requirement of easy access to the inner parts of the detector for maintenance. For this purpose, the iron yoke, and hence the detector, is sectioned into five barrel-wheels, each 2.5m wide, and three end-cap disks at each end, for a total weight of 12,500 tonnes (see Figure 2.3). The field is returned through the iron yoke which houses four muon stations to ensure the robustness of measurement and full geo-metric coverage. The sectioning enabled the detector to be assembled and tested in a large surface hall while the underground cavern was being prepared. The sections, weighing between 350 and 2000 tonnes, were then lowered sequentially between October 2006 and January 2008. The central barrel wheel, which is the only sta-tionary part around the interaction point, is used to support the superconducting coil.

We give below details on some of the challenging design features and the tech-nology selection. Techniques developed for the construction of large solenoids used in previous LEP detectors such as ALEPH (Apparatus for LEP Physics), DELPHI (Detector with Lepton, Photon and Hadron Identification), at HERA particle accel-erator and the detector H1 at DESY, Germany, were introduced into the design of the CMS solenoid.

The main features that led to the high quality and reliability of these large magnets were the use of a high purity Aluminium stabilised conductor and indirect cooling. However, the large increase in some parameters such as magnetic field, Ampere-turns, forces, and stored energy density, necessitated some changes from the previous designs. In particular a four-layer winding was adopted. For the previously cited solenoids the radial and axial forces were low enough to hold the coil on an external and the cooled mandrel. For CMS the 20 kA state of the art conductor had to share with the mandrel the outward force. The refrigeration scheme followed in many ways a classical scheme: indirect cooling by thermosiphon.

CMS' inner tracking aims to efficiently reconstruct high transverse momentum isolated charged tracks using a smaller number of measurements each with a high precision. The requisite fine strip pitch therefore led to a highly granular system of almost 10 million channels[2] based on silicon microstrip detectors with three layers of pixel detectors close to the beam pipe.

---

[2] For details see section 10.3.4 Data acquisition and processing, HCAL Technical Design Report (CERN/LHCC 97-31, CMS TDR 2, 20 June 1997) http://uscms.fnal.gov/uscms/Subsystems/HCAL/hcal_tdr/ch10/.

The high number of channels, with low occupancy, helps with efficient reconstruction of charged particle trajectories, and precise measurement of their momenta. The high momentum precision is a direct consequence of the high magnetic field and the point-precision.

In a solenoidal field—since the bending takes place in the transverse plane—the interaction vertex, whose precision is set by the transverse size of the proton beams (15 μm), is usable. Careful attention was paid to the robust identification of muons and 'complementary' measurements of their momenta. In the solenoidal configuration, centrally produced muons are measured three times; in the inner tracker, after the coil, and in the return flux. Each muon is identified and measured in four identical muon stations inserted in the return yoke. The four stations ensure that there are always three stations recording the muon trajectory, leading to full geometric acceptance. All the stations include fast Resistive Plate Chambers (RPC) triggering planes that also identify the bunch crossing.

The two types of chambers each enable a cut on the muon transverse momentum at the first trigger level, providing the desired robustness. The large bending power is the key to very good momentum resolution even in the so-called 'stand-alone' mode especially at high transverse momenta.

The coil radius was chosen to be large enough to install essentially all the calorimetry inside, so as to avoid placing the coil in front or behind the electromagnetic calorimeter. In addition, a special attention was paid to the ability of the experiment to search for, and discover, a low mass SM Higgs boson.

CMS considered four options: lead-scintillator sampling calorimeter ('shashlik'), and three fully active dense scintillating materials: cerium fluoride, lead tungstate ($PbWO_4$) crystals, and hafnium fluoride glass. After much R&D, prototyping and beam tests, and considering other aspects such as physics performance, radiation tolerance, manufacturability and cost, CMS chose $PbWO_4$ crystals. These dense scintillating crystals offer very good energy resolution for electrons and photons. The scintillation light is detected by then-novel silicon avalanche photodiodes in the barrel region and vacuum phototriodes in the endcap region (Bell et al., 2004).

The electromagnetic calorimeter is followed by a brass/scintillator sampling hadronic calorimeter. The blue scintillation light is captured by wavelength shifting fibres, embedded in the scintillator plates, and fused to clear fibres and channelled to then-novel hybrid photodiodes. These photodiodes can provide gain and operate in high axial magnetic fields. Coverage up beam pipe is provided by a Cu/quartz fibre calorimeter. The Cerenkov light emitted in the quartz fibres is detected by photomultipliers.

To the extent possible, CMS decided to take advantage of the anticipated future industry trends concerning the computing power (CPU), data transmission speeds, and storage capacity. In a break from the past, custom triggering hardware would be used only at the first level, in anticipation that the link transmission speeds and CPU power would be adequate to send and analyse full events in commercial CPUs. The trigger and data acquisition comprised four parts: the front-end on-detector electronics; the calorimeter and muon first level trigger processors; the readout network; and an online CPU farm. The first two parts are synchronous and pipelined with a pipeline depth corresponding to $\approx 3$ µs. The latter two are asynchronous and based on industry standard data communication components and commercial processors. Being on the surface, the whole of the switch network and CPU farm has already been upgraded several times using the latest available commercial components.

## 2.7 Upgrades for the Future High-Luminosity Phase of the LHC

In July 2012, the ATLAS and CMS experiments announced the discovery of a Higgs boson, confirming the conjecture put forward in the 1960s. Further results from the two experiments show that, within the current measurement precision, the Higgs boson has the properties predicted by the Standard Model (SM). However, several theories of physics beyond the SM (BSM) predict the existence of more than one Higgs boson, and one of these would only be subtly different from that predicted in the SM one with signal strengths differing by between 0.5% and 5%, depending on the model in question, indicative of the required level of sensitivity to distinguish it from a SM Higgs boson.

In Run 2 (2015–2018), the LHC provided proton–proton collisions at $\sqrt{s}$=13 TeV with a peak instantaneous luminosity of $2\times10^{34}$ cm$^{-2}$s$^{-1}$, a factor of two beyond the design value. Initially it was intended to operate the collider at $\sqrt{s}$=14 TeV, the full design energy, after the second long shutdown (LS2), and to integrate a luminosity corresponding to some 300 fb$^{-1}$ by the end of Run 3. Finally, the Center of Mass energy for Run 3 is 13.6 TeV, in order to achieve optimal performances. The Run 3 started in spring 2022 and is scheduled to last until the end of 2025.

More precise measurements of the properties of the new boson will be made, as well as a more extensive exploration of physics beyond the SM, for which many possibilities are conjectured including supersymmetry, extra dimensions, unified theories, superstrings, etc.

However, the results will still be mostly dominated by statistical errors. Much more data needs to be collected to enable rigorous testing of the compatibility of the Higgs boson with the SM and to get clues to physics lying beyond the SM, in case of a significant deviation in results. This is one of the main motivations for the high luminosity LHC project, labelled the HL-LHC.[3]

Europe's highest priority in particle physics calls for the exploitation of the full potential of the LHC, including the high-luminosity upgrade of the accelerator and detectors in view of collecting 10 times more data than in the initial design. It is planned to increase the instantaneous luminosity of the LHC to $5\times10^{34}$ cm$^{-2}$s$^{-1}$,

---

[3] Details of a scientific case for HL-LHC can be found in The High-Luminosity LHC (HL-LHC) Project, https://cds.cern.ch/record/2199189/files/English.pdf.

and record, by around 2035, an integrated luminosity corresponding to ~3000fb$^{-1}$ (10 times larger than the original design value).[4] Such an integrated luminosity also requires very substantial technological upgrades of the ATLAS and CMS experiments, to allow a very precise measurement of the properties of the Higgs boson and the study of its rare decay modes and self-coupling, in addition to the search for physics beyond the SM. Many theories beyond the SM make different predictions for the properties of one or more Higgs bosons. These very ambitious upgrades to cope with the next level of challenges are a further leap into the forefront of technologies, demonstrating the continuous evolution of detector technologies in ATLAS and CMS where this is possible.

## 2.8  Select Cases of Success and Limitations and the Lessons Learnt

There are many interesting stories that could be told about the many exciting years of detector component construction all over the globe. Just a few illustrative examples can be given here, others are documented (Butler, 2018; ATLAS Collaboration, 2019b).

### 2.8.1  CMS

#### 2.8.1.1  An Example of Complexity in Construction

The engineering design, the construction at various sites around the world, and subsequent installation at CERN of the elements of the LHC experiments were complex tasks. One of the most illuminating examples of this is CMS' superconducting solenoid coil.

Early in the design process, thought was given to possible sites of construction and the method of delivery of the coil to CERN. The transport was likely to be through Marseilles up the Rhone river in France and then by road transport from Macon to CERN, Geneva. In order not to destroy and rebuild existing buildings or bridges, the diameter of the coil had to be less than 7m and the cylindrical length of the individual coil units less than 3m. To give the overall length of the cylindrical coil of 13m, it was sub-divided into 5 units.

The CMS solenoid has several innovative *state-of-the-art* features compared with previous magnets used in particle physics experiments. The most challenging feature is the four-layer coil winding, reinforced to withstand the huge forces at play. The challenge was to design the superconducting 'cable' that could run 20kA and handle large outward forces (corresponding to an outward pressure of roughly 60

---

[4] Luminosity is an important indicator of the performance of an accelerator. It is proportional to the number of collisions that occur in a given amount of time. The higher the luminosity, the more data the experiments can gather to allow them to observe rare processes.

atmospheres). The design of the coil was carried out at the Saclay laboratory near Paris, France, and CERN. CMS' solenoid magnet is the most powerful ever built.[5]

The manufacture of the solenoid coil started around 1998, with the superconducting wire made by the Outokompu company in Finland, comprising fine (~130 µm diameter) Nb-Ti filaments embedded in a copper matrix in the ratio of about 1:1. The wire was sent to the Brugg Kabel AG company near Zurich, Switzerland. Thirty-two wires were wound into a flat 'Rutherford' cable, with a twisting wire pattern to cancel out the inevitable Eddy currents that otherwise would render the coil non-superconducting. This flat cable (size of 21mm x 2.3mm) was co-extruded in the Nexans company near Neuchatel, Switzerland with ultra-high purity aluminium from the Sumitomo company in Japan. In order to assure perfect continuous lengths of over 2 km the control of the quality of this 'insert' (size 30mm x 21.6 mm) was defined by and carried out under the supervision of the Swiss Federal Laboratory for Materials Science and Technology (EMPA). The insert was then sent to the Techmeta company near Grenoble, France where two 'blocks' of high-strength aluminium alloy from Alcan (Switzerland) were electron-beam welded on each side of the insert.

These blocks enable the conductor to take up the above-mentioned pressure. Twenty perfect lengths were needed and 21 were produced with one spare which did not need to be used. The superconducting 'cable', now more of a plate measuring 64 mm x 21.6 mm, was sent to the ANSALDO company in Genoa, Italy to be wound into five coil modules. The winding was carried out on the inside and against a high-strength aluminium drum. The five coil modules were individually shipped from Genoa to Marseille, then up the River Rhone and finally by truck to CERN.

The five coils were stacked vertically at CERN, rotated into the horizontal position using a platform made in Korea, and the inner and outer cylinders that would form the vacuum tank 'sleeved' over the coil. The magnet was successfully energised to full field in 2006, having taken a journey of nearly ten years. Several challenges were overcome. The most important feature was the implementation of 'obsessive' assurance and control of quality in every stage of the manufacturing process to avoid any backward step or excessive spares that would have been costly in time and financial resources.

### 2.8.1.2 An Example of Technological and Economic Evolution during the Long Period of Construction

CMS chose dense lead tungstate scintillating crystals for its electromagnetic calorimeter. On a 'collaboration-building' visit to Ukraine in 1993, the CMS team was shown some results from a new dense scintillating crystal, lead tungstate ($PbWO_4$). This crystal has several advantages for a precision calorimeter, but also has several drawbacks. The light yield was low and had strong temperature dependence. A further drawback, unrelated to the crystal itself, was that the scintillation had to be

---

[5] For the CMS detector, the Saclay laboratory was responsible for the design, manufacture, and commissioning of the calibration system, by laser light injection, of the electromagnetic calorimeter (ECAL) with lead tungstate crystals and its permanent online monitoring (see details: https://irfu.cea.fr/en/Phocea/Vie_des_labos/Ast/ast_technique.php?id_ast=2292).

measured inside CMS' very large magnetic field. That ruled out the use of photomultipliers. It was presumed that by detecting the light with existing silicon photodiodes that could operate in magnetic fields, a good energy resolution could be achieved. However, when tests were carried out in a high energy electron beam at CERN, a large 'tail' was observed for mono-energetic electrons. Electrons and positrons at the end of the electromagnetic shower were producing a signal in the photodiodes that was larger than that generated by the scintillation light. A photo-device that could amplify the light-signal, and attenuate/eliminate the charged particle signal would be needed.

The solution to this problem turned out to be in a presentation given at an instrumentation conference earlier that summer. Results were presented from some novel photo-devices called silicon avalanche photodiodes (APDs) that could work in a magnetic field[6]. A few phone calls located some of these devices at the Paul Scherrer Institute in Villigen, Switzerland. They were brought to CERN the following day— just in time before CERN's accelerator complex shut down for the winter break that year. The tests were extremely encouraging: it seemed that the high-side tail had disappeared. However, there remained the 'small' matter of going from a few crystals available to the 75,000 $PbWO_4$ crystals needed by CMS, and from the few available APDs to the 130,000 needed. After much hard work by CMS members working closely with industry, mass-production of both items was established while satisfying the strict quality and performance criteria required.

R&D was carried out between 1993 and 1998. For APDs, this involved amelioration of radiation tolerance, decreasing temperature dependence, and the suppression of signals arising from the passage of charged particles.

For crystals, the transparency and the radiation-hardness of the lead tungstate crystals had to be improved. This involved optimising the fraction of lead oxide and tungsten oxide, the purity of the raw materials, and compensating for remaining defects by specific doping. More tests in beams showed the need for a powerful laser monitoring system to track precisely the changes in performance due to radiation damage. The crystals then had to be inserted into specially developed, light mechanical structures and integrated with high-performance electronics and a cooling system that can keep the temperature of some 100 tons of crystals constant to within 0.1°C, as the amount of light emitted is a strong function of temperature.

A remarkable backstory was how CMS managed to get 75,000 crystals grown in time for the start of collision data-taking at the LHC in 2008, and 130,000 APDs delivered, both starting from the handful available in the mid-1990s.

A round-the-clock crystals production line was set up in the small town called Bogoroditsk, near Moscow in Russia. The local people's livelihoods depended on the continued operation of this factory which had previously been deployed in the Russian military-industrial sector. CMS was able to use funds from the International Science and Technology Centre set up in 1992, with contributions from the USA,

---

[6] For details of Silicon Avalanche Photodiodes, see: https://indico.hep.caltech.edu/event/11/attachments/38/51/apd_intro_jra.pdf.

Europe, Japan, and Canada to convert such factories to non-military industrial production. After years of research, development, and production, CMS finally received the last consignment of crystals in March 2008, taking a total of 15 years from a novel idea to its realisation. It was, and still is, a great achievement for CMS.

CMS' lead tungstate crystal electromagnetic calorimeter was finally ready to record proton–proton collisions. And in July 2012 it was in these very same crystals that CMS' signal for the Higgs boson was the strongest, making the long and arduous crystals journey well worth it. It is inevitable that projects with such long durations such as CMS and ATLAS, will be buffeted by difficulties. The crystals project was no exception.

In the mid-1990s when the procurement contracts were negotiated, CMS agreed on a reasonable price per unit volume, in US dollars. However, as Russia's economy began to pick up, and with Russia's initiative to join the World Trade Organisation requiring de-regulation and discouragement of state subsidies, the raw material and energy prices started to increase sharply. By 2004, the factory in Bogoroditsk declared that the unit price had to triple and informed CMS that they could no longer continue the production of lead tungstate crystals.

A period of intense negotiation ensued, led by the Russian Minister of Education and Science on one side and the CERN Director-General (2004–2008), Robert Aymar, on the other. A few months later, and to CMS' relief, a new mutually acceptable price was agreed upon. Furthermore, given the evolving world economic environment, all later contracts (for the remaining almost half quantity of crystals) were placed in Russian roubles as the rouble was now considered by the factory to be a more stable currency than the US dollar.

The evolving economic situation was felt first-hand by CMS scientists involved. Looking back to the early 1990s, on their visits to Russia they took with them bottles of water, bars of chocolate, packets of cheese, biscuits, dried fruits, and other preserves, as there were no real restaurants in town. Much has changed since then, and for the better, however, mostly at a high cost. For much of that time some members of CMS working on the crystals project had open visas to visit Russia.

Even with production back on track in Russia, CMS had to bring in another supplier to be ready in time for the first LHC beams that eventually came in September 2008. The second supplier was a known manufacturer of crystals in China that eventually provided about 10% of the crystals needed, just enough for CMS to be ready in time. However, this supplier wanted CMS to provide platinum, used to line (make non-stick) the insides of the ceramic crucibles, operating at 1200°C, in which the melt was contained to grow lead tungstate ingots.

Since Switzerland's UBS bank holds considerable reserves of precious metals in its vaults in Zurich a loan of some 10M$ worth of platinum was arranged. Some members of CMS, aided by key members of the purchasing group at CERN, had to quickly learn about negotiating deals for the loan of precious metals, shipping them to China and then getting it back to Switzerland, getting it purified and returned to the UBS vaults.

The culture of Quality Assurance (QA) and Quality Control (QC) had to be introduced into the company in Russia, chosen for the production of crystals. The company required new equipment to substantially increase the volume of production (production capacity was quadrupled). Technically they were very strong but their managerial and quality control practices had to be improved. Much effort was spent on facilitating the procurement of the equipment needed for the increased production, e.g. 150 ovens were purchased using funds granted by the International Science and Technology Center (ISTC), and installed and commissioned at the factory. Achieving repeatable quality was essential. More sophisticated quality control instrumentation was installed on each oven. Another example was the procurement of platinum from the Russian state reserves that had to be authorised by the Russian President. These examples demonstrate the complexity of technological and political factors in negotiating Big Science procurement processes.

## 2.8.2  Two examples from ATLAS

### 2.8.2.1  Distributed Construction of the Complex Magnet System

The construction of the ATLAS magnet system was a great effort of many collaborators worldwide. In normal circumstances, a tendering process would have preceded such a megaproject, resulting in a consortium of companies that would have taken care of production design, procurement issues, integration, and testing. However, this required all financial resources to be present and collected in a bank account in Geneva, from which the deliveries could be paid for. This was not so in ATLAS. Since cash was only available for some 15% of these deliverables, the decentralised resources had to be used by the funding agencies (FAs) interested to support their laboratories and industries working on the magnet system.

The solution found was to split the magnet system up into smaller units, which could then be covered by the budgets available at the various FAs over the construction period. At first impression, this looked like a recipe for disaster, as many of the additional technical and managerial interfaces between collaborating companies and institutes and their workshops had to be controlled, which in principle could lead to more risk and potential delay due to inefficiency. But on the other hand, this manner of system procurement turned out to be a good solution, as it allowed funding for less cost since financial reserves were decentralised and remained under the responsibility of the FAs. It also guaranteed a fully motivated engagement of all partners to succeed.

This is illustrated in the case for the Barrel Toroid by showing the procurement circumstances for only the main parts since presenting the details in full is well beyond the scope of this book. As explained above, in a conventional project tender, a single leading contractor in a consortium would take responsibility and manage all procurements and deliver the system to CERN, ready-made for installation. Not in this case. In the project's set-up all manufacturers were only responsible for their own deliveries; the ATLAS magnet team at CERN only handled the final financial and technical responsibility. When going from inside out, from the conductor to assembled toroid, one sees the following procurements, with the percentage of the system cost in brackets:

- Production of superconductor (18%) was a 50–50 effort of FAs from Germany and Italy, using many companies for the delivery of the main ingredients comprising superconducting wire and billets of pure aluminium, conductor cabling and co-extrusion at two sites, and quality control by yet other companies and the supervising magnet laboratories, the French Alternative Energies and Atomic Energy Commission (CEA), Istituto Nazionale di Fisica Nucleare (INFN). and CERN.
- The manufacturing of 16 coil winding packs (17%) was all funded by and made in Italy.
- The eight cold mass structures called coil casings (22%) were funded by Switzerland and Germany. The 16 coil packs and eight cold masses were delivered to CERN for integration, which was performed by a company delivering eight cold masses ready for integration with the eight cryostats. A cryostat comprises an outer vacuum vessel, a radiation screen and MLI (multi-layer) superinsulation, and cold supports locking the cold mass in the vacuum vessel.
- The cold mass supports have two families, 17 stops and 8 tie rods per cryostat. The stops (1%) were made at a small company in France, while the special and expensive titanium tie rods (2%) were delivered in-kind by Russia, benefiting from a project funded by the ISTC.
- The thermal radiation shields for all eight cryostats (3%) are in-kind from Italy and were made by an Italian company. The superinsulation blankets and installation (1%) were made in-kind by an institute in Russia.
- The vacuum vessels (10%) were in-kind from Spain and manufactured in Spain, However, the stainless steel used was delivered in-kind from Sweden by a Swedish company. Eight cryo-ring sections and a cryogenics valve box 3%), which guides bus bars for interlinking coils and helium cooling lines, interconnect the eight cryostats.
- When the eight coils were completed, bolting the parts together into a toroid took place in the cavern using the so-called warm structure comprising an inner and an outer ring of very special aluminium alloy struts (8%), delivered in-kind by Russia and Belarus through forging and H-profile extrusion and welding of connection flanges at a company in the Netherlands.

The coil parts listed above were delivered to CERN where all integration and coil testing on the surface prior to installation (4%) in the cavern took place. It was a deliberate and conscious choice to do all integration activities (12%) at CERN under the direct responsibility of the magnet team, in order to minimise technical risk and to control cost.

Indeed, there were some failures of industrially fabricated components that could be efficiently recovered by in-sourcing of the work to CERN. Figure 2.6 shows the vast building at CERN used for the barrel toroid coil integration.

The team was supported in these works by the engineering laboratories involved such as CEA, INFN, and the Joint Institute for Nuclear Research (JINR) and they provided most of the manpower, manufacture capabilities, large-scale tooling, as well as connecting with several companies for mechanical integration services and others specialising in welding repairs and quality control.

Without going into too much detail, a similar procurement organisation for the manufacturing of conductors, coils, cold masses, cryostat parts, integration, and testing was set up for both Endcap Toroids. However, it was supported by other funding agencies, in particular from the Netherlands and Israel, but again leaving the integration of coils and cryostats at CERN for the same good reasons. Again, the integration activities were performed by the ATLAS magnet team, which was supported by the magnet engineering laboratories Rutherford Appleton Laboratory (RAL) in the United Kingdom and the National Institute of Subatomic Physics (NIKHEF) in the Netherlands. JINR in Russia provided most of the tooling and integration manpower. An exception was the procurement of the central solenoid. Its complete cold mass and proximity cryogenics were in-kind contributions from Japan with engineering

**Figure 2.6** The ATLAS barrel toroid cold mass and cryostat integration area in building 180 at CERN, showing the several stages in coil integration from cold mass (bottom) to cryostats (upper right)

*Source:* © CERN

leadership coming from the High Energy Accelerator Research Organisation (KEK) laboratory. The solenoid was manufactured almost entirely and pre-tested in Japan, thereafter, it was shipped to CERN and integrated with the barrel LAr calorimeter. In this configuration, an ultimate test was performed before its installation in the experimental cavern in 2005.

### 2.8.2.2 Construction and Installation of the Main ATLAS Detector Support Structure

A second example from the ATLAS construction is the story of the heavy support structure carrying the entire detector system (with the exception of the outer end-cap muon chambers). One of the basic elements of the detector is a system of supports on which the most massive components—the central solenoid and inner tracking detector, the calorimeter system, the toroid magnet system, and the muon detectors of the barrel and the small wheel—are mounted. The total weight of these elements is about 7000 tonnes. This support system, called 'Feet and Rails' (F&R), had to satisfy the following requirements:

- Minimum magnetic permeability, so as not to spoil the characteristics of the magnetic field;
- Sufficient strength;
- High machining accuracy;
- Ability to withstand emergency loads, for example, rapid cooling with cryogenic liquids;
- Convenience for mounting and maintenance of the muon barrel detectors; and
- Acceptable cost.

The main components of the F&R are the bedplates, which provide an inclination of ATLAS equal to that of the LHC ring, and nine pairs of feet with rails on top, for the longitudinal movement of detectors during the installation phase and the maintenance periods.

In 2000, the ATLAS Technical Coordination was nearing the completion of the definition drawings, and the Collaboration was about to decide on placing a manufacturing order in the industry. It should be noted that F&R manufacturing deadlines were very tight since the system had to be first mounted before any other parts of the installation could follow. This project, along with other infrastructure components, was a Common Project (CP) item; thus neither a Funding Agency (FA) nor a responsible institution wereinitially defined. The Russian participants of ATLAS were then invited to perform this work at one of the Russian enterprises as an in-kind contribution to CPs. The Institute for High Energy Physics (IHEP) Protvino was appointed as the main Russian institute responsible for the project.

After considering possible manufacturers, 'Izhorskiye Zavody' (IZ), a firm in St Petersburg that could perform all the above work, plus the production of materials, was selected. IZ is one of the oldest industrial enterprises in Russia, founded in 1722 by a decree of Peter the Great. It is a large enterprise with its own metallurgical plant, rolling mills, and advanced metal processing. In its portfolio are equipment for nuclear power plants (including vessels of nuclear reactors) and petrochemical equipment, metallurgical castings from steels with special properties, and bridge and road metal structures.

For this plant, the F&R order was quite small, 'only US$4 million for 500 tonnes of stainless-steel products', and non-standard, so one could expect a variety of hard-to-predict problems. It was not surprising that this proposal was not met with much enthusiasm. However, after studying technical documentation and further communications with colleagues from CERN and IHEP, it was found that this project was not only feasible, but also interesting for the development of the company's technological potential and prestige. One of the interesting tasks was the production of the necessary stainless steel. CERN's designers relied on European standards, which differed from typical Russian ones. It took a thorough analysis of the technical requirements and the manufacturing capabilities of the plant to achieve the parameters for the metal that would satisfy all the magnetic and strength characteristics needed.

To prepare the fabrication drawings, the IHEP made a separate agreement with the IZ design bureau. However, the main issue with the drawings done then was the uncertainty of the welding deformations, and at the same time, the attainability of the required accuracy in the subsequent machining of the parts. Another contract was concluded to produce a trial batch of steel plates and for the production of welded joint samples from these plates. The contract for the manufacture of the F&R stipulated the supply of six batches of components for the ATLAS support structure during 2002 and 2003.

The quality of welds was controlled by several methods, including ultrasonic inspection, and when any defects were found, the seams were repaired. For an independent control of welding works, IHEP attracted the St Petersburg Maritime Register under a separate contract. The flatness of the bedplates was controlled by the independent GRADAN company, which specialises in optical measurements. The list of control operations was summarised in the Quality Control plan for the project. It is worth acknowledging the readiness of IZ to carry out and document all necessary checks and measurements, as well as their liberal attitude towards the changes requested by the customer in the drawings, including some significant ones. The cost of the project did not increase in comparison with the initial contract. The ATLAS F&R was successfully and timely assembled in the pit from October 2003 until February 2004.

## 2.9  General Challenges and Lessons Learnt

A substantial challenge for the ATLAS and CMS experiments was to keep within the cost ceiling specified in a letter of approval in January 1996—the same 475 million CHF in 1995 prices for both detectors. The experiments were completed in 2009 marking the start of collision data taking. A lot of the funding and the expenditure took place in currencies other than the Swiss Franc. Using the US dollar as a proxy, folding the purchasing power with the spending profile, the 1996 approval ceiling transformed to an equivalent of about 570 million CHF in 2009. The accounts were closed on the experiments, for the configuration that discovered the Higgs boson, at around 550 million CHF each.

There are many lessons to be learnt from the construction of the two unique scientific instruments that are ATLAS and CMS. Foremost, one cannot stress enough the importance of careful engineering and quality control considerations right from the onset of the projects. Sufficient resources should be invested in these aspects already at the start of the projects, preferably even more than this was possible in the cases of ATLAS and CMS, to minimise corrective actions to be taken at later stages that can turn out to be costly both in money and time.

Of paramount importance are strict quality assurance (QA) and quality control (QC) for all components of the detectors. Quality assurance is a way of preventing mistakes and defects in manufactured products and avoiding problems when delivering products or services to customers. Quality control is a process by which entities review the quality of all factors involved in production. Much thought had to be put in considering what could go wrong, and to introduce mitigation measures.

Specifications were defined for all critical elements. Prototyping of most of the critical elements was crucial, e.g. a half size coil module was manufactured and tested for the CMS solenoid as well as for the ATLAS toroid coils. Testing and measuring instrumentation were developed and used to maintain manufacturing quality within the strict specifications laid down for successful operation of the magnet system at the design value, plus a small margin. Careful monitoring of the progress of the manufacture of each element was carried out, requiring an understanding of how to smoothly transition from one operation to another or one company to another. This was vital to keep to the required and desired schedule.

Intensive R&D and prototyping have to be successfully carried out before the launch of production. This applies to all aspects of the detector, ranging from heavy mechanical structures to highly integrated electronics components. Much attention has to be paid to selecting the companies that are able to carry out mass production within the desired specifications that are almost always stricter than usual for the industry concerned.

It is interesting to note further lessons learnt, among others:

i) judicious changes of design specification, in consultation with industry, without compromising much in the way of physics performance, so as to ease manufacture and consequently lower costs and risks;

ii) 'ride the technology wave' as much as possible and be prepared to adopt newer technology:

    a. for elements that gain in performance at the same cost (e.g. computing power);

b. for elements whose costs decrease as technology develops (e.g. cost of micro-electronics used in the front-ends as the feature size of the technology decreases);

iii) assist industrial companies in figuring out ways of lowering costs while maintaining the specification (cost benefit v/s knowledge transfer); and

iv) an open and judicious purchasing policy (e.g. allowing for in-kind contributions, world-wide tendering instead of the policy of *'juste retour'*,[7] expenditure of financial contributions locally, as much as possible, from participating countries).

## 2.10  Human Factors Led to the Success of ATLAS and CMS

Big Science projects rely heavily on the contributions from individual scientists who in turn rely on intellectual contributions and talents from many. Human factors refer to how people do their work and bring in those social and personal skills such as communication and decision making which complement technological skills. Human factors are defined as the science of people at work. It is primarily concerned with understanding human capabilities and then applying this knowledge to the design of equipment, tools, systems, and processes of work (ARPANSA, 2021) As discussed in this chapter, human participation in these experiments requires not only scientific and technological intellectual capacity, but also more subtle factors such as resilience, desire to participate, motivational power, and working under very difficult conditions that requireimmense self-determination and personal sacrifices.

Complex particle physics megaprojects like ATLAS and CMS are only possible with the cooperation of a large spectrum of talents, deeply motivated by the physics goals. The ATLAS and CMS experiments could not have been built as designed without the talents and resources from institutions the world over. The remarkable growth of both collaborations, by almost a factor of three in the number of scientists and Institutions since the Letter of Intents (LoIs) in 1992, was marked by specific events in addition to a steady influx of new collaborators (ATLAS, 1992). Finding new collaborators was a high priority for the leaders of the experiments in order to secure enough resources to realise their challenging aims. This recruiting activity included many visits and discussions in European and non-European countries, to motivate and invite physicists and institutes to participate in and contribute to the experiments. Much value was placed not only on material contributions but also on intellectual ones.

[7] The 'juste retour' mentality refers to budgetary decisions taken on the basis of highly misleading indicators result in poor policies as they are biased towards programmes with monetary backflows into Member States.

In the early days, what was striking was the great interest in participating in these experiments and the desire to contribute successfully, sometimes under very difficult conditions, to the construction of the parts of the detectors. An example of this spirit is the contribution of the scientists from the former Soviet Union states who made very substantial early contributions to the design, development, and construction of both ATLAS and CMS. It was also of great mutual benefit to the experiments, and the Institutes concerned, that this cooperation could profit from special inter-governmental programmes such as ISTC and the International Association for the promotion of cooperation with scientists from the independent states of the former Soviet Union (INTAS), to revitalise and convert industries for peaceful applications in these countries.

A particularly significant event that occurred soon after the submission of the LoIs was the unfortunate discontinuation of the Superconducting Super Collider (SSC) in 1993. It led to the sudden growth of ATLAS and CMS with several tens of US Institutions joining the experiments. The US scientists and engineers strengthened almost all the sub-detector teams in ATLAS and CMS, integrating the experience gained from the SSC-related R&D and experiment designs. This wide area of contribution enabled a smooth integration of a large new community into the collaborations, a spirit that also characterises today's common work. The US teams were already well integrated into the experiments, years before the formal CERN—US DOE/NSF LHC agreement was signed in December 1997.

The collaborations continue to be open to new institutions. The early endeavours of the leaders in building up the collaborations are explained below.

In ATLAS, strong and fruitful collaborations were extended to and established with many Institutions from Japan, Canada, Israel, Australia, and several other countries, followed notably by several Chinese institutes in the second half of the 1990s. The participation of South American groups was, for a long time after the submission of the LoI, limited to Brazil, but got a real boost through the European Union sponsored exchange programmes in the mid-2000s, enabling several other Latin American countries to become members of ATLAS. Participation from the African continent is still sparse: Morocco joined in the mid-1990s, but it took another 15 years before universities from South Africa joined.

The ATLAS Collaboration today comprises some 180 Institutions (230 Institutes) from 38 countries as basic constituents with a formal voting right at the Collaboration Board. Institutions can be a cluster of Institutes (universities or laboratories), typically from a given country or region. The count of scientific authors is close to 3000, but the ATLAS detector would not exist and run efficiently without an additional large number of engineers, technicians, and administrative personnel.

In CMS in the early 1990s, enlarging the CMS collaboration resembled a world-wide grand tour for some of its leaders. In looking for collaborators they travelled to numerous countries including Brazil, China, India, Iran, Ireland, Korea, New Zealand, Pakistan, Russia, various Eastern European states, Taiwan, Turkey, and the US, not to mention essentially all of the CERN member states. The enthu-siasm encountered at all levels, from students and university rectors through to science ministers, was gratifying, placing the LHC project at the heart of their scientific programme. Watching the growth of international scientific collaboration in countries that previously had little experience of such endeavours was particularly heart-warming. Today the CMS collaboration comprises over 3000 scientists and engineers from about 200 Institutes from about 40 countries.

One of the most difficult challenges that the high energy particle physics community is facing now is how to retain the expertise for their instruments and how to transfer to the younger generations the know-how for the design, construction, and operation of such very complex experiments.

Nowadays one observes an increasing specialisation of knowledge as the experiments stretch over longer and longer time scales. Twenty or 30 years ago, the gap between two experiments was about 10 years, and young people were able to follow through all the phases of the experiment, from construction to operation and finally the analysis, learning from each of these phases and preparing themselves to lead the next steps of experimentation. Now that the gap between two (phases) of experiments can reach several decades, this natural evolution of learning and then leading the next experiment is lost.

The conceptual designs of ATLAS and CMS were made in the early 1990s, followed by construction in the 2000s with exploitation starting in 2009 and scheduled to go on up till the end of the 2030s. The first phase of upgrades began around 2011 and these upgrades are expected to continue and be fully functional in 2022. The second phase of upgrades is under construction. The people who learned (through experience) how to manage a detector construction project, have in many cases not continued in the field, and those who remained were not able, or did simply not have a chance, to pass on their expertise because of the lack of projects in the construction phase for a long period. This created a generational and cultural gap that is now very difficult to bridge.

Another difficulty, which adds to the concerns mentioned above, is the fact that in general, physicists who are more interested in the work on instrumentation and technologies (R&D on new detectors, construction, operations) tend to be less likely to be awarded long term positions in academia. This is one of the reasons why the people who devoted most of their time to the construction of the large experiments were not able to continue in the field. However, there has been in recent years, growing awareness in the community of the need to counteract this situation and to find ways to offer a future for these vital competences and to ensure their longevity in the field.

## 2.11 Conclusions

The ATLAS and CMS experiments at CERN's Large Hadron Collider (LHC) are two prime examples of Big Science projects that illustrate many aspects, ranging from their fundamental scientific motivation, to sophisticated technical developments and challenges for laboratories and industry. The development of impressive international scientific collaborations with their distinctive collegial management structures took decades of careful planning and perseverance, strong organisational backing, and the dedication of thousands of physicists, engineers, and technology experts.

The science motivating the projects was the quest to elucidate long-standing fundamental questions of modern elementary particle physics, the basis for understanding the subatomic world as well as the universe. Much in the focus of the experimental physicists in the early 1990s was the search for the famous Higgs boson, hypothesised in 1964, or in its absence, whatever else was responsible for the spontaneous breaking of the electro-weak symmetry, as the key element of the Standard Model of Particle Physics.

The prospect of the future LHC, with its high proton collision energy and luminosity, made it conceivable that conclusive experiments could be conducted. It took 20 years of design and construction of two instruments (detectors) that were finally successful in discovering the Higgs boson, reported jointly by ATLAS and CMS in 2012. The two complementary detectors were built and operated independently by the two separate collaborations CMS and ATLAS, thereby underlining the highest standards for scientific discoveries.

As the chapter exhibits in detail, the technological challenges for the detectors were unheralded and could only be met by innovative developments and engineering innovation, with concepts based on extensive R&D networks and the continual development of prototypes for various components of the instruments, ranging from semiconductor pixel devices to huge superconducting magnet systems. Various technologies had to be tested in cooperation between universities, research laboratories, and industry, and often difficult decisions had to be made for the final choices. These decisions can be a real challenge for the leaders of the experiments, as beside the technical performance issues they also included important aspects of costs and the human pride of the inventors which all had to fit the overarching common physics goals of the projects.

All that has been mentioned above for detector components applies as well to the giant steps that had to be made in terms of micro-electronics, data handling, and computing. Enormous collision rates, from LHC bunch crossings occurring every 25 ns, are registered in the typically 100 million channels of the detector sensors. Furthermore, a lean engineering of the overall concepts of the ATLAS and CMS detectors was of paramount importance. In that sense one can say that each one of them was its own prototype.

After more than 10 years of operation, one can proudly state that the ATLAS, CMS, and the LHC are successes. But they are not at the end yet, by far, of their journey. Based on the experience gained, major parts are being upgraded to operate for another two decades at a much higher intensity, providing a tenfold increase in sensitivity with an equivalent increase in registered collisions. These will allow the collaborations to search for physics clues that point beyond the Standard Model, so called New Physics beyond the Standard Model for which there exist solid indications as discussed in Chapter 5.

There are many lessons that can be learned from ATLAS and CMS, that can help guide future projects on a world-wide collaborative scale. Their nature covers both technical and human aspects and is implicit in Sections 2.9 and 2.10 of this chapter. Most importantly, building the experiments required extensive resources, both financial and human, that was only possible by a world-wide collaboration, involving the talents of a large number of scientists, engineers, and technicians, all of whom were motivated by the prospect of making significant advances in the knowledge and understanding of nature, and achieving unique technological breakthroughs.

# 3
# A Machine with Endless Frontiers
## The Large Hadron Collider (LHC)

*Lyn Evans, Frédérick Bordry, and Shantha Liyanage*

## 3.1 Introduction

The Large Hadron Collider (LHC), at the European Organization for Nuclear Research (CERN) in Geneva Switzerland, is heralded as the world's biggest particle physics accelerator. The construction of the LHC was a massive undertaking, spanning almost 30 years from its conception, construction, and commissioning to its first operation.

To appreciate the significance of the LHC at CERN, it is important to understand the role of accelerators or colliders in particle physics experiments. It was not until as recently as the 1970s that two particle beams were colliding to study high energy collisions to discover subatomic particles, as explained in Einstein's $E=mc^2$ (energy equals mass times the speed of light in a vacuum squared) relationship to determine the collision energy E required to produce a particle of mass m.

Some of the early accelerators such as cyclotrons and electrostatic generators operated in the range of millions of electron volts (MeV) from the 1930s to 1950s. Then came the Betatrons, Synchrotrons and Proton LINACs, which were operating at over a billion electron volts (GeV). More powerful accelerators such as the Large Electron Positron (LEP) in the late 1980s and the Large Hadron Collider (LHC) in the late 2000s are part of the most recent generation of colliders. LEP was the largest circular electron-positron collider; the energies reached 209 GeV (Giga electron volts). LEP ceased operation in November 2000 to be replaced by its successor LHC (Aymar, 2014).

In the construction of the LHC, it had to fit into the former LEP tunnel imposing considerable civil engineering challenges for the LHC accelerator developers. Several research studies and technical innovations were necessary to achieve the successful construction and commissioning of the LHC. The LHC will attain a proton particle beam energy of 7 TeV (7 trillion electron volts). With two beams operating, the LHC provides proton beams of 7 TeV for proton–proton collision at the centre of mass energy of 14 TeV. The LHC accelerator (also known as the LHC machine) delivers heavy ion (lead–lead) and even lead–proton collisions.

Lyn Evans et al., *A Machine with Endless Frontiers*. In: *Big Science, Innovation, and Societal Contributions*.
Edited by: Shantha Liyanage, Markus Nordberg, and Marilena Streit-Bianchi, Oxford University Press.
© Lyn Evans et al., (2024). DOI: 10.1093/oso/9780198881193.003.0004

For a high energy accelerator, beam energy is not the only beam quality factor in its development. The type of particles, beam intensity, beam trajectory, and luminosity are all important factors. Luminosity gives a measure of how many collisions are happening in a particle accelerator. Luminosity is measured in $fb^{-1}$ (femto barn −1). Each accelerator tries to increase its luminosity. Luminosity is not necessarily the collision rate but it measures how many particles can be squeezed through a given space at a given time. Although this does not guarantee that all of those particles will collide, the more particles that can be crammed into a given space, the more likely they will collide (Gillies, 2011). Indeed, the LHC has exceeded its design luminosity by a factor of two and delivered an integrated luminosity of almost 200 fb−1 in proton–proton collisions.

During the early 1980s, the accelerator builders faced many scientific, technical, and financial challenges. The pioneering role of the Tevatron, which began operations in 1983, as the first large superconducting machine was important for accelerator builders. For example, the Relativistic Heavy Ion Collider (RHIC) at Brookhaven in the USA and the electron–proton collider HERA at DESY in Germany derived directly from the experiences of building the Tevatron.

A giant step in the detector concepts was achieved in the Tevatron experiments at Fermilab, USA with respect to the physics signatures, sophistication, and granularity of the detector components (Fermilab, 2017). The Tevatron collider reached a final collision energy close to 1 TeV (exactly 0.98 TeV). At that time, the Tevatron was the world's highest-energy proton–antiproton collider. It was shut down in September 2011 (Fermilab, 2014). The development of the LHC began in 1984, giving an added advantage to CERN in accelerator technology (Evans and Jenni, 2021).

The LHC can be regarded as a penultimate success in the development of particle physics accelerators with the discovery of the Higgs boson in 2012.

A comprehensive review of the LHC machine design can be found in the LHC Design Report (CERN, 2004), which gives a detailed description of the machine as it was built and comprehensive references. A more popular description of the LHC and its detectors can be found in Evans (1998; 2010; and 2018).

This chapter aims to provide an overview of the organisation of accelerator technology that has led to various applications in particle physics, nuclear medicine, new materials, and other medical fields. In doing so, it aims to provide the personal accounts and experiences of those who were involved in the building of the LHC. To explain, how trials and tribulations managed to bring this truly remarkable machine functional and valuable, it is necessary to understand and appreciate those personal insights.

## 3.2  Genesis of the LHC—Science and Diplomacy

The concept of the Large Hadron Collider (LHC) emerged in 1976 when the European particle physics community began to discuss the building of a Large Electron Positron (LEP) collider at CERN. The LEP was formerly approved in 1981 and civil construction started in 1983. The LEP was eventually completed in 1987 and began its pilot runs in 1989. The LEP was installed in a circumference of 27 km (about 16.8 miles) and 3 metres (about 10 feet) tunnel that runs below 100 metres (328 feet), underground. The tunnel now houses the current LHC.

The LHC was not the only kid in town. It was among several other particle accelerators in the world. One of the obstacles to the approval of LHC had been the Superconducting Super Collider (SSC) in the United States, which was approved in 1987 (Wojcicki, 2008). The SSC was planned to have 20 TeV (trillion electron volts) with a centre-of-mass energy of 40 TeV much more than 7 TeV per beam proposed by the LHC. The SSC was almost three times more powerful than what could ever be built at CERN at that time (Smith, 2014).

It was only the resilience and conviction of Carlo Rubbia, former Director General of CERN and 1984 Nobel Prize winner in physics for the discovery of the W and Z bosons that kept the project alive. Carlo Rubbia, who became Director General of CERN in 1989, argued that, despite its energy disadvantage, the LHC could be competitive with the SSC by having a luminosity of an order of magnitude higher than what the SSC could achieve at a fraction of the cost. It was argued that, in addition to colliding protons, the LHC would be able to accelerate and collide heavy ions at no extra cost.

As explained in an article by one of CERN's former Director Generals, Christopher Llewellyn Smith (Smith 2014), on the genesis of the Large Hadron Collider, the LHC had a scientific, technical, and political genesis. From a political perspective, the eventual cancellation of the SSC in 1993 strengthened a case for the development of the LHC (Wojcicki, 2009). However, the financial climate in Europe at the time was not conducive to the approval of such a large and expensive project in Europe. CERN's biggest financial contributor, Germany, was struggling with the costs of reunification and many other countries were trying to get to grips with the problem of meeting the Maastricht Treaty (European Monetary Union, 1992) for the introduction of the single European currency.

The cost of the LHC was a major concern. During the course of 1993, an extensive review was made in order to reduce the cost as much as possible, although a detailed cost estimate was particularly difficult to establish since much of the research and development on the most critical components was difficult to determine (Smith, 2014). In December 1993, a plan (CERN, 1993) was presented to the CERN Council to build the machine over a ten-year period, while reducing CERN's experimental programme to the absolute minimum, with the exception of the full exploitation of the LEP collider.

Although the plan was generally well received, it became clear that two of the largest contributors, Germany and the United Kingdom, were very unlikely to agree to the budget increase required. They also managed to get the CERN Council voting procedures changed from a simple majority to a double majority, where increased weight was given to the large contributors so that they could keep better control of the budget. On the positive side, after the cancellation of the SSC, the US panel on the future of particle physics (Drell, 1994) recommended that 'the government should declare its intentions to join other nations in constructing the LHC'. Positive endorsements were also being received from India, Japan, and Russia.

In June 1994, the proposal to build the LHC was made once more. The CERN Council adopted a very unusual procedure in which the vote on the Resolution was opened so that countries in a position to vote could do so, but neither the vote nor the Council Session was closed (CERN, 1994b). Seventeen-member states voted to approve the project.

In 2019, the CERN council adopted new rules of procedure which allowed each Member State to appoint one or two delegates to represent it. This is also known as the double vote method, the approval for the LHC was thwarted by Germany and the UK. They wanted significant additional contributions from the two host governments, France and Switzerland, claiming they received disproportionate returns from the CERN budget. They also requested that financial planning should proceed under the assumption of 2% annual inflation, with a budget compensation of 1%, essentially resulting in a 1% annual reduction in real terms, making the LHC project difficult to take off.

To address this new constraint and for cost savings, CERN was forced to propose a 'missing magnet' machine, in which only two-thirds of the dipole magnets required to guide the beams on their quasi-circular orbits would be installed in the first stage, allowing the machine to run with reduced energy for a number of years before upgrading to full energy.

The proposal was put before the CERN Council in December 1994. After a round of intense discussions between France, Switzerland, Germany, and the UK, the deadlock concerning extra host-state contributions was settled, when France and Switzerland agreed to make extra voluntary contributions. In the 100th Session of Council, the project was finally approved (CERN, 1994) for two-stage construction, to be reviewed in 1997 after the size of the contribution offered by non-member states interested in joining the LHC programme would be known.

Negotiations with France and Switzerland were couched in diplomatic language in the *Considerata* of the Council Resolution: '(The CERN Council) and it noted with gratitude, the commitments of France and Switzerland to make voluntary contributions to help and accelerate the LHC Project'. The negotiations and political wrangling for the LHC project were difficult and they were hard fought. This was to be expected for a large-scale and long-term project like the LHC, which required financial investments from several countries.

There followed an intense round of negotiations with potential international contributors. The first country to declare a financial contribution was Japan, which became an observer to the CERN Council in June 1995. The declaration from Japan was quickly followed by India and Russia in March 1996 and by Canada in December.

A final sting in the tail came in June 1996 from Germany, which unilaterally announced that, in order to ease the burden of reunification, it intended to reduce its CERN subscription by between 8% and 9%. It proved impossible to confine this reduction to Germany. The UK was the first to demand a similar reduction in its contribution in spite of a letter from the UK Minister of Science during the previous round of negotiations stating that the conditions were 'reasonable, fair and sustainable'. The only way out was to allow CERN to take out loans, with repayment to continue after the completion of the construction of the LHC. In December 1996, single-stage construction of the LHC was approved by the CERN Council, accompanied by a large cut in the budget (and also a one-year reduction in CERN salaries), on the basis of the contributions offered by non-Member States (Smith, 2014).

Ultimately, the LHC was built from 1998 to 2008 in the circular tunnel of the former Large Electron-Positron (LEP) collider at CERN with a 27-km circumference, 50–175 metres (average 100 m) deep underground beneath the French–Swiss boarder northwest of Geneva. The total cost of the LHC was reported to be 4332 million Swiss Francs (CH), out of which 3756 million of CH was for the machine and areas, 493 CH million for detectors and detector areas and 83 million was CERN's share of LHC computing (CERN, 2021). Figure 3.1 outlines the key components of LHC.

Following a recommendation from the US panel, and in preparation for a substantial contribution to LHC, the US Department of Energy, responsible for particle physics research, carried out an independent review of the LHC project (Department of Energy (DOE), U.S, 1996). DOE found 'the accelerator-project cost estimate of 2.3 billion in 1995 Swiss francs, or about $2 billion USD, to be adequate and reasonable'.

Moreover, DOE declared that 'Most important of all, the committee found that the project has experienced and technically knowledgeable management in place and functioning well.' The DOE noted that 'the LHC will be the latest and largest in a series of colliders built with superconducting magnets, beginning with the Tevatron at Fermilab. Support from the USA was followed by HERA in Germany,

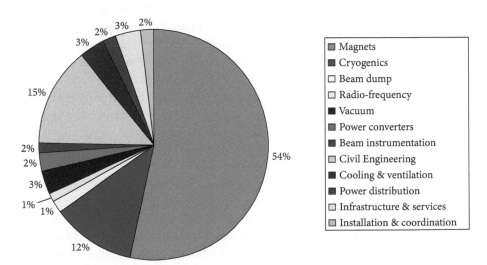

**Figure 3.1**  Key technology components of the LHC

*Source:* 'The Large Hadron Collider: a Marvel of Technology', 1st ed. Evans Lyn (ed.), Lausanne: EPFL Press, (2009).

soon to be joined by RHIC at Brookhaven National Laboratory' (DOE, 1996: ii). The DOE report also concluded that the strong management team, together with the CERN history of successful projects, gives the committee confidence in the successful completion of the LHC project. In December 1997, at a ceremony in Washington in the splendid Indian Treaty Room of the White House Annex, an agreement was signed between the Secretary of Energy and the President of the CERN Council.

After a shaky start and a mid-term hiccup, the LHC project has proceeded reasonably smoothly to completion. The LHC is a fine example of international collaboration with European leadership in high energy physics.

Although the LHC is CERN's largest and most powerful accelerator, it should be noted that CERN has a system of accelerators with a complex network of beam lines that feed particles from one accelerator to the next in order to ramp up their energy along the way. Before reaching the LHC, protons travel from the source down a linear accelerator known as Linac2 (Linac2 had been replaced later by Linac4) and then through a series of more accelerators known as Proton Synchrotron Booster (PSB), the Proton Synchrotron (PS) and the Super Proton Synchrotron (SPS) and finally the LHC.

After the operation of the LHC in 2008, there were nine major experiments installed/performed. ATLAS and CMS experiments were the largest and others smaller experiments included: ALICE, LHCb, TOTEM, MoEDAL, LHCf. The newest two LHC experiments are FASER and SND@LHC, which are situated close to the ATLAS collision point in order to search for light new particles and to study neutrinos. In addition, the CERN Control Centre (CCC) is an integral part of the LHC accelerator chain. The experiments have their own control rooms. The CCC combines the control rooms of the accelerators, cryogenics, and technical infrastructure necessary to provide high energy particle beams for the experiments. Figure 3.2 outlines the accelerator system including the LHC and experiments. All these work as a complex system in which all parts are well connected and operate harmoniously.

## 3.4  The Design of the LHC

The design of the LHC is a synchrotron storage ring that consists of two evacuated pipes passing through a ring of magnets where the magnetic field can be kept constant. The different types of magnets of the LHC accelerators are listed in Table 3.1. It allows charged particles to circulate in the ring indefinitely. In the LHC these storage rings can operate at energies up to 7.7 TeV per beam (LHC Study Group, 1991). The LHC has several innovative design features.[1] It uses superconducting magnets to create fields to guide particles in a circular path; it is designed to accelerate protons (almost at the speed of light), allowing high energy particle collisions; it uses high

---

[1] For a detailed discussion of accelerators and detectors, see M. Stephen and S. Herwig (eds), *Particle Physics Reference Library, Volume 3: Accelerators and Colliders*, Springer Open, https://link.springer.com/book/10.1007/978-3-030-34245-6.

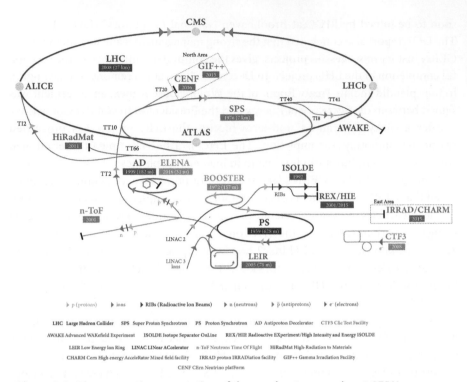

**Figure 3.2** Diagrammatic presentation of the accelerator complex at CERN
*Source:* © CERN

**Table 3.1** Magnet types

| Type | Number | Function |
|------|--------|----------|
| MB | 1232 | Main dipoles |
| MQ | 392 | Arc quadrupoles |
| MBX/MBR | 16 | Separation and recombination dipoles |
| MSCB | 376 | Combined chromaticity and closed orbit correctors |
| MCS | 2464 | Sextupole correctors for persistent currents at injection |
| MCDO | 1232 | Octupole/decapole correctors for persistent currents at injection |
| MO | 336 | Landau damping octupoles |
| MQT/MQTL | 248 | Tuning quadrupoles |
| MCB | 190 | Orbit correction dipoles |
| MQM | 86 | Dispersion suppressor and matching section quadrupoles |
| MQY | 24 | Enlarged-aperture quadrupoles in insertions |
| MQX | 32 | Low-beta insertion quadrupoles |

precision sensors to monitor the position and quality of the particle beam; it uses compact design to minimise the distance particles need to travel and collide with the right particles and allowing innovative thinking for future upgrades, expansion and uses of the LHC.

## 3.4.1  Machine Layout

In parallel with the approval of the LHC machine, proposals for the experimental programme were being examined by the LHC Experiments Committee (LHCC), whose job it was to give advice to the CERN management and through it to the Council. Unlike the LHC machine, the detectors have considerable independence. Only 20% of their funding comes through CERN. The rest comes from collaborating institutes all around the globe. However, it is the responsibility of CERN to provide the infrastructure, including the caverns in which the experiments are housed. Eventually, the LHCC proposed the approval of two large general-purpose detectors: ATLAS and CMS; as well as two smaller, more specialised detectors, ALICE for heavy-ion physics and LHCb for the study of matter-antimatter asymmetry (see Chapter 2 for more about the LHC experiments).

## 3.4.2  Civil Engineering Issues

The first job was to decide where these detectors were to be located. The LHC ring is segmented into eight identical arcs joined by eight 500-m Long Straight Sections (LSS) labelled from 1 to 8 (see Figure 3.3). Four of these LSS (at Points 2, 4, 6, and 8) already contain experimental caverns in which the four LEP detectors are located. These caverns are big enough to house the two smaller experiments.

ATLAS and CMS required bigger caverns, where excavation had to start while the LEP was operational; therefore, the four even points were excluded. Point 3 lies in a very inhospitable location deep under the Jura Mountains and for various reasons, Point 7 could also be excluded. There remained Point 1, conveniently situated opposite the CERN main campus and diametrically opposite to Point 5, the most remote of all. There was considerable pressure from both ATLAS and CMS collaborations to get the more convenient Point 1.

In the end, geology prevailed. Sample borings showed that Point 1 was much better suited for the larger cavern required for ATLAS. CMS was allocated Point 5. ALICE re-used the large electromagnet from one of the former LEP experiments at Point 2 and LHCb was assigned the cavern at Point 8. Therefore, ATLAS detector is located at Point 1 and CMS at Point 5, which also incorporates the small angle scattering experiment TOTEM. Two more detectors are located at Point 2 (ALICE) and at Point 8 (LHCb), which also contain the injection systems for the two rings. The beams only cross from one ring to the other at these four locations.

The excavation of the large caverns at Points 1 and 5 posed different problems and complexity. At Point 1, the cavern is the largest ever excavated in such ground conditions. At Point 5, although the exploratory borings showed that there was a lot of ground water to be traversed when sinking the shaft, the speed of the water flow took the project team by surprise. Extensive ground freezing was necessary to produce an ice wall around the shaft excavation.

An additional complication at Point 5 was that during the preparation of the worksite, the foundations of an ancient Roman farm (fourth century AD) were discovered.

## LHC LAYOUT

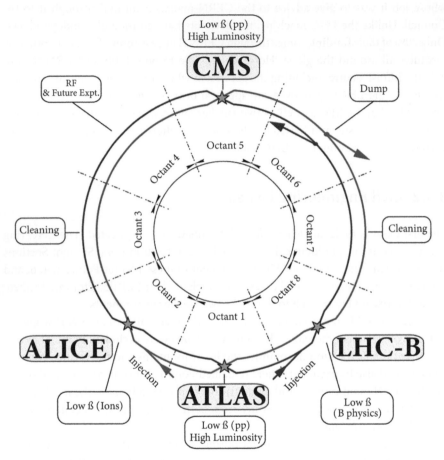

**Figure 3.3** Schematic layout of the LHC
*Source:* © CERN

Work was immediately stopped so that the mandatory archaeological investigation could be conducted (Evans, 2018). A third civil engineering work package was the construction of two 2.6 km long tunnels connecting the SPS to the LHC and the two tunnels leading to the beam dump caverns.

### 3.4.3  Machine Utilities

Once the four straight sections were allocated to the detectors, the other four could be assigned to the essential machine utilities.

Figure 3.4 shows a schematic layout of the LHC ring. The two beams cross from one ring to the other at the four collision points 1, 2, 5, and 8; elsewhere, they travel in separate vacuum chambers. They are transported from the SPS through two 2.6 km long tunnels, TI2 and TI8. Due to the orientation of the SPS with respect to the LHC, these tunnels join the LHC ring at Points 2 and 8. It was therefore necessary to integrate the injection systems for the two beams into the straight sections of the ALICE and LHCb detectors.

Clockwise from Point 2, the long straight section at Point 3 lies deep below the Jura Mountains. It contains no experimental cavern from the LEP days and, moreover, it is known from the experience of excavating the LEP tunnel that the geological conditions in this region are very bad. Cracks and fissures in the rock allow water to percolate from the very top of the mountain, more than 1000 m high, producing a large static water pressure.

In view of this it was decided that no additional civil engineering for tunnel enlargement would be allowed in this region. It was therefore assigned to one of the two-collimation systems, which could be fitted into the existing tunnel.

The LHC was divided into eight octants, each including a straight section (insertion region), used for the purposes indicated in Figure 3.4.

In most of the LHC, the two beams are located side by side, with a separation of 194 mm. Detectors are foreseen at four of the insertion regions, where the

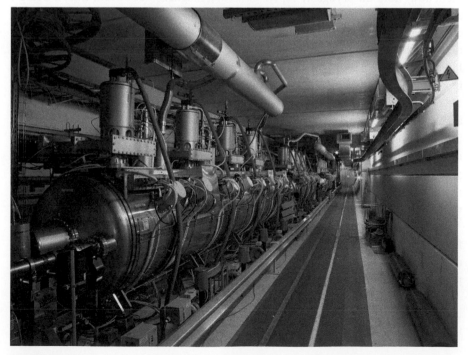

**Figure 3.4**  The superconducting LHC radio frequency cavities
*Source:* © CERN

beams cross in a symmetric fashion, such that the two beams go the same distance around the ring.

A perfect collimation of beams is essential in a collider. As the beams are stored for many hours, a halo of particles slowly builds up around the core, mainly due to nonlinearities in the magnetic field or by the interaction of one beam with the other. If it were left uncontrolled, eventually particles would hit the vacuum chamber wall, producing an unacceptable background in the detectors, and risking a *quench* (a transition from the superconducting state due to the accompanying temperature rise) in some of the magnets.

Collimators are specially designed motorised blocks that can be driven into the machine aperture to 'clean' the beam by removing the halo locally. The collimators constitute the primary aperture restriction in the machine. When they are in their operating positions, the machine aperture is just a few millimetres.

Two counterrotating proton beams are circulating around the 27 km ring more than 11,000 times per second. Each is accelerated up to a top energy of 7 TeV, the energy at which they are brought to collision at four points to generate showers of particles that are recorded by ATLAS, CMS, ALICE, LHCb, and other detectors of smaller experiments like TOTEM.[2]

After a few hours of operation, the colliding beams need to be disposed of to allow a new fill for physics. Operators in the CERN control centre instruct beam-transfer equipment to shunt the circulating beams into external trajectories that transport them away from the cryogenic superconducting magnets. Each beam exits the ring and travels in a straight line for 600 metres before arriving at a compact cavern containing a large steel cylinder approximately 9 m long, 70 cm in diameter, and containing approximately 4.4 tonnes of graphitic material in its centre (Calviani, 2021). Injection of the two beams of protons into the LHC from the existing, so-called LHC injector accelerator complex takes about 7 minutes per beam (then a total of 14 minutes for the two beam injectors both clockwise and anticlockwise), followed by an acceleration to full energy taking about 20 minutes. The beams will then be brought into collision and made to collide for several hours while the detectors record selected interactions. After several hours, the interaction rate is significantly reduced due to proton–proton collisions, beam cleaning by collimators, or beam rest-gas interaction. At that time, the remaining protons, will be ejected (dumped) and the LHC refilled.

Including the time to tune up the injection system, the interruption of data-taking for refilling may take as little as two hours. The beams are stored at high energy for about 10 hours, the so called 'beam lifetime', and particles will have made about 400 million revolutions around the machine.

The energy stored in the superconducting magnet system exceeds 10 GJ and each beam has a stored energy of 362 MJ. This total beam energy at top energy may cause

---

[2] TOTEM—TOTal cross section, Elastic scattering and diffraction dissociation Measurement at the LHC is a physics programme dedicated to the precise measurement of the proton–proton interaction cross section.

major damage to accelerator equipment in the case of an uncontrolled beam loss. The safe operation of the LHC therefore relies on a complex system for equipment protection, the so-called machine protection system. The systems for the protection of the superconducting magnets in case of a quench must be fully operational before powering the magnets (Schmidt, 2016).

Point 4 is assigned to the all-important Radio Frequency (RF) acceleration system. Acceleration is obtained by a longitudinally oscillating electric field at a frequency of 400 Megahertz (MHz) in a set of resonant cavities. The electric field in the cavities is very high, in excess of 5 million volts per metre.

Superconductivity came to the rescue. The cavities are made of copper but there is a thin film of niobium deposited on the inside surface. When cooled with liquid helium, this film becomes superconducting, enabling currents to flow through the cavity walls without loss.

With each revolution, the beam is given a small increase in energy as long as the field is pointing in the right direction. To achieve this, the frequency of the RF must be a precise harmonic of the revolution frequency so that each time a particle comes around, the field is pointing in the same direction. As the energy slowly increases, the magnetic field must also rise to keep the beams in the centre of the vacuum chamber since the magnetic field required to bend a particle on a constant radius is proportional to its energy. The RF system needs considerable infrastructure and profits fully from the space available in the old LEP cavern at Point 4 (see Figure 3.4).

As mentioned before, at 7 TeV with nominal intensity, the stored energy in one of the beams is 362 Mega Jules (MJ), equivalent to more than 80 kg of TNT. For any reason, if this beam is lost in an uncontrolled way, it can do considerable damage to machine components, resulting in months of down-time. Beam-intercepting devices have to withstand extremely high mechanical and thermally induced stresses. It is therefore essential to have a system that can reliably extract the beams very quickly and deposit them on special absorber blocks. This 'beam-dump' system is located at Point 6. A set of special magnets can be pulsed very rapidly to kick the whole beam out of the machine in a single turn. In general, the energy deposited in beam-intercepting devices is directly proportional to the beam energy, its intensity, and the beam-spot size, as well as to the density of the absorbing material (Calviani, 2021).

The LHC Beam Dumping System is meant to ensure a safe beam extraction and deposition under all circumstances. The system adopts redundancy and continuous surveillance for most of its parts. The beam dumping action is performed to reduce the risk of a faulty operation at the subsequent dump trigger. In order to secure safety, LHC processes carry out redundancy, surveillance, and diagnostics to achieve the required safety level (Carlier et al., 2005). Many sources can trigger the beam dump, for instance if an excessive beam loss on the collimators is detected or if a critical power supply fails. It is also used routinely during operation; when the intensity in the beams falls too low the beams are 'dumped' by the operators in order to prepare the machine for the next filling cycle. The LHC was designed to

**Figure 3.5** The LHC 'inner triplet' in the long straight sections left of Point 1 (ATLAS)
*Source:* © CERN

withstand some 20,000 such cycles in 20 years lifetime, as well as 20–30 full thermal cycles.

### 3.4.4 The Inner Triplet

The long straight sections on each side of the four detectors house the magnets needed to bring the beams together into a single vacuum chamber and to focus them to a small spot with a radius of about 15 microns at the collision points inside the detectors (Figure 3.5). This requires special elements and is a prime example of international collaboration in the machine construction. The superconducting magnets required to focus the beams were built in the USA and Japan, with the Japanese magnets shipped to the USA for integration into their cryostats before delivery to CERN.

The special dipoles used to bring the two beams into the same orbit were built in Brookhaven in the USA and the current feed boxes for all superconducting elements in the straight sections come from FERMILAB. Other equipment in these long straight sections comes from India and Russia.

As shown in Figures 3.5 and 3.6, the orange cryostats contain quadrupole magnets, which focus the beams to a 30-micron spot at the interaction point.

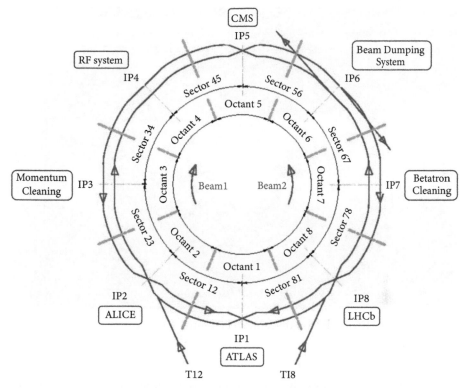

**Figure 3.6** Illustration of the LHC accelerator layout and beam collision
*Source:* © CERN

## 3.5 Superconducting Magnets and Cryogenics

The 27-km circumference of the former LEP tunnel (Figure 3.7) was a major challenge to the machine builders. The maximum energy attainable in a circular machine depends on the product of the bending radius of the dipole magnets that have to keep the particles on their trajectory at their end energy and the maximum field strength attainable. Since the bending radius is constrained by the geometry of the tunnel, the magnetic field should be as high as possible. The field required to achieve the design beam energy of 7 TeV, is 8.3 Tesla, which is about 60% higher than that achieved in previous machines. This pushed the design of superconducting magnets and their associated cooling systems to a new frontier.

The next constraint was the small (3.8 m) tunnel diameter. The LHC is not designed as one but two machines (Figure 3.8). Superconducting magnets occupied a large space. To keep it cold, it must be inserted into a vacuum vessel and well insulated from external sources of heat.

Due to the small transverse size of the tunnel, it would have been impossible to fit two independent rings into the space. Instead, a novel and elegant design with

**Figure 3.7** The LHC installed in the former LEP tunnel
*Source:* © CERN

the two rings separated by only 19 cm inside a common magnet yoke and cryostat was developed. This was not only necessary on technical grounds but also saved a considerable amount of money, some 20% of the total project cost.

Finally, the re-use of the existing injector chain governed the maximum energy at which beams could be injected into the LHC. Commenting on the ingenuity of scientists involved in the LHC design, the OECD (2014: 32) reported:

> CERN engineers and administrators drew up all of the specifications, purchased raw materials, delivered them to selected manufacturers, and received the resulting components which they then provided to other contractors for further processing or assembly. For example, once the specifications for the all-important niobium titanium cables were finalised, CERN placed orders for the necessary raw materials (including the very special variety of copper which makes up a significant fraction of the cable mass), delivered them to the cable manufacturing company, and then provided the finished cable to the three main contractors who wound the coils and assembled the 'cold masses' i.e. (coils+collars+yokes+numerous smaller magnet components) that were delivered to CERN.

There were many components that were made outside of CERN with the close involvement of CERN, delivered, and installed at various stages. At the heart of the LHC is the superconducting magnet system and the associated cryogenics. The magnet and cryogenics systems constitute 66% of the estimated LHC accelerator project cost, with the 1232 main bending magnets (a '2-in-l' design with twin-tore

## LHC DIPOLE : STANDARD CROSS-SECTION

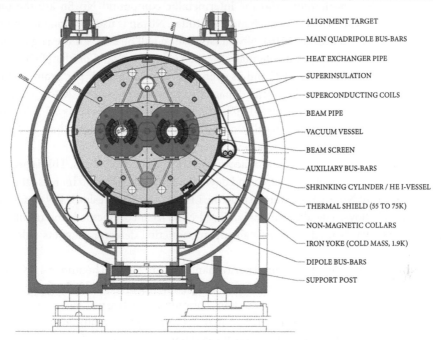

ALIGNMENT TARGET

MAIN QUADRIPOLE BUS-BARS

HEAT EXCHANGER PIPE

SUPERINSULATION

SUPERCONDUCTING COILS

BEAM PIPE

VACUUM VESSEL

BEAM SCREEN

AUXILIARY BUS-BARS

SHRINKING CYLINDER / HE I-VESSEL

THERMAL SHIELD (55 TO 75K)

NON-MAGNETIC COLLARS

IRON YOKE (COLD MASS, 1.9K)

DIPOLE BUS-BARS

SUPPORT POST

**Figure 3.8** A cross-section of the two-in-one LHC bending magnet. The two rings are concentrated inside a single vacuum vessel to save space (and money)
*Source:* © CERN

magnets) representing over half the magnet systems cost. The overall main bending magnet design was well established with the scope completed and the cost estimate was adequate. However, several time-consuming iterations were necessary to finalise the engineering details of the main bending magnets before production. The main dipoles need to operate at a much higher field (8.3 Tesla for 7 TeV energy) than in any previous machine. Besides the 1232 main dipoles there are about 5400 smaller magnets installed (quadrupoles, sextupoles, insertion quadrupoles, etc.) for orbit correction, insertion, separation, and recombination of the beams.

The first development stage of the collider was an energy level of 5 TeV per beam, ready for experiments in 2004. This was upgraded in the second stage to 7 TeV per beam in 2008. At 7 TeV, the proton beams had over seven times the energy of the world's present highest energy accelerator, the Tevatron at Fermilab. Unlike the Tevatron collider, which had a single beam pipe containing both a proton beam and an antiproton beam, the LHC needed two separate beam pipes. This was because both of the two LHC beams had a positive charge and the two counter-rotating beams needed oppositely directed magnetic fields. While this required separate magnetic channels for the two beams, it allowed for a higher beam intensity, and thus, higher interaction rates that can be obtained with protons compared to antiprotons.

This high field level can be achieved with two types of superconductors. The ductile alloy niobium-titanium and the intermetallic compound $Nb_3Sn$ are the only materials that can be used for such magnets today. $Nb_3Sn$ is used for magnets above 9 Tesla (Barzi and Zlobin, 2019). $Nb_3Sn$ fabrication will be used for the High Luminosity LHC (HL-LHC) and potentially for the Future Circular Collider (FCC). The 'future LHC' could be the FCC and its energy could reach 50 TeV per beam. A study was undertaken to install higher field magnets (between 16 ton and 20 ton magnets) in the present LHC tunnel (HE-LHC: High Energy LHC; towards 27 TeV collision energy). However, the installation of magnets weighing more than 16 ton in the LEP-LHC tunnel (3.8 m) was never an easy task.

Nb-Ti is a mature technology for accelerator magnets up to 9 Tesla. This is because a very high investment in research and development is necessary to increase the current density of $Nb_3Sn$. If we were to redo the LHC today at 8.3 Tesla CERN would still use Nb-TI alloys. The aluminium-stabilised Nb-Ti/Cu conductor is the traditional workhorse that is used for nearly all superconducting detector magnets (Mentink et al., 2023).

$Nb_3Sn$ could reach the required performance in supercritical helium at 4.5K, but it is mechanically brittle and costs at least five times as much as Nb-Ti. It was therefore excluded for large-scale series production. The only alternative was Nb-Ti, but it must be cooled to 1.9K, below the lambda point of helium to get the required performance. This requires a very innovative cryogenic system.

Superconducting cables are required to provide the magnets with the necessary electrical current while avoiding any heating of the magnets. The superconducting cable is made of strands of wire, about 1 mm in diameter and composed of one-third superconducting material and two-thirds copper. The Nb-Ti filaments are 6–7 μm in diameter and precisely positioned with a 1 μm separation in the copper matrix. They are produced by multiple co-extrusions of Nb-Ti ingots with copper rods and cans. The strands and multi-strand cables are shown in Figure 3.9.

It is of interest to make the dipoles as long as possible in order to:

- to reduce the number of units and interconnects, and therefore the cost; and
- to maximise the filling factor to reduce the magnetic field required for a given energy.

For a given circumference, the energy of the accelerator is given by the field of the dipoles and the integral of their length. Hence, the interest is to minimise the number of interconnections and their length.

A number of practical factors, including the road transport of magnets and the facility of installation put an upper limit on their length. The maximum length of 15 metres of the main dipole was determined by the transport. It is not possible to transport objects longer than 15 metres on a massive scale on the roads (for the same reason, the FCC magnets are also designed with a length of 15 metres).

The final magnets have a magnetic length of 14.3 metres with a physical length of 15 metres. The arcs of LHC lattice are made of 23 regular arc cells, each of which is 106.9 metres long. This regular lattice period is made out of six dipoles and two 3 metres long quadrupoles per period (Figure 3.10). The ends of the dipoles contain the small octupole and decapole correctors to control unwanted multipoles in the dipoles, especially in the 'snapback' regime at the start of acceleration when persistent currents cause strong nonlinearities.

The mechanical forces in the dipole are very large, up to 300 tons/metre pushing the coils outwards at full power. These forces are contained by strong non-magnetic steel collars surrounded by an iron yoke and stainless-steel cylinder. Several other technical options are possible but this option was the best after an optimisation process. Series production of dipoles and quadrupoles has been a monumental task. It

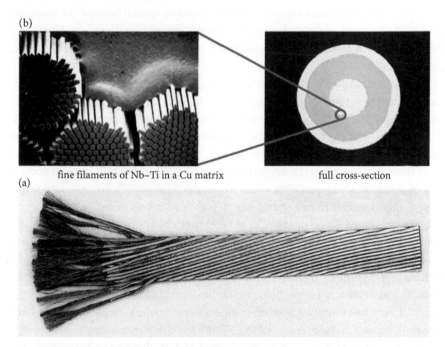

(b)

fine filaments of Nb–Ti in a Cu matrix                    full cross-section

(a)

**Figure 3.9** LHC superconducting cables

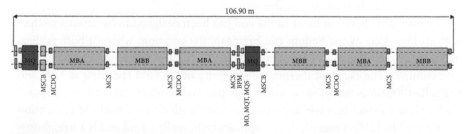

**Figure 3.10** Schematic layout of the LHC arc corrector magnets
*Source:* © CERN

seems obvious that the production of 1232 dipoles of 15 metres in length weighing 35 tons and 400 quadrupoles is a monumental task with all strict specifications.

All superconducting cables and many mechanical components were supplied to the cold mass assemblers (three for the dipoles and one for the quadrupoles) by CERN in order to ensure uniformity of production and to allow control of the distribution of contracts between countries. The cold masses were assembled into their cryostats at CERN.

All magnets were tested at 1.9K before installation in the tunnel. The total number of cryogenic magnet assemblies, also known as cryogenic magnets, consists of 1232 dipoles with correctors, 360 short straight sections (SSS) for the arcs with quadrupoles and integrated high-order poles, and 114 special SSS for the insertion regions (IR-SSS) with magnets for matching and dispersion suppression. All of these magnets had to be tested at low temperatures before they could be installed in the tunnel.

The collaboration with India resulted in the completion of cold testing of 1706 superconducting magnets for the LHC in 2013. Superconducting magnet testing had several aspects. For each magnet the tests had to verify the integrity of the cryogenics, mechanics, and electrical insulation; qualify the performance of the protection systems; train the magnet up to the nominal field or higher; characterise the field; ensure that the magnet met the design criteria; and finally accept the magnet according to its performance in quenches and in training (CERN, 2007). From start to finish, production, from cable to fully tested magnets took about six years.

The magnets are cooled by eight large helium refrigerators, each with a nominal rating of 18 kW at 4.5 Kelvin. This was claimed to be the largest cryogenic system in the world, not just the largest with superfluid helium. The design of this system was based on past experience with accelerators, as well as the Tore Supra tokamak. At 1.9 Kelvin, liquid helium is a superfluid, which has highly efficient heat transfer properties and very low viscosity, providing an excellent cooling medium for the LHC magnets. The total mass to be cooled to 1.9K is 42,000 tons, requiring approximately 130 tons of superfluid helium to be maintained at 1.9K during the entire period of operation. The main reason for operating in a superfluid was to extend the operating range of the Nb-Ti superconductor. However, operating below the lambda point brings its own advantages and challenges. The rapid drop in the specific heat of the conductor at low temperatures makes it imperative to use the special properties of superfluid helium in the best possible way.

The insulation between turns in the coil has been designed to be porous so that, with its low viscosity the helium can permeate the windings where it buffers thermal transients thanks to its high specific heat (2000 times that of the conductor per unit volume). The excellent thermal conductivity of the fluid (peaking at 1.9K and typically 1000 times that of oxygen-free high thermal conductivity (OFHC) copper) enables it to conduct heat without mass transport with no need for fluid circulation or pumps. The LHC cryogenic design appears technically sound and is a straightforward extrapolation of past experience with collider technology. The cryogenic costs were estimated to be 15% of the accelerator project.

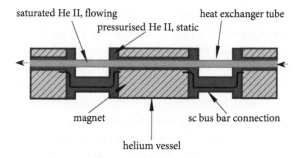

saturated He II, flowing

pressurised He II, static

heat exchanger tube

magnet

sc bus bar connection

helium vessel

**Figure 3.11**  Schematic of LHC magnet cooling scheme

*Source:* Collier, Paul (2015). The Technical Challenges of the Large Hadron Collider, *Philosophical Transactions of the Royal Society A.* 3732014004420140044. https://doi.org/10.1098/rsta.2014.0044. ©

The magnets operate in a static bath of superfluid at atmospheric pressure using an unconventional cooling scheme. The bath is continuously cooled through a linear heat exchanger made out of cryogenic grade copper and extends the full 107 m length of each cell (Figure 3.11). The pressure inside the heat exchanger is 15 mbar. Helium expands into the tube through a Joule-Thomson valve and is cooled to 1.8K. The static helium in the magnets is then cooled by latent heat from the vaporisation of the small quantity of superfluid inside the heat exchanger. This scheme worked beautifully, keeping the LHC temperature stable for long periods.

## 3.6  Vacuum System in the LHC

The vacuum is crucial to maintaining an airy and dust free environment for the beam of particles to travel unobstructed. Electromagnets steer and focus the beam of particles while it travels through the vacuum tube. A very high vacuum is needed because collisions of beam particles with gas molecules remaining in the beam tubes would reduce the beam's lifetime. This must be accomplished in spite of the effect of the synchrotron radiation emitted by the protons—this radiation strikes the walls and desorbs gas molecules, which must be given a means of migrating to an area shielded from the radiation.

The three primary vacuum systems required for the LHC consist of the cold beam vacuum in the cold bore tubes of the arc magnets, the cryostat insulating vacuum system for the cryogenic magnets, and the warm straight section vacuum in the straight sections.

With the first start-up of beams in 2008, the LHC became the biggest operational vacuum system in the world. It operated at different levels of pressure and used an impressive array of vacuum technologies. With a total of 104 kilometres of piping under vacuum, the vacuum system of the LHC is regarded as the largest in the world. The insulating vacuum, equivalent to some $10^{-6}$ mbar, was made up of an impressive 50 km of piping, with a combined volume of 15,000 cubic metres. Building this vacuum system required more than 250,000 welded joints and 18,000 vacuum seals.

The remaining 54 km of pipes under vacuum are the beam pipes, through which the LHC's two beams travel. The LHC's vacuum systems are fitted with 170 Bayard-Alpert ionisation gauges and 1084 Pirani and Penning gauges to monitor the vacuum pressure (CERN, 2023).

At 7 TeV, even protons start to produce synchrotron radiation. The power emitted was about 4 kW per beam. It was much too low to provide useful synchrotron radiation damping but is quite a nuisance since it must be absorbed on the cold surface of the beam pipe. The most important multi-bunch effect in the LHC is transverse resistive wall instability. Its growth rate is proportional to the square root of the resistivity of the beam pipe and to the inverse cube of its radius.

The instability exhibits no threshold behaviour but its growth rate can be reduced by coating the inside of the beam 5 screen with a 50 mm layer of copper and cooling it to below 30K where its resistivity is further reduced. One watt at 1.9K corresponds to a kilowatt at room temperature, which cannot be accepted by the refrigerators.

Therefore, the beam vacuum chamber contains a liner cooled to 20K in order to intercept the heat load with better thermodynamic efficiency. At this temperature, the cryo-pumping capacity is strongly reduced and it has been shown that gas, particularly hydrogen, desorbed from the body of the liner by the synchrotron radiation, accumulates on the surface and gradually deteriorates the vacuum.

This function is provided by the beam screen, a perforated tube centred within each magnet beam tube. The perforations allow the desorbed molecules to pass through the screen to the cold bore of the magnets, where they firmly adhere to the lower-temperature surface. By running at a higher temperature than the magnets (5 to 20 Kelvin, compared to 1.9 Kelvin) this screen also allows for more efficient operation of the helium refrigerator system. Other vacuum systems will provide the insulating vacuum required for the superconducting magnets, as well as the beam-tube vacuum in the warm straight sections of the accelerator.

The total cost of all vacuum systems was estimated at 79 million Swiss francs. There should be little cost risk for conventional insulating vacuum and warm vacuum systems since they are based on standard components. The cold beam vacuum system is the largest and most challenging, comprising over half of the total cost.

Even though the cold bore tube has a great capacity for cryo-pumping, the synchrotron radiation power that the beams emit and the resistive wall power loss of the beam tube could result in a significant heat load for the 1.9 Kelvin cryogenic system. Therefore, an intermediate beam screen—a perforated copper-lined tube centred within each magnet cold bore tube—was provided to intercept this power at a higher temperature; the screen is maintained at a temperature between 5 Kelvin and 20 Kelvin by gaseous helium flow. An analysis of the beam screen system was carried out to ensure the vacuum properties and mechanical properties could be maintained. A large-scale manufacturing capability was necessary for producing approximately 45 km of cold beam tubes and screens. Both systems, although very large in scale, were modelled on extending designs similar to those used at LEP and other accelerator systems throughout CERN.

## 3.6.1  Magnet Power

A large number of power converters are needed to provide current to the various magnets that guide and control the beams. The largest of these are eight current-source power converters located in pairs at the even-numbered insertion regions. Each of these converters powers a sector of 154 bending magnets connected in series. A ramp rate of 10 amperes per second while accelerating the beam requires 185 volts. At 7 TeV, about 10 volts will be needed to maintain the operating current of 11,500 amperes. These eight converters must track one another precisely. LHC is unique in the world for having eight independent sectors of 154 magnets in series.

It was a challenge and at the beginning, few people believed it possible. The challenge was to measure current up to 13,000 amps with an absolute precision of a few parts per million (ppm) and to design high precision current loops. The LHC's superconducting magnets are the pinnacle of high technology. The LHC's power converters are very different from those of the LEP or the SPS since the new accelerator's magnets are mostly superconducting. That means that they require much higher currents at a lower voltage since superconductors have no resistance to current flow (CERN, 2001).

However, to work, magnets need the help of high-precision power converters to supply them with extremely stable DC current. Perfection with an accuracy of just 1–2 parts per million (ppm) is required. The LEP, for the sake of comparison, could live with 10–20 ppm.

The main focusing and defocusing quadrupole magnets are separately powered. Many of the insertion magnets and correction magnets are powered in series as families of magnets, but some must be powered separately. In total, about 1,550 power converters are needed, supplying a total current of about 1,750 kA. They require a steady-state input power of about 19 MW, with a peak of 41 MW.

## 3.7  First Commissioning

By 10 September 2008, seven of the eight sectors had been successfully commissioned to 5.5 TeV in preparation for a Run[3] at 5 TeV. Due to lack of time, the eighth sector had only been taken to 4 TeV. Beam commissioning started by threading beam 2, the counter-clockwise beam around the ring, stopping it at each long straight section sequentially in order to correct the trajectory. In less than one hour, the beam had completed a full turn.

Very quickly, a beam circulating for a few hundred turns could be established. The decay in intensity is due to the debunching of the beam around the ring since the radiofrequency system was not yet switched on. Without the Radio Frequency (RF) capture, the beam debunches as it should in about 250 turns, or 25 msec.

The first attempt was made to capture the beam, but the injection phase was completely wrong. Adjusting the phase allowed a partial capture, but at a slightly

---

[3] The LHC's first run (Run-1) was between 2009 and 2013 at centre-of-mass energies ($\sqrt{s}$) between 900 GeV and 8 TeV and this was the first data taking period of LHC operation.

wrong frequency. Finally, adjusting the frequency resulted in a perfect capture and the closed orbit could then be corrected.

## 3.8 The Incident

In a sophisticated machine with various complex components which operate at high energy levels, hiccups and occasional breakdowns can be expected. It is important to note that, apart from the LHC engineers and physicists who work intensively to construct the LHC machine, there are several safety related services to ensure proper safety not only for the people working on the machine but also for the civilians living close by. The LHC has to have a plan for troubleshooting and recovery and must also have the necessary spare parts in stock. It is also important to reduce equipment downtime and get the machine into operation soon.

All these components have their own physics. If there is at least one component in the chain malfunctioning, there will be no beam. There are so many systems to work through—the vacuum pump, the cryogenic system, exchangers, the power converter, and so on.

Commissioning proceeded rapidly with the circulating beam in the other ring until 18 September 2008 when a transformer failed at Point 8, taking down the cryogenics in that sector. Since it was impossible to circulate the beam, attention turned to bringing the last remaining sector up to 5 TeV like the others.

On 19 September 2008, the last remaining circuit was being ramped to full field when, at 5.2 TeV a catastrophic rupture of a busbar occurred causing extensive damage in Octant 3–4 (refer to Figure 3.3). These busbars are connected by induction brazing with three layers of tin/silver solder in a copper box. Initially it was foreseen to clamp these busbars as well as the solder mechanically, but this was discarded on the grounds that it would increase the hydraulic impedance in the interconnect region and would therefore reduce the effectiveness of conduction cooling in the superfluid helium.

The investigation into this incident found a large helium leak into Octant 3–4 of the LHC tunnel. It was confirmed that the cause of the incident was a faulty electrical connection between two of the accelerator's magnets. This resulted in mechanical damage and the release of helium from the magnet's cold mass into the tunnel (CERN, 2008).

A fact-finding commission was established, where it was concluded that the most probable cause of the accident was too high a resistivity in one of the 10,000 superconducting busbar joints due to the omission of the solder. In a normal machine, this would have caused minor damage. However, the joint rupture resulted in an arc piercing the helium vessel. The resultant high pressure in the insulating vacuum and the volume of helium gas were too high for the rupture discs to take, resulting in overpressure and the displacement of magnets off their jacks. In total, 14 quadrupoles and 39 dipoles needed replacing.

One of the leaders recalled the incident as follows:

The most striking example in my career was the famous LHC incident at the very beginning of the operation. On 19th September 2008, during powering tests of the main dipole circuit in Sector 3-4 of the LHC, a fault occurred in the electrical bus connection in the region between a dipole and a quadrupole, resulting in mechanical damage and the release of helium from the magnet cold mass into the tunnel. Proper safety procedures are in force, the safety systems perform as expected, and no-one is put at risk.

   A big incident, ending up with almost a kilometre of the molten accelerator, no clue as to why the safety systems and protection measures did not work, and ten thousand physicists waiting to start to take data: it could have turned into a disaster. The construction teams were already dismissed, and the tooling to make magnets was already dismantled together with production lines: projects nowadays cannot afford margins and once construction ends, all is discontinued. And the temptation of a top-down approach was very well present also at CERN: myself at the centre of the trouble, I strongly defended the collaborative approach: no blaming and hunting for the faulty people, rather uniting the energy to find the fault. And thanks to the collaborative approach very large resources could be mobilised, both internally from collider teams and also getting help from experimental teams. It was so clear that we were in the same boat. If it sank, no one could survive. And with a real common effort a solution was found despite all difficulties so that we could finish the magnet repair in nine months, and one year after the incident the collider was operating again and in November 2009 LHC gained the record energy, taking the baton from the highest energy accelerator from Tevatron at Fermilab. And all with a modest extra-cost of about 2–3% of the total cost of the collider. The approach of particle physicists to large projects, seems more expensive at a first view, because of the overheads embedded in collaborations: however, it is a much more resilient model and avoids propagation and amplification of problems, which is so easy when each team feels responsible only for its own part, while charging cost and time for any supposed mishap coming from peer teams.

(Lucio Rossi, University of Milan, former HL-LHC Project Leader
Interviewed by the author, May 2021)

## 3.9  The Search for Further Defects

An urgent priority after the incident was to go through post mortem data to see if any precursors of the accident could be detected in particular, any anomalous temperature increase in the affected area.

   Detecting a temperature rise in the superfluid helium is difficult for two reasons. The first is the enormous thermal conductivity of superfluid helium. This provides good cooling of joints initially, but the thermal conductivity is a function of flux density, so as the heating increases, the cooling capacity quickly collapses, especially in the region of the splices with high hydraulic impedance.

Every magnet is equipped with a 'post mortem' card containing an ADC (Analogue to Digital Converter) and a buffer memory opening up the possibility of making ohmic measurements across each splice. All the joints in the dipole chains of Sectors 67 and 78 were powered during a stepwise current ramp to 5 kA. In Octant 6–7, there was one anomaly visible with a resistance of 47 nano-Ohms. It was possible to locate exactly which splice was responsible. Both the 100 nano-Ohm splice previously mentioned and the 47 nano-Ohm splice were inside magnets, which had already been tested to full current. They have both been removed and the bad splices have been confirmed. No other such splices have been detected anywhere else in the machine.

However, during the removal of damaged magnets it was discovered that in some instances, solder had been leaking out of the interconnect joints during brazing, weakening the joints in the event of a (very unlikely) busbar quench. Consequently, it has been decided to operate the LHC at reduced energy until additional consolidation be made during a shutdown. This consolidation consisted of strengthening the interconnects, increasing the number of rupture discs in sectors where it has not already been done and reinforcing the jacks at the vacuum barriers so that they can take higher differential pressure in case of a very unlikely further incident of this kind.

## 3.10  Recommissioning

The repairs and hardware recommissioning took until November 2009. In the short time available until the end of the year, beams were accelerated to an energy of 1.18 TeV, equivalent to a dipole field of 2 kA, and a small amount of physics data taking was done. On 30 March 2010, first collisions were obtained at a centre-of-mass energy of 7 TeV. Since then, operating time has been split between machine studies and physics data taking.

In view of the very large stored energy in the beams, particular attention was focused on the machine protection and collimation systems. More than 120 collimators are arranged in a hierarchy of primary, secondary, and tertiary collimators to ensure tight control of the beam orbits.

## 3.11  Operations

The machine's performance at this early stage was very impressive (Lamont, 2013). A beam lifetime of more than 1000 hours had been observed, an order of magnitude better than expected, proving that the vacuum was considerably better than expected. Moreover, the noise level in the RF system was very low.

The LHC operations spans across so many components, including the delivery of beams to the experiments. The operation of some of the large LHC experiments such as ATLAS and CMS experiments alone involves over 3500 institutions and closer to 6000 physicists and engineers and many postgraduates and technicians. The LHC has been designed to follow a carefully set out programme updates.

The LHC produces colliding subatomic particles, such as protons or heavy ions, for a period of time known as 'Run'. This run is followed by a period of shut down which is known as Long shut down (LS). Each Run and Shut down has a particular set of physics goals and objectives.

Run 1 of the LHC (2010–2013) took place first at 7 TeV at the centre of mass and then at 8 TeV (centre of mass, cm) in the final year of the Run (2012). This gradual increase in energy allowed us to learn to manage the energies stored in the magnets and to gradually increase the number and intensity of the proton bunches.

The long technical stop 1 or Long Shutdown 1 (LS1) in 2013 and 2014 allowed the interconnections between the dipole magnets to be consolidated. Nearly one-third of the people involved in the consolidation work were conducting only quality control and quality assurance work. During the long shut down in 2013–2014, the motive was 'safety first, quality second, and schedule third'. Afterwards the energy of the LHC could be safely increased to 13 TeV (centre of mass, cm).

LS1 also allowed in-depth maintenance to be carried out on the equipment after the long Run 1. Based on the training tests of the dipole magnets and in order to be able to restart quickly in 2015, it was decided that Run 2 (2015–2018) would take place at 13 TeV (centre of mass, cm).

The LHC Run 2 was very successful and allowed a rich harvest of collisions in the LHC experiments. The following two curves show the integrated luminosities during Run 1 and Run 2 in the two multipurpose ATLAS and CMS experiments (Figures 3.12 and 3.13).

As discussed above, the first Run for physics started in 2010 at 3.5 TeV per beam with 248 bunches per beam and a maximum luminosity of $10^{32} \mathrm{cm}^{-2}.\mathrm{s}^{-1}$. The integrated luminosity for the year was 50 pb$^{-1}$. In the second year of operation in

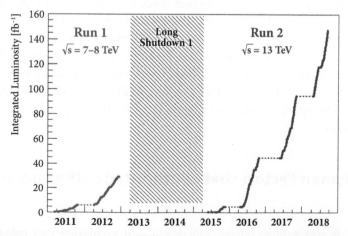

**Figure 3.12**  Run 1 and Run 2 energy levels after long shutdown 1
*Source:* © CERN

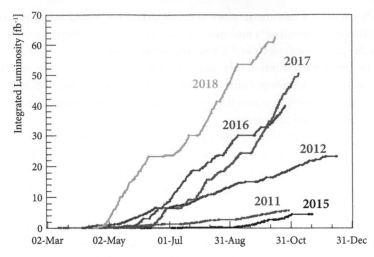

**Figure 3.13** Integrated luminosity increases during different phases
*Source:* © CERN

2011, still at 3.5 TeV the number of bunches per beam was increased to 1380 and the peak luminosity increased to $3.8 \times 10^{33}$ cm$^{-2}$s$^{-1}$, well beyond expectations.

The integrated luminosity in ATLAS and CMS was 5.6 fb$^{-1}$, enough to give the first hint of a signal of the Higgs boson (Figure 3.13).

In 2012, the energy was increased to 4 TeV and the number of bunches per beam to 1380, with a bunch separation of 50 ns. By the time of the summer conferences, a further 6.7 fb$^{-1}$ was accumulated, enough to confirm the existence of a new boson, announced in a seminar at CERN on 4 July 2012. By the end of the year, a total integrated luminosity of 23.3 fb$^{-1}$ was accumulated in both ATLAS and CMS, enough to measure the spin of the new particle and to confirm that it was indeed the Higgs boson.

The versatility of the machine has been amply demonstrated. In addition to the proton operation, dedicated runs have been made with heavy ion (Pb-Pb) collisions and even a run with lead on protons, a tricky procedure since the two beams must rotate on different orbits in order to have the same revolution frequency.

The LHC went into a shutdown in 2019 for an improvement of the machine and the detectors. With Covid-19 coming into play, the shutdown lasted until 22 April 2022. The collision energy for the Run 3 in July 2022 had been 13.6 TeV (i.e. 6.8 TeV per beam). This energy was chosen after an intensive training period for the main dipole magnets.

## 3.12 Human Factors that Contributed to the Success of LHC

The LHC is also a huge human experiment and a collision and collaboration of human minds. The LHC experiment brings together an unprecedented number of scientists, engineers, and technicians to work towards a common purpose.

A complex machine like the LHC is the culmination of human ingenuity, the synthesis of human vision and a technological marvel. Without human factors working seamlessly, such a complex machine would not have been invented. In Big Science organisations and experiments, the technical competence paired with an outstanding enthusiasm and team spirit of all work forces in combination with competent, transparent, and respectful leadership were the keys to success for making such a complex machine like LHC real. The project involved not only CERN employees where the LHC was built but also many teams across international organisations and research laboratories. Indeed, various parts, modules, and components were perfectly aligned and bought together as a system operating with perfect efficiency. Many minds from various nations were willing to collaborate and contributed. The thorough testing of all the LHC magnets by the Indian teams is just one of the numerous examples of the enthusiasm and endurance of all LHC collaborators.

To lead such an international work force that comprises engineers for work planning and logistics, quality control, civil engineering as well as engineers and physicists for accelerator technology and safety requires a strong, unanimously recognised personality, appreciated by colleagues on all hierarchical levels.

Leaders responsible for LHC worked tirelessly to build a common work culture based on technical competence, mutual respect, transparency, and fairness.

Effective communication among scientists, engineers, and technicians with a common language is a key feature of this LHC experiment. For this purpose, various forums for open discussions and information sharing had been introduced.

**Figure 3.14** Inside the LHC tunnel—a member of staff peddles inside the tunnel during maintenance in 2020

*Source:* Getty Images

This complex information and decision-making process culminated in the early set up of the LHC initiative, which took place in the Chamonix Workshop. Several hundred engineers and physicists met over the course of a week to exchange ideas and information. Each team had been invited to present the status and challenges of its work package to the Project Leader, the CERN Directors, and the Machine Advisory Committee. Even in carrying out routine maintenance and constant monitoring of the beam quality, humans have to be on alert and resort to basic human efforts (Figure 3.14).

All these contributions were published in the various proceedings of the LHC workshops. The Chamonix workshops were always a very efficient team building event where Project Leaders and Directors mingled with colleagues on all hierarchical levels while at the same time everybody was brought on the same level of information. Everybody felt they belonged to the same big LHC team, contributing, and committing to an outstanding, unique endeavour: the construction of the LHC machine.

On 10 September 2008, the endless joy of the entire team broke out when the first beam had been successfully circulated around the 27 km LHC accelerator. As several scientists admitted, it gave many of the direct witnesses the creeps, with weird feelings of a first moon landing. The construction, installation, and operation of a complex machine like the LHC required careful and strict management processes so that nothing was left to guesswork or laissez-faire approaches. The HL-LHC Project Management Structure is outlined in Figure 3.15.

Within the project, a Work Breakdown Structure (WBS) had to be applied to ensure that all tasks were performed in accordance with the work plans and the

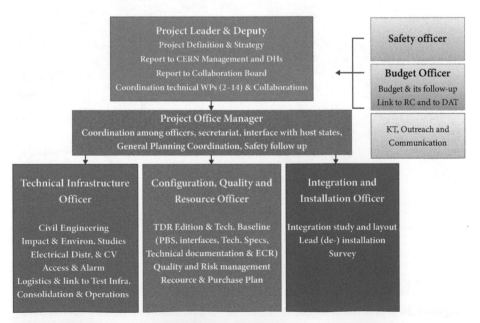

**Figure 3.15** HL-LHC project management structure
*Source:* Created by author

agreed schedule. The applied work package management was similar to the one used for HL-LHC and which is sketched in Figure 3.16. It was created using personal experience and ingenuity in combination with full Computer Assisted Design (CAD). The 3D designs and the following engineering specifications for different parts were conducted accordingly. Any changes had to go through an engineering change request (ECR) and all members of the team were consulted before the management team decided whether to carry out such changes or not.

Ultimately, the smooth operation of the LHC without incidents is the responsibility of the Project Leader. Soon after the first beams had been successfully circulated, the LHC Project leadership as well as the enthusiasm and commitment of the entire work force were soon subject to the next stress test: due to the incident on 19 September 2008 the operation of the machine had to be stopped for a long period. The identification of the root cause of the problem and its solution was a great challenge, the decision-making process needed to be well managed to minimise machine down time and cost overrun.

There were many excellent engineers who were capable of resuming operation of the machine; however, it needed to be done in a well-structured manner led by the Project Leader and his project team. The Project Leader had to face a difficult knowledge management process as he had to consider budget and costs besides the technical feasibility. Dealing with people and their opinions has always been a difficult but rewarding aspect of human resource and knowledge management.

CERN leaders in various situations have to have the knowledge and skills to analyse technical, economical, and operational realities, take decisions, and move on. In a complex machine like the LHC, operations are challenging and one cannot anticipate

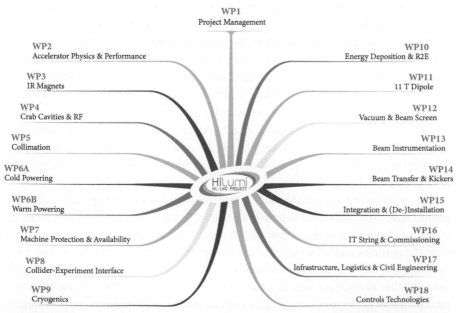

**Figure 3.16** The organisation of work packages in HL-LHC project management
*Source:* © CERN

all the complex challenges and unforeseeable problems that may arise. One has to anticipate the unexpected. Big Science is inherently risky and uncertain and the ability to build such a complex LHC machine is truly remarkable. Its success is squarely attributable to a large number of scientists and technical community, engineers, and technical people.

## 3.13  Conclusions

The Large Hadron Collider (LHC) is the world's largest and most powerful particle accelerator built over nine years to accelerate two beams of protons to almost the speed of light and then collide to smash particles allowing scientists to study the fundamental properties of matter. The LHC design work commenced back in 1984 and the first circulating beam in 2008 was a major achievement (Smith, 2014). The proton beams were smashed for the first time at a record high energy of 3.5 TeV on 30 March 2010. The discovery of the Higgs boson in 2012 is the most significant achievement of the LHC up to now. It is a significant scientific and engineering achievement that demonstrates how humans can collaborate to solve the world's most complex, unanswered questions and produce cutting-edge knowledge that benefits human understanding of the universe around us. The discovery of the Higgs boson in 2012 is the most significant achievement of the LHC up to now paving the way for revolutionary scientific discoveries.

The LHC is a complex machine, which operates in a relatively narrow 27-km circumference circular tunnel and 50–175 metres (average 100 m) deep underground. The tunnel criss-crosses the Swiss-Franco borders. The LHC contains more than 7000 superconducting magnets ranging from the 15 m long main dipoles to the 10 cm long octupole/decapole correctors inside the dipole cold masses to direct proton beams. These magnets are cooled with pressurised superfluid helium.

During the construction, design, and operation, the LHC team faced numerous engineering and technical challenges, that required innovative problem-solving. Effective communication and collaboration among accelerator physics scientists, designers, and engineers was vital. Industry inputs were essential to develop numerous components such as magnets, electronic and communication devices. The construction of the two caverns for the ATLAS and CMS detectors was a complex task having to navigate through civil engineering, archaeological sites, and deal with environmental and safety issues.

The engineering supervision and coordination of such a large-scale scientific project have to be performed to perfection. The technological complexity of the accelerator was extensive, as many powerful superconducting magnets to accelerate and steer the particles had to be placed in the tunnel with millimetre precision. The magnetic fields of the LHC dipoles represent almost twice the strength of those previously used in other particle accelerators. Industries were involved in the development and production of numerous components such as magnets, electronics, and communication devices, and organisational management of these relations was not easy.

The LHC accelerator faced many design, technical, and organisational challenges. One of the most important challenges of a circular accelerator is to bend a high-intensity proton beam circulating around the circumference of the machine, where the magnets and two beams have enormous amounts of stored energy in the magnet system at 7 TeV that exceed 10 GJ (giga joules). Such stored energy must be managed under all circumstances with a highly reliable magnet protection system. To achieve high collision rates in the LHC experiments (high luminosities), the key word is scientific precision. This applies to all systems for beam guidance and focusing (magnets and power converters), acceleration (RF system), beam instrumentation, vacuum, and cryogenics.

The LHC is a complex knowledge system as it involves many interacting components that operate independently as well as collectively as a biological system. For example, the LHC operates in a vacuum system which needs careful design and the ability to withstand extreme temperatures, in order to avoid particle collisions with an air and dust-free environment for the beam of particles to travel unobstructed. Moreover, for the superconducting magnets that direct and focus the proton beams to remain in a superconductive state, the entire 27-kilometre LHC ring must be cooled down to 1.9K (−271°C), which is colder than the 2.7K (−270.5°C) of deep space. The LHC cryogenics system uses a combination of liquid helium and liquid nitrogen and uses the most complex system of compressors, heat exchangers, and storage tanks.

In the development of the LHC, one of the important lessons was the continuous effective communication among scientists, engineers, and technicians and making the right decisions. Human commitment is just as crucial to any Big Science experiment. The LHC machine would not have been possible without the dedicated involvement of everyone involved. Human input from large teams assembled around the world was fundamental to the design and construction of components often at a nanoscale. Dedicated teams were working around the clock for the operation, control, data analysis, and modelling, as well as routine inspections for the safety of the LHC.

The lessons learnt were considerable and the knowledge gained will be applied to the design and development of future accelerators like the High Luminosity LHC (HL-LHC) and the Future Circular Collider (FCC).

The LHC is truly a remarkable technological masterpiece which was possible due to the and collective efforts of scientists, science administrators, and communities who are willing to share intellectual resources to create and generate next generation scientific facilities. These facilities are legacies of previously unattained scientific investigative forces, components, and modules that work both systematically and serendipitously. They are new materials and technology capable of withstanding adverse conditions known to human beings and operate in a very high energy environment that recreates conditions closer to those after the Big Bang. The scientific achievements of the LHC also opens up many unanswered questions: What will life be like after the LHC and heading towards the Future Circular Collider? Subsequent chapters will delve deeper into these questions.

# 4
# Innovating Accelerator Technologies for Society

*Amalia Ballarino, Tim Boyle, and Shantha Liyanage*

## 4.1 Introduction

Science and technology are inseparable companions in high energy physics (HEP). The scientific effort that tries to answer fundamental questions about nature and the universe relies on incremental confirmations of theoretical predictions via experimental work. Conducting experimental work requires complex instruments, which are the result of years of challenging research and experimental development (R&D) that leads to the maturity of innovation and innovative technologies (see Chapters 2 and 3 for details). Science and technology have been evolving and expanding over the years in highly interdependent ways.

Science defines the goals in the evolution of fundamental knowledge whereas technology provides the means for exploration and limits the attainable upper limits of such exploration on a realistic timescale. The technological advancement of a perfected refracting telescope enabled the father of modern astronomy, Galileo Galilei, to observe Jupiter's moons in 1610. This discovery challenged popular beliefs about the bodies of our solar system at the time and provided evidence for the Copernican theory of the universe.

Since then, astronomers have constantly pushed the boundaries of increasingly detailed observations at a variety of wavelengths in the electromagnetic spectrum and developed increasingly complex telescopes seeking to understand the universe (see Chapter 8).

This chapter illustrates the social, economic, and innovation impact of Big Science by using accelerators and superconductivity as drivers, both of which play critical roles in the scientific and technological missions of Big Science projects.

## 4.2 The Role of High Energy Research

Technological breakthroughs in high energy physics are driven by the demands of particle physics. When the already available technology becomes insufficient to provide the experimental means for an agreed-upon physics programme, innovation

Amalia Ballarino, Tim Boyle, and Shantha Liyanage, *Innovating Accelerator Technologies for Society*. In: *Big Science, Innovation, and Societal Contributions*. Edited by: Shantha Liyanage, Markus Nordberg, and Marilena Streit-Bianchi, Oxford University Press. © Amalia Ballarino, Tim Boyle, and Shantha Liyanage (2024). DOI: 10.1093/oso/9780198881193.003.0005 begins. Fundamental research is the driver, and the genesis is within the corresponding research environment. Ideas and innovative concepts are followed by feasibility studies. These studies aim to highlight challenges, R&D requirements, intermediate milestones to be met to prove feasibility, reliability aspects (vitally important for systems that operate in an accelerator), economic affordability, and cost-benefit analysis.

Preliminary concepts call for a complex multidisciplinary approach, and the R&D phase requires flexibility for optimising the final performance as a function of the achieved R&D targets. Innovation may entail basic research in one or more fields, and it always implies technological development. Success relies on a rigorous approach that leaves nothing to chance and critically analyses intermediate results to steer the project. The typical timescale for completing complex physics systems, from R&D to commissioning, can be 15–20 years.[1]

Industrialisation follows, but this is facilitated by addressing the associated constraints already during the development phase. Big Science experiments extend well beyond the research and experimental development phase. They differ from the standard industrial, technological, and organisational innovations. Big Science leads to innovation that involves rational creativity, focused design work, and advanced engineering.

Fundamental research has a cost to society, so it is natural to wonder what benefits it will have in terms of technology that can solve societal issues as well as basic knowledge. Science and technology provide significant benefits to society through the production and accumulation of useful knowledge in a wide range of fields, as well as the resulting innovation. Innovative developments based on fundamental science that impact society may not initially be conceived with specific benefits in mind. Serendipity comes into play. Fundamental science tackles problems that would be unthinkable in a less multidisciplinary environment, and it transforms that knowledge into mature technology and innovation that benefit society. Once a technology has been proven and demonstrated to have potential, it can be attractive for application in other fields.

As discussed in this chapter, accelerator technology is widely used in a variety of industries, but is particularly used in medical applications, which are covered in detail in Chapter 9. These applications include diagnostic and therapeutic uses for radioisotope production, cancer treatment, and medical imaging.

The challenges and scale of large scientific accelerator projects have grown significantly over time. Solving these challenges will necessitate the collaboration of thousands of people from various nationalities, cultures, and educational backgrounds in order to share and generate complex multidisciplinary knowledge in the field. Such projects necessitate not only scientific and technological knowledge, but also human efforts, with motivation derived from common goals to overcome cultural diversity. Scientific and technical personnel from international organisations such as CERN work on-site in a multi-ethnic environment where almost everyone

---

[1] See for example, Evans, 2010; CERN 2010; and Zhu and Qian, 2012.

comes from different countries. De facto, this aids in the process of local knowledge assimilation and creative talent integration. Furthermore, contributions from national laboratories, universities, and industries around the world add a new dimension to these vibrant organisational cultures and give the projects an international flavour.

## 4.3 Superconductivity: An Accelerator Technology for Society

Particle accelerators are the major instruments for HEP research. By probing the smallest scales, over the past 80 years these accelerators have contributed to the understanding of fundamental physics. They have been powerful tools for new discoveries and precision measurements that extended existing knowledge and laid the groundwork for new discoveries. Circular colliders have challenged the energy frontier, producing through the years of collisions between particles with increasingly high centre-of-mass energy (Ballarino, 2019; see also Figure 4.1).

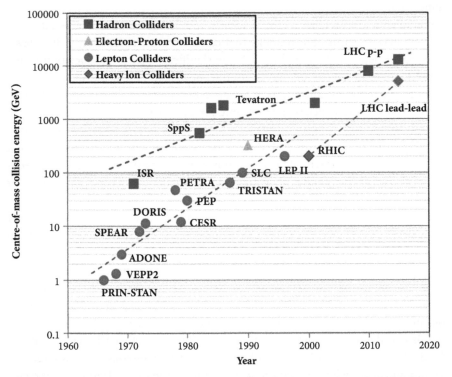

**Figure 4.1** Main circular colliders constructed during the years (superconducting colliders are highlighted in orange)

*Source:* Zimmermann, Frank (2018). Future Colliders for Particle Physics—"Big and Small". *Nuclear Instruments and Methods in Physics Research Section A: Accelerators, Spectrometers, Detectors and Associated Equipment.* 909. https://doi.org/10.1016/j.nima.2018.01.034.

Superconductivity enables us to break the magnetic field limits of conventional magnets, and thus exceed the approximately 300 GeV centre-of-mass energy in colliders of affordable size and cost. The Tevatron at Fermilab in the USA, HERA at DESY in Germany, RHIC at Brookhaven in the USA and finally the LHC at CERN in Switzerland, with up to 14 TeV design collision energy, have challenged the development of Nb-Ti magnets (see Chapter 3 related to the LHC).

Nb-Ti superconducting magnets generate high magnetic fields, which reach 8.3 T at 1.9 K in the large series of LHC dipoles. Superconductivity is an enabling technology for accelerators, and its complexity has been largely rewarded by the successful results achieved in high energy physics. In accordance with the historical trend that attributes to hadron colliders the role of discovery machines and to lepton colliders that of precision measurements, the Tevatron announced, in 1995, the discovery of the top quark (Fermilab, 1995), and the LHC, in 2012, that of the Higgs boson (ATLAS Collaboration, 2012 and CMS Collaboration, 2012) and Della Negra et al., 2012). These discoveries have served to confirm the adequacy of the so-called Standard Model of particle physics (Figure 4.2).

Particles in white circles were discovered by circular colliders (Figure 4.2): SPEAR for the charm quark and the tau lepton, PETRA for the gluon, SPS-AA for the Z and W bosons, Tevatron for the top quark, and LHC for the Higgs boson (Fermilab, 2010; Ballarino, 2019).

The recognition of superconductivity's application to experimental particle physics resulted in an intense development of superconducting materials suitable for use in magnets as early as the mid-1960s. The industry of Magnetic Resonance Imaging (MRI) is a direct spin-off of this intensive work. MRI is a non-invasive diagnostic method for capturing images of internal organs and tissues. It requires strong and uniform magnetic fields, stable both in space and time, which span the range from about 0.5 T to 11.7 T. About 25,000 MRI systems are operating in the world, and the volume production of Nb-Ti superconductor for MRI is today the main reason for its comparatively low cost (and thanks to which the LHC was affordable!).

The pay-off of a technology in the medical field, where innovation plays a crucial role in sustaining health, is immediate social acceptance. At the time of the Covid-19 pandemic, in 2020, it was also found that existing technology can play a role in supporting emergency situations. For instance, at CERN, accelerator technology and knowledge were proposed for a novel streamlined ventilator, which was shown to be easily manufactured and integrated into a hospital environment to support patients (Buytaert et al., 2020; Abba et al., 2021). Similarly, CERN's computing resources were made available for the global research effort against Covid-19 (CERN, 2020).

Nuclear Magnetic Resonance Spectroscopy (NMR) uses large magnets to determine the structure of organic compounds by probing the intrinsic spin properties of atomic nuclei. A strong magnetic field is needed for the polarisation of the magnetic nuclear spin. The first superconducting NMR magnet (200 MHz) also dates back to the mid-60s (Nelson and Weaver, 1964). Today NMR operating close to 1 GHz, i.e. ~ 23.5 T, is commercially available (Bruker, 2009, 2019). These very high magnetic fields exceed the capability of Nb-Ti, adequate for operation in magnetic fields of not more than about 10 T, and require the use of Niobium-Tin ($Nb_3Sn$) and High Temperature Superconductors (HTS), which have remarkable high field performance when used at low temperatures. Ongoing studies on future circular hadron colliders, that should reach centre-of mass-energy of the order of 100 TeV (Benedikt et al., 2020), rely on high-field magnet technology based on $Nb_3Sn$ superconductor, for fields up to 14 T–16 T, and HTS for even higher fields. To achieve this goal, R&D on superconductors (Ballarino et al., 2019) and on magnets (Tommasini et al., 2018) is needed. It should be noted that the only superconductors to date used in accelerators are Nb-Ti for magnets—about 1200 tons in the LHC (Evans, 2010)—and HTS Bismuth Strontium Calcium Copper Oxide (BSCCO 2223), for the electrical transfer in the LHC (Ballarino, 2002). Performance requirements for superconductors suitable for accelerator technology are specific and very demanding.

In high energy physics, electro-magnetic fields generated by radiofrequency (RF) cavities are used to accelerate charged particles. Superconducting cavities excel in applications requiring Continuous Wave (CW) or long-pulse accelerating fields above a few million volts per metre ($MV \cdot m^{-1}$). RF cavities are key components of linear colliders, which are of interest for industrial applications and life sciences.

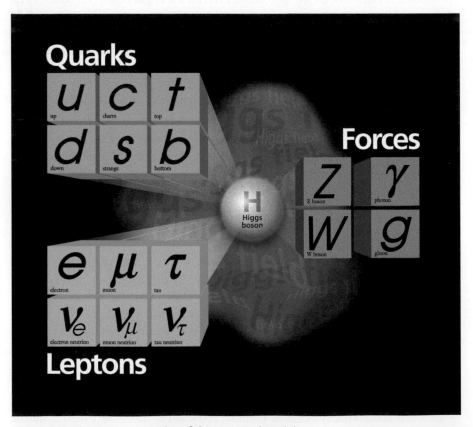

**Figure 4.2** Elementary particles of the Standard Model

*Source:* Fermi National Accelerator Laboratory, Public domain, via Wikimedia Commons

Spallation sources, for instance, use neutrons, emitted from metal targets by the bombardment of accelerated, high intensity, proton beams as material probes. The associated research relates to materials, with applications in the fields of technical sciences, industrial developments, energy, health, and the environment. Light sources use electrons to generate beams of light with unique properties. RF cavities in light sources accelerate the electrons to boost their energy. The resulting X-rays are used for imaging materials on the micro and macro-scales, studying chemical processes, studying bonds and surfaces. They enable research in diverse fields, including biology (e.g. protein crystallography); pharmaceutical and chemical science (e.g. X-ray crystallography to develop new drugs); energy (e.g. development of advanced materials such as those involved in the new generations of solar cells). There are many facilities in operation worldwide, which serve tens of thousands of users every year. Five Nobel prizes were awarded from 1997 to 2012 to scientists whose research had been made possible by light sources (www.nobelprize.org/prizes/lists/all-nobel-prizes/).

Superconductivity is an enabling technology for the production of energy via nuclear fusion. ITER, the international Tokamak reactor in under construction at Cadarache, in France, is the world's largest fusion experiment. It relies on strong superconducting magnets to confine, shape, and control the plasma (Mitchell et al., 2012). The development of high-field magnets for accelerators, in particular for the LHC, has provided the fusion community with a solid ground of accumulated know-how and available technologies. The powering of the ITER Tokamak magnets, for instance, will use HTS current leads of the type developed for the LHC machine (Ballarino, 2002; Bauer et al., 2020). After the development and adoption of the LHC, HTS current leads became the standard choice for transferring high currents from room temperature to superconducting systems, thereby reducing the power consumption of the associated cryogenics.

## 4.4  Origins of Innovation and Technical Challenges

Innovation starts with an initial concept and, if successful, it moves to large-scale applications. Initially, a concept is born in the mind of an individual, often derived from discussion among a few experts on how to approach the solution of a recognised problem. However, this is only the beginning—most of such ideas evolve under the light of reality: the next steps are vital in the genesis of what can eventually become ground-breaking technology. Big Science is a collective force of such individual thoughts applied to problems that no individual alone can solve.

In a Big Science innovation, the research process usually comes first, with initially theoretical and then an experimental demonstration of the concepts that can address specific scientific challenges and demonstrate benefits that justify the replacement of conventional technologies with new ones. The development phase follows and undertakes prototype activities that aim at demonstrating the feasibility of the most critical and innovative aspects of the project. This stage provides essential rounds of feedback and iterations on the initial concepts. New research topics can also arise in parallel and need to be agreed upon among scientists. The last step of the

innovation chain, if carried through, is the construction of the final products or outcomes and, in some areas, the industrialisation for large-scale production. During
this last phase, technological diffusion and the transfer of knowledge are required.
Technology transfer is a challenging and risky process where success may be determined by disciplined development and understanding of concepts. Often scientists
and researchers have to be involved to further refine innovation until it is industrialised. This implies the participation and continuous effort of experts involved in the
previous phases of the innovation process to ensure technical guidance, refinement,
and assistance. When the required expertise cannot be found in a single industrial partner due to the complex multidisciplinary nature of science projects, it is
necessary to carefully follow-up with additional partners. Industry aims at cost optimisation. The choice and selection of technologies can be done while keeping in
mind the final requirements of the system being developed.

Electrical transfer from a room temperature power source to a superconducting
system today can be done via conventional or superconducting current leads. For
the development of the HTS current leads carried out by CERN for the LHC, for
instance, the benefit derived from the first adoption of the new HTS technology
in a large system was studied in detail (Ballarino et al., 1996) and R&D on materials and prototype current leads followed (Ballarino, 2002). After a successful R&D
programme, which included an extensive qualification of prototypes constructed at
CERN, the technology was adopted for the powering of the LHC magnet circuits. The
Budker Institute of Nuclear Physics (BINP) in Novosibirsk and industry both participated in the construction of the series of more than a thousand current leads of three
different types. CERN provided the technology transfer, procurement, and assembly
services for the HTS BSCCO 2223 superconductor. The last qualification process
included national laboratories. HTS current leads were commercially available after
the LHC.

The challenges associated with an innovation project are many. This is because of
the extreme and specific requirements of high energy physics that demand technological research and development in a complex multidisciplinary setting. Shortcuts
are not possible, which may affect performance. If a system fails while the accelerator is running, it could cause a significant delay in the physics programme and have
serious financial repercussions. Access to the accelerator's components is not always
easy or possible; reliability issues must be considered early in the design process.

## 4.5  Challenges of Contracts with Industry

Big Science organisations like CERN rely on industry for the production and supply
of a series of components and equipment. Some of this equipment is unique and
cannot be purchased 'off the shelf'.

Typically, research institutions like CERN design such equipment, including
the essential designs, specifications, and early prototype work. Then, industry is
involved in the replication of goods that are acquired based on specifications via

competitive bidding. Since purchasing regulations typically award contracts to the lowest conforming bidder, it is crucial to make sure that the specification is clearly communicated and accepted by potential industrial suppliers in order to satisfy the technical requirements. Furthermore, once the contract is signed, professional technical follow-up is implemented to guarantee that the requirements are being followed, potential derogations are negotiated, and a timely delivery is guaranteed.

If inadequacies do occur, Big Science laboratories like CERN can intervene to help put production back on track due to their experience and technical expertise. Such an industrial production process is a joint venture with industry partners and CERN. The goal is to ensure a timely delivery that meets performance and reliability criteria while keeping within the estimated costs.

Some equipment incorporates technology that is unfamiliar to industry, and a laboratory like CERN can take responsibility for the delivery of related performance, only requiring the guarantee of strict adherence to the specification and specified procedures. With this approach it is possible to reduce the risk taken by industry, allowing it to reduce margins and thereby making the product more affordable. Technology transfer is critical. Several detailed studies have shown that, on average, working with large facilities like CERN in this manner benefits industry (Autio et al., 2011). It should be noted that such an approach is dependent on CERN having a reliable source of expertise, and that the organisation must actively promote the necessary training.

When a project would benefit from extending one or more of these technologies, it is obvious that corresponding collaborations should be established. Industry experts (and patents) protect important aspects of some technologies. In such cases, CERN as a host laboratory can approach companies to enter into agreements that address corresponding aspects of the project, with the aim of extending the state-of-the-art technology and purchasing the resulting components to incorporate in the system under development. Such partnerships require careful negotiation because the cost will vary depending on how much can be allocated to estimated benefits, both present and future, and how much must be paid directly. Such partnerships, however, depend on a thorough specification of the requirements and are transactional in nature. Follow-up from experts is mandatory on a regular basis.

## 4.6 Challenges of Collaborations with Industry and Universities

Collaboration can be defined as the practice of working together across different functions and across different locations. Making it work efficiently is a serious challenge, but the scale of the effort required for Big Science projects is such that one must rely on various forms of collaboration and an in-kind supply of equipment. The experience gained with building a machine like the LHC shows that for a collaboration-based complex project to work it is essential to maintain a strong, dedicated, and skilled home team to accompany the process—and step in with

expertise if any issues arise. This implies two things: (i) the home team must be sufficient in terms of number and expertise, and be assigned sufficient financial resources,
and (ii) the tasks assigned to the various collaborating bodies must be clearly defined.

The approach to collaboration depends on whether it is with an industrial partner,
an institution, or a university. In the case of collaboration with other institutions, the
task of elaborating such specifications is shared and agreed upon. Clearly, to arrive
at a consensus, this implies a great deal of discussion. It is crucial that the collaborating parties have common goals and priorities. In the case of collaborations with
universities, one must accept that an important aspect is to provide subjects for scientific studies often carried out by doctoral students. The derived privilege consists of
engaging more effort to work on a given project. The effort required to ensure proper
coordination and efficient communication—functions that should be performed in
parallel with the direct technical work—must also be considered, and it is imperative to assign suitably experienced staff. It is important to develop ways to make
collaborations work.

The next important question is how to drive collaborations with institutes and
industrial partners towards a successful innovative development. This is not always
easy as innovation is inherently a risky process (Klitsie et al., 2019). As discussed
in the previous section, managing Big Science innovation can be quite a challenge,
but there can also be great satisfaction in making it work smoothly. First, to share the
work efficiently one must evaluate the domains and levels of expertise of the potential
collaborating partners.

Collaboration with industry plays a key role in two broad cases: (i) fabrication
of prototypes that may be beyond what is feasible in the home laboratory, with
the view of developing procedures and tooling for the eventual manufacture of a
series; (ii) development of materials or components in view of applications that reach
beyond the current state-of-the-art. Marketing the potential benefit to the company
concerned and building mutual trust are critical. This includes an assurance that
proprietary technology will be fully respected, as even the suspicion of information
leakage to competing firms could put an end to the collaboration.

Case 1 is relatively straightforward because it relies on a thorough specification,
professional follow-up, and evaluation of the outcome, much like in the case of a
regular order. Case (ii) is more delicate because an eventual series production based
on its results is more distant and almost certainly outside the scope of future activities as the company management has them planned. Concerns over the disclosure of
trade secrets are also more likely to arise. It is also more susceptible to worries about
the leakage of trade secrets. However, in Big Science projects, such R&D is critical,
so it is worth the effort to pay special attention to establishing and maintaining the
collaboration.

Industry requires deep experiential knowledge and insights from scientists working on Big Science projects to verify technical and scientific efficacy and reliability.
After all, it is the role of scientists to verify to technical rigour and performance. Such
collaboration is a win–win opportunity for both industry and the host laboratory.

Not all collaborations are the same. Collaborations with institutions and
universities should be approached in a different manner. Consensus must be sought

from the outset, with agreement on deliverables and how to monitor progress. If the assignment is to procure a batch of equipment, the collaborating entity acts as an extension of the project team, effectively providing additional staff, and the job can be defined as a work package, with a corresponding allotment of project funds. Effective collaboration on R&D projects is more difficult to arrange due to the uncertainty of research outcomes, and it is customary to agree on a series of intermediate steps, defined by milestones, and comprehensive reporting. Careful documentation of progress and test results is vital, especially in the case of collaboration with universities where students performing the work may move on before the project is completed.

## 4.7  Human Factors Contributing to Success in Big Science

The success of large science projects relies on contributions from individuals who work together—on the same site or via collaborations—in an effective and constructive way. The management of human resources in complex R&D multidisciplinary projects requires identification and allocation of the different skill levels, adaptation of personnel to the various phases of the project possibly also via a continuous learning process, and the capability of motivating colleagues. In a motivated team, different ethnic, cultural, and political backgrounds are not obstacles to the achievement of the common goal.

Creativity is an integrated part of the R&D process and it calls for a unique style of leadership to foster and cultivate a conducive environment that encourages creative talents. Creativity fuels innovative ideas and solutions to come up with novel scientific and technical knowledge. The leaders are respected for technical and managerial competence, approachability, and fairness. It is necessary to create an environment where team members can identify with a project, relate to milestone results, and enjoy (or lament!) the outcome. R&D calls for passion as well as skill and dedication (see also Chapter 7). Managing human resources and recruiting talented scientists is a constant battle for Big Science organisations.

In Big Science operations, technological advances have in most cases overtaken human interventions. With the rapid growth of computers, the use of robots, algorithms, and artificial intelligence, scientists have taken on the role of supervisors and monitors rather than that of direct interventionists and controllers (De Winter and Hancock, 2021).

## 4.8  Case Study on Accelerator Technology: Superconducting Electrical Transmission

Electrical transmission lines are an application of superconductivity in the field of energy transmission. The potential of HTS for this purpose was already identified in the mid-90s. Very low total electrical losses and high-capacity transmission, along with several other technological advantages (e.g. low environmental impact), motivated studies, and demonstrations of power transmission lines, some of which operated in the grid, initially using BSCCO 2223 superconductors and then REBCO HTS superconductors. The high cost of HTS materials has, however, made acceptance and wide adoption unaffordable up to this point. The challenge of creating high-current power transmission lines, known as Superconducting Links, based on Magnesium Diboride (MgB2) superconductor, was taken up as part of the LHC High Luminosity upgrade (HL-LHC) (Ballarino, 2014).

This superconducting material was discovered in 2001. It can be effectively cooled by helium gas, as opposed to liquid helium, as is needed for Nb-Ti, or liquid hydrogen, and has a critical temperature of 39 K, meaning it can be used at up to 25 K to 30 K. It transfers high currents at these temperatures and can withstand peak magnetic fields of up to 1.5 T. The Powder-In-Tube industrial process, which has already been created and used for other superconducting materials, can be used to produce it in long unit lengths at a potentially low cost.

The power converters for the HL-LHC superconducting magnets will be installed in new radiation-free galleries about 100 m away from the magnets. Following the LHC's successful development of HTS current leads, it was decided to research using a similar technology for the cables connecting the power converters to the magnets. Each of the eight final HL-LHC systems, collectively referred to as cold powering systems, includes a superconducting link.

The Superconducting Links are superconducting transfer lines that will electrically connect, in the LHC underground areas, the power converters to the HL-LHC superconducting magnets. They will transfer DC currents of up to |120| kA at 25 K, and the physical length will be about 120 m. The advantages of the $MgB_2$ superconductor are its low cost with respect to HTS and its high operating temperature (~ 25 K) with respect to Nb-Ti. These two aspects make $MgB_2$ power transmission lines economically affordable, when compared to HTS, and able to operate at higher temperatures with reduced complexity of the cryogenic system and a significantly greater temperature margin, when compared to Nb-Ti.

Initial studies done at CERN on concepts relying on $MgB_2$ proved their suitability and identified the advantages and feasibility of this solution. At the time of the study, the superconducting material was only produced commercially in the form of tape, not suitable for cabling. CERN contributed with its experience with other types of superconductors to suggest alternative wire layouts, with regular testing to qualify wire performance. After a few years of development, the company industrialised the process and launched the first large-scale production of $MgB_2$ wire. The feasibility of cabling was investigated in depth at CERN via the conception and testing of cable layouts. This cabling is special in that the multi-filament superconductor in the wire is brittle, and a strain of about 0.3% is sufficient to cause permanent damage. For large-scale cabling, a collaborative effort was launched with a specialised company. After the successful completion of this collaboration, a series production of long and complex cables was launched in industry. Prototype cold powering systems were developed and tested concurrently at CERN. To create a Superconducting Link, the MgB2 cables are housed in a lengthy, semi-flexible cryostat; orders for such cryostats were placed with several manufacturers to confirm their viability. Via a dedicated testing campaign at CERN, it was shown that thanks to the relatively high operating temperature of $MgB_2$ such a cryostat does not require a gas-cooled thermal screen. The design of the system was therefore simplified, and series production was launched.

The cryostats of the Superconducting Link's terminal at the magnet end were designed in partnership with the University of Southampton. The university provided the series components through a collaboration agreement. These cryostats of the terminal at the power converter ends were further developed at CERN and will be supplied through a collaboration with Uppsala University in Sweden. The HTS Rare-earth Barium Copper Oxide (REBCO) current leads are based on the ReBCO-CORC technology.[2] CERN developed prototypes for this technology and went through product development and construction at CERN. This project is an example of an innovation generated at CERN, which went through research, experimental development and commercialisation phases involving industry and academic partners.

The R&D on the Superconducting Links started with the development of the first $MgB_2$ round wire, industrialised in kilometre long length, and continued for about ten years with development and design work to conceive and construct prototype cold powering systems. It was delivered in early 2020 with the first successful demonstration of a very high transmission capacity, and culminated in late 2020 with the first successful demonstration of a high current—|120| kA—$MgB_2$ superconducting transmission line system (Pralavorio, 2020).

For other applications, including those for electrical utilities, Superconducting Links cooled with liquid hydrogen, at about 20 K, are an option that efficiently combines the transfer of fuel and electricity. Studies were launched by the Institute of Advanced Sustainability Studies, IASS, (Thomas et al., 2016) and a collaboration agreement between IASS and CERN demonstrated the first feasibility via the successful qualification of an ad-hoc designed system (Del Rosso, 2014).

A European collaboration, involving both industry and a European Transmission System Operator, continued the effort in this direction (Ballarino et al., 2016). The study demonstrated the potential of Superconducting Links in future grids for efficient and powerful electricity transmission (Best Paths, 2018). Short-term applications were identified in existing networks in urban areas where space is limited or civil engineering work for the installation of conventional lines is expensive, e.g. when crossing rivers. In the longer term, the use of High Voltage Direct Current (HVDC) long Superconducting Links transferring Giga Watts of remote renewable energy to consumption centres in large urban areas is considered an enabling technology. The potential of sustainable electric transfer via Superconducting Links in this application for society lies primarily in their small size, with potential benefits in terms of efficiency, low environmental impact, and widespread public acceptance.

---

[2] For details of the development of ReBCO-CORC technology—see https://indico.cern.ch/event/760666/contributions/3390620/attachments/1885467/3107852/20190724-Mulder-TenKate-CEC-ICMC_2019-CORC-Development-CERN.pdf and https://indico.cern.ch/event/760666/contributions/3390620/contribution.pdf.

The adoption of superconducting electrical transfer is also being studied for other applications, including aircraft propulsion (Green Car Congress, 2021). In this case, superconductivity may be a technology that makes it possible to scale up electric propulsion for larger aircraft because it allows for the distribution system to be controlled and weighted efficiently while still delivering high power density.

## 4.9   Future and Present Applications of Accelerator Technology to Society

Both accelerators and detectors for experiments have been, and will continue to be, rich sources of innovative technology for the medical field. This is highly visible and easy to appreciate.

Additionally, it is likely that superconducting technology will be used in motors, generators, and power transmission systems as infrastructure addressing energy and climate change issues receives more attention. The work done for large accelerator projects will also serve as a rich source of information and a solid foundation for the development and deployment of innovative superconducting devices.

The application of particle physics technologies to the general benefit of society in areas as diverse as climate change, archaeology, and art is discussed below.

## 4.10   Links of Accelerator Technology to Society: Climate Change, Archaeology, and Art

Starting with early experiments conducted by the Cavendish Laboratory, in Cambridge, particle accelerator technology has evolved over a period of over 100 years. During this time, there have been several fields where accelerator technology development has diverged from fundamental research towards applications that impact society more directly.

It is estimated that there are currently more than 40,000 accelerators operating globally, ranging from large bespoke machines like the LHC at CERN to machines that are virtually ubiquitous in the clinical setting (Chernyaev and Varzar, 2014). The societal impact of accelerator technologies is illustrated below through examples drawn from real-world situations.

### 4.10.1   Climate Change

Climate change is the gradual change in environmental conditions such as temperature, weather patterns, and rainfall over a period of time. The effects of anthropogenic climate change can already be seen in our ecosystems, bushfires, food security, health, and infrastructure (Australian Academy of Science, 2020). One of the key data-driven mechanisms for characterising this change is through the measurement of atmospheric greenhouse gas concentration (GHG). The most comprehensive dataset of GHG concentrations was published in Geoscientific Model Development in early June 2017 and tracked changes in all 43 greenhouse gases that have contributed to human-induced climate change over the past 2000 years.

While a dataset detailing the measurements from the past 100 years has been curated via direct measurement, there is a technical challenge in obtaining a representative dataset for the previous 1900 years. The historical data in this landmark publication were obtained by analysing ice cores and compacted snow from the most remote places on Earth such as Antarctica. Every year that it snows in Antarctica, the current snow layer weighs on the previous layer, compacting over hundreds or thousands of years to eventually form layers of ice. These ice layers contain air bubbles, which are like tiny time capsules (see Figure 4.3). Using sophisticated sample preparation techniques, researchers are able to extract the ancient air contained within these bubbles. To accurately reconstruct an atmospheric record, these gases are analysed using accelerator technologies such as accelerator mass spectrometry and radiocarbon dating.

Researchers at the Australian Nuclear Science and Technology Organisation (ANSTO) contributed to this research by using the unique capability for precise

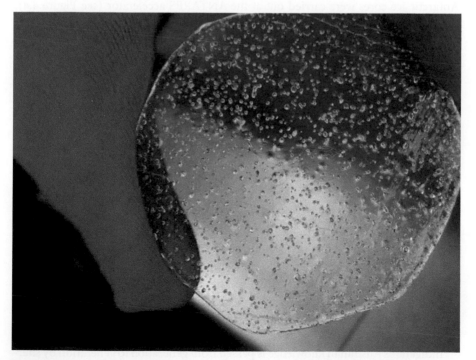

**Figure 4.3**  Air bubbles trapped in a section of ice core from an Antarctic glacier
*Source:* ANSTO

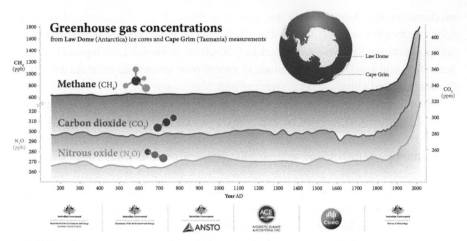

**Figure 4.4**  GHG levels over time as determined by accelerator techniques
*Source:* ANSTO

measurements of the radiocarbon content in minuscule carbon samples, such as those derived from carbon dioxide, methane, and carbon monoxide extracted from ice core air bubbles (ANSTO, 2017). Measurements of the carbon-14 content of methane ($14CH_4$) and carbon monoxide ($14CO$) in ancient air extracted from Antarctic ice cores were carried out at the ANTARES accelerator (see Figure 4.4), operational since 1991, at ANSTO's Centre for Accelerator Science (Petrenko et al., 2017).

The determination of historical GHG concentrations highlighted the anthropogenic impact on the climate since the industrial revolution. Popularised by the documentary *An Inconvenient Truth*, societal awareness of climate change and GHG levels increased as pressure was exerted on governments globally to act. This has resulted in several global policy initiatives such as the Kyoto Protocol and the Paris Agreement, which seek to reduce GHG emissions over time as a unified global climate effort.

## 4.10.2  Early Human Culture

Australian indigenous culture is recognised as the world's oldest continuously existing culture. With evidence suggesting that the ancestors of indigenous Australians first migrated to the Australian Continental Landmass between 65,000 and 120,000 years ago (Clarkson et al., 2017), the early inhabitants of Australia existed in isolation from the rest of the world until European colonisation in the late 1700s.

Indigenous Australian knowledge is passed from generation to generation by sharing events through storytelling, music, art, and ceremony. Application of accelerator-based techniques to surviving artefacts such as artwork offers a unique opportunity for society to understand early human knowledge and culture.

### 4.10.3  Australian Rock Art

Indigenous Australian rock art is the oldest surviving human art form. Across all of Australia, this form of art represented an integral part of pre-colonial Australian life and culture and a multigenerational knowledge exchange.

The age of indigenous Australian rock art cannot be directly determined using accelerator-based techniques like radiocarbon dating because the medium of painting is ochre, an inorganic mineral pigment that does not contain carbon. A new method which dates calcium oxalate from mineral crusts deposited within the art was applied to accurately date nine different artworks from Arnhem Land, in the Northwest of Australia, by using the STAR and ANTARES accelerators at the Centre for Accelerator Science at ANSTO.

The dating indicated that the rock artwork originated in the Pleistocene and early Holocene periods, ranging from at least 3500 to 9400 years ago across the nine artworks (see Figure 4.5).

### 4.10.4  Culturally Modified Trees

Indigenous Australians have a long history of using trees as a resource to make canoes, containers, shields, tools and implements, and weapons (Cooper, 1981). Many Australian trees contain artefacts such as carvings, scratches, and scars from

**Figure 4.5**  Indigenous Australian rock art
*Source:* ANSTO

indigenous Australian interaction. These trees are often referred to as culturally modified trees. Recently a culturally modified tree was discovered in the Wiradjuri country, near Orange in central New South Wales. This tree is unique as it has an indigenous stone tool embedded within regrowth surrounding a large scar in the tree. This discovery represented the first example of an ingenious Australian tool to be found in a tree.

Radiocarbon dating undertaken at ANSTO's Centre for Accelerator Science indicated that the tree was relatively young having begun growing at the start of the twentieth century and died during Australia's millennium drought at the age of approximately a hundred years. Radiocarbon dating also indicated that the tool was embedded in the tree sometime between 1950 and 1973—which was an unexpected result.

This unprecedented discovery indicates the resilience of Aboriginal culture in Australia: the Wiradjuri culture continued even during the active discouragement and assimilation policies that prevailed in Australia in the twentieth century.

## 4.10.5  Accelerator Technology for Cultural Preservation

Accelerator technology plays a key role in the study of art objects, e.g. authentication or identification of the artist, and in their conservation and restoration. The IBA (Ion Beam Analysis) facility, installed in the Centre for Research and Restoration of the Museums of France, is devoted to the study of cultural heritage. Among the techniques adopted, Particle-Induced X-ray Emission (PIXE), based on photon detection, is used to identify constituent materials of art works, derive techniques adopted, and eventually attribute authors (Dran, 2002).

About 15 metres under the glass pyramid at the Louvre there is a 27 metre-long accelerator weighing 5 tons called AGLAE (Accélérateur Grand Louvre d'Analyse Élémentaire) (AGLAE, 2021). It uses ion beams and Pixe technology as non-invasive techniques to quantify composition and attribute the place of origin and age of art works. The famous Egyptian head in blue glass, which was for a long time considered an ancient portrait of Tutankhamun, was via AGLAE dated as an object of the eighteenth century, thanks to the identification of the presence of lead and arsenic.

MACHINA (Movable Accelerator for Cultural Heritage In-Situ Non-destructive Analysis) is a novel compact and transportable accelerator, based on the radio-frequency quadrupole technology developed at CERN (CERN, 2017). It is a collaborative effort between CERN and the Italian National Institute for Nuclear Physics (INFN) and it will be based at the laboratories of the Opificio delle Pietre Dure in Florence. It is designed to be used to analyse in-situ large immovable art works, such as frescoes, or works too fragile to be transported. It is less than 2 m long and weighs about 300 kg.

# 4.11  Conclusions

The core scientific mission of Big Science is to deliver fundamental knowledge. Closely associated with this scientific mission is the ensuing technological development that benefits society at large. It is always difficult to assess all the social and economic benefits of conducting Big Science experiments. As demonstrated in this chapter, the development of accelerators in Big Science has many social, medical, and economic benefits for a variety of stakeholders. The translation of Big Science knowledge into practical and tangible benefits is a complicated process that can take years of hard work and dead ends and departures. It is important not only to understand how fundamental knowledge can be produced but also how such knowledge can be decodified and translated into industrial applications.

In this chapter, we have used superconductivity as one example, and then briefly described the use of accelerators in diverse fields such as studying climate change, early human culture, and archaeology. Superconducting electrical transmission has the potential to bring social benefits as this technology can have a big impact on energy efficiency, greenhouse gas emissions, and the efficiency and reliability of the electric grid to support different industries. We have also talked about how accelerator technology can help with environmental and societal issues like climate change, archaeology, and the arts, as well as how superconductivity, a key technology for the LHC, can be used in fields like energy production, biomedicine, and medical imaging.

What can we learn from the experience of translating fundamental knowledge of Big Science?

- Translating Big Science knowledge is a step-by-step process, following many feedback loops and iterations;
- Interactions are complex while engaging in joint innovation—very often there are disparities in knowledge levels, and differences between scientific and industry cultures. These need to be overcome for effective knowledge transfer. Organisational and administrative processes must be put in place to facilitate these processes;
- Evidence suggests that in some complex technological areas, such as superconducting materials, it is essential to elaborate long-term strategies and work with partners over many years using different forms of collaborative formats. Radical innovation takes time; and
- Accelerator technology plays a key role in the study of art objects, e.g. authentication or identification of the artist, and in their conservation and restoration.

The high energy physics is one of the main drivers of particle accelerators and associated technologies. The next generation of accelerators is faced with new challenges like extremely high energy and intense magnetic and radiation energy. Affordable and well-integrated R&D is necessary for the next generation of the HEP community to advance into the next stage of discovery. This is the systematic nature of the never-ending search to extend the frontiers of human knowledge.

# 5
# Leapfrogging into the Future

*Michael Benedikt, John Ellis, Panagiotis Charitos,
and Shantha Liyanage*

## 5.1 Introduction

The primary objective of particle physics is to understand the underlying structure of matter and its role in the history and structure of the Universe. As discussed in the previous chapters of this book, much progress has been made in recent decades, particularly with the LHC hadron–hadron collider and the previous LEP electron–positron collider, housed in the existing 27 km tunnel straddling the Franco-Swiss border in the Geneva region. Nevertheless, despite this progress there remain many open questions in particle physics and open cosmological issues that future colliders may be able to resolve.

The most effective and the most comprehensive approach to explore thoroughly the open questions in modern particle physics is research infrastructures offering a staged research programme that combines precision measurements with direct exploration at previously uncharted energies. This vision lies at the heart of the Future Circular Collider (FCC) study that integrates a lepton collider (FCC-ee) (FCC Collaboration, 2019) as a first step followed by a hadron collider (FCC-hh) (FCC Collaboration, 2019b) in a manner reminiscent of the complementarity between the LEP and the LHC.

Today, there is overwhelming consensus on the research agenda of particle physics for a lepton collider that could operate as a Higgs factory, producing copious Higgs bosons, yielding precise knowledge of this unique particle. The novelty of the Higgs boson, and thus the great interest in studying its properties and interactions with the other known particles of the Standard Model (SM), derives largely from its scalar nature. It is the only fundamental particle without spin.

Four Higgs-factory designs are presently being considered. Two are based on linear accelerators, namely the International Linear Collider (ILC) under consideration in Japan and the Compact Linear Collider (CLIC) proposed at CERN, which have been studied since 1975 (Amaldi, 1976) as they are considered to be the most mature approach towards high energy lepton collisions. The advantages of linear accelerators are that they can be extended to higher energies, though this would require additional

Michael Benedikt et al., *Leapfrogging into the Future*. In: *Big Science, Innovation, and Societal Contributions*. Edited by:
Shantha Liyanage, Markus Nordberg, and Marilena Streit-Bianchi, Oxford University Press. © Michael Benedikt et al., (2024).
DOI: 10.1093/oso/9780198881193.003.0006

civil engineering work, and the beams can be polarised longitudinally. The other two concepts are circular: the lepton option of the Future Circular Collider (FCC-ee) at CERN; and as discussed in detail in Chapter 13, the Circular Electron Positron Collider (CEPC) in China. A circular collider can provide higher luminosities and better performance for energies up to 400 GeV, while the same infrastructure can be used to host energy-frontier proton colliders like the proposed FCC-hh.

While one of the main motivations for a future lepton collider is the precise study of the interactions of the Higgs boson, seeking answers to open questions in particle physics requires many high-precision measurements of the other three heaviest SM particles, namely the W and Z electroweak bosons and the top quark. The proposed operation models for the circular colliders comprise data taking at and around the Z pole (90 GeV), at the WW threshold (180 GeV), at the ZH cross-section maximum (240 GeV) and, for FCC-ee, an extension up to 365 GeV at and above the top pair threshold. With the highest luminosities at the Z pole, the WW threshold, and the top-pair threshold, and with transverse polarisation to precisely calibrate the beam energies, precision electroweak measurements are the realm of FCC-ee. The designs are sufficiently flexible to allow for operation at other centre-of-mass energies, if justified by compelling physics arguments.

The experience from the FCC-ee would be valuable for the next step: a future high energy collider (FCC-hh), which could be the hadronic successor to the LHC. The FCC-hh would be a circular proton collider housed in the same tunnel as the FCC-ee. It could reach energies of some 100 TeV (approximately seven times higher than the 14 TeV of the LHC) and luminosities 50 times higher than at the LHC, using new high-field magnets reaching 16 T (fields twice as high as the 8 T magnets of the LHC). Exploring the multi-TeV regime is the only way to study how the Higgs interacts with itself. Experimental searches at the FCCs will offer an exhaustive understanding of the SM and guide our theoretical understanding as we face the pressing questions (FCC Collaboration, 2019c) that we discuss in the next sections.

The Conceptual Design Reports (CDRs) of the FCC-ee and FCC-hh projects were published in January 2020, in time to inform the update of the European Particle Physics Strategy. At present, as recommended by this 2020 update, a feasibility study for the FCC (including both FCC-ee its subsequent hadron-collider stage, FCC-hh) is ongoing, with the goal of presenting an updated conceptual design report in 2026, in time for the next strategy update.

## 5.2 What We Know

The visible matter in the Universe is described very accurately by the so-called Standard Model (SM) of Particle Physics. Ordinary matter is built out of molecules, which are made out of atoms that contain nuclei surrounded by clouds of electrons. The nuclei are bundles of particles called protons and neutrons that are themselves composed of apparently fundamental constituents called quarks. The SM prescribes how molecules and atoms are held together by photons, particles that produce light

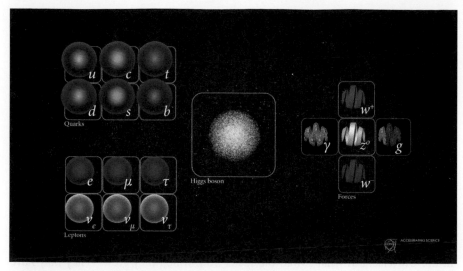

**Figure 5.1** Constituents of the SM of particle physics
*Source:* © CERN

and radio waves when they escape these bound states. Similarly, the quarks are held together inside protons and neutrons by particles called gluons, though these are never detected directly, because they are confined inside nuclear matter.

In addition to the electromagnetic interactions mediated by photons and the strong interactions mediated by gluons, there are weak interactions that cause radioactive decays of heavier particles into lighter ones. These weak interactions are mediated by massive particles, the W and Z bosons.

The particles introduced above and shown in Figure 5.1 are the fundamental building blocks of Nature and through their interactions they make up the visible matter that we observe around us. The SM describes all these physical phenomena in a framework that is consistent with quantum mechanics and Einstein's Special Theory of Relativity and has been used to make many very accurate and successful predictions.

## 5.3  What We Do Not Know

Nevertheless, the SM is deeply unsatisfactory, for several reasons: Why these specific particles, rather than others? Why not more? Why not less? These questions are frequently labelled collectively as the problem of 'flavour'.

We also ask why these specific interactions? Perhaps there are others? Can we find a more unified description of all the fundamental interactions, perhaps including gravity, which is currently left outside the SM? These questions are often grouped as the problem of unification.

Then there is the problem of mass: the SM accommodates particle masses via a mechanism whose physical manifestation is the Higgs boson. However, nothing within the SM explains the magnitudes of these particle masses, nor the vast hierarchies between their measured values.

Beyond these intrinsic shortcomings of the SM, our observations of the Universe around us pose several other problems that are extrinsic to the SM.

How did the matter in the Universe originate? One would have expected the numbers of matter and antimatter particles produced by the Big Bang to be almost identical but, somehow, it produced significantly more matter than antimatter, and the latter all annihilated with matter, leaving behind the excess of matter that surrounds us today, and no significant quantities of antimatter. The SM is unable to explain the magnitude of the matter–antimatter imbalance.

And what is the nature of the unseen dark matter that has formed massive halos around galaxies, holding their stars together? The SM contains no candidates for dark matter, which might be composed of one or more unknown species of particle. Dark matter is essential for the formation and existence of galaxies and other structures in the Universe, but what sowed the seeds from which they grew? They may have originated from quantum processes in the very early Universe within some extension of the SM or Einstein's general theory of relativity.

Finally, cosmologists tell us that the majority of the density of matter and energy in the Universe is in the form of dark energy, which does not cluster, but is spread universally and is causing the expansion rate of the Universe to accelerate. Here the problem is not so much the existence of dark energy, but rather why it is so small. The SM suggests a density with a magnitude far greater than the measured value.

## 5.4  What the FCC Integrated Programme Offers

The FCC programme offers a comprehensive, multi-pronged approach to these outstanding problems beyond the SM. Experiments at FCC-ee, an intensity-frontier lepton collider, lay the basis for offering unparalleled precision in measurements of the SM, including the Higgs boson, the electroweak gauge bosons Z and W, and the top quark, opening indirect windows on new physics. Experiments at FCC-hh, on the other hand, will directly explore possible new physics at the highest accessible energy scales, and will also produce vast numbers of SM particles, providing opportunities for more precision measurements that will enable further indirect probes of new physics (Biscari and Rivkin, 2019).

The different phases of the FCC project depicted in the planned plot include: administrative steps, infrastructure development, the FCC-ee schedule, and the FCC-hh schedule (see Figure 5.2).

SM particles, provide opportunities for more precision measurements that will enable further indirect probes of new physics (Biscari and Rivkin, 2019).

The integrated programme for the FCCs, combining FCC-ee and FCC-hh, extends over 70 years in time, as illustrated in Figure 5.2. The capabilities offered by the combination of a lepton circular collider (FCC-ee) with a hadron circular collider (FCC-hh) are illustrated in the following sections for the examples of the Higgs

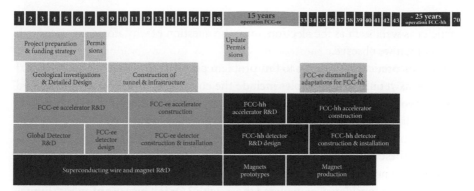

**Figure 5.2** The FCC project extends over 70 years from its starting date (year 1)

*Note:* The different phases of the FCC project depicted in the plot include: administrative steps, infrastructure development, the FCC-ee schedule, and the FCC-hh schedule.

*Source:* Created by author

boson and dark matter, which are among the most mysterious puzzles in particle physics and cosmology, respectively.

The different phases of the FCC project depicted in the plot include: administrative steps, infrastructure development, the FCC-ee schedule, and the FCC-hh schedule.

## 5.5  A Puzzling Particle

The discovery of the Higgs boson, the last particle in the SM to be detected, leaves many questions unanswered while also raising new ones (ESPPU, 2019). It is the first and only example so far of a novel type of elementary particle, one without any spin. Is it truly elementary, or is it a composite object made out of more fundamental constituents? The latter possibility was considered actively before the discovery of the Higgs boson, but the LHC experiments have found no evidence in its favour. The best way to explore this possibility may be to measure its properties as accurately as possible, a task at which FCC-ee will excel (Blondel et al., 2019). If it is indeed composite, it is likely to be accompanied by other, heavier particles in which its constituents are arranged in different ways, a possibility that will be explored comprehensively at FCC-hh. Whether the Higgs boson is elementary or not, it may well be accompanied by other spin less particles, such as scalar particles, whose existence can be indirectly confirmed at the FCC-ee or directly at the FCC-hh through various experiments.

If the Higgs boson is indeed elementary, many more questions arise. What determines its mass and those of other elementary particles? The existence of the Higgs boson is a manifestation of the mechanism that gives masses to elementary particles but does not explain how large they are. The sizes of atoms depend on the mass of the electron, and the strengths of radioactive decays depend on the mass of the W particle that generates them, so understanding their magnitudes would give important insights into major features of the Universe. This issue is particularly problematic because the Higgs has no spin which makes the measured value of its mass seem unnaturally low and, by extension, the masses of other elementary particles as well, such as the electron, raise the question of why atoms are not much smaller than we observe.

Many theoretical approaches to this problem postulate the existence of additional particles, as yet unseen. Examples include the composite Higgs models mentioned above, theories with additional dimensions of space, and theories that partner particles of different spins in which the mass of the Higgs boson would be protected by its spinning partner and other new particles, an idea called supersymmetry. But where are these additional particles? The LHC has found no evidence of additional particles beyond the Higgs boson. Is this because they behave in ways that were not experimentally considered, or explored thoroughly? Or is it because of energy limitation or because the LHC has not simply collected enough data or analysed such data to find rare particles? Or is it beyond our current understanding and experimental capabilities? In that case the very clean experimental conditions and high collision rates provided by FCC-ee may enable us to find them. Or is the absence of additional particles so far simply because they are too heavy to have been produced by the LHC? In that case the very high collision energies provided by FCC-hh offer the best chances of finding them.

There are other issues, which concern the way in which the Higgs boson interacts. The SM controls the possible forms of its interactions but does not specify their strengths. For example, the mechanism for fixing the overall density of the Higgs field in the Universe today requires that it has self-interactions. What determines their strengths? A priori, they could have been strong, but present data suggest that they are rather weak, though the LHC is unable to measure them directly. FCC-ee could provide a first indirect measurement by studying the production of the Higgs boson very precisely, but an accurate, direct measurement will require studies of pairs of Higgs bosons at FCC-hh.

The Higgs particle is a quantum manifestation of a field extending throughout space, much as the photon is a quantum of the electromagnetic field. If the self-interactions of the Higgs boson are indeed weak, the energy of the Higgs field whose quantum is the Higgs boson does not depend strongly on its value, and other questions arise. How was the present value of the Higgs field determined during the evolution of the Universe, and could it change in the future?

Within the SM, the answers to these questions depend on the interactions of the Higgs boson with the top quark and their masses, and calculations of their effects are subject to considerable uncertainties. However, they indicate that the present configuration of the Higgs field may be unstable in principle, though on a time-scale longer than the present age of the Universe. Accurate measurements at FCC-ee and FCC-hh will resolve this issue, which has interesting implications at the frontier between physics and philosophy. What if FCC measurements and SM calculations confirm the Higgs instability problem? Would this mean that the Universe as we know it is doomed? Or does it rather suggest that there must be some physics beyond the SM that restores stability? If the latter, FCC-hh would be the most powerful instrument to search directly for any such new physics.

The interactions of the Higgs boson with matter particles also pose many puzzles that are linked to the problem of flavour (de Blas et al., 2019). In the SM the strong and weak interactions are similar for different flavours of matter particles with identical electric charges but varying masses. On the other hand, the interactions of the Higgs boson do not share these universality properties. Instead, the SM predicts that they are proportional to these different masses, which range over several orders of magnitude. Will this prediction hold up under the scrutiny of FCC-ee and FCC-hh? A corollary question is, what is the origin of the big differences between the masses of different matter particles? Will FCC studies of the interactions of different matter species find deviations from the universality predicted by the SM?

The Higgs boson is the most recent particle to have been discovered, and it is possible that it may have other interactions beyond those predicted in the SM, that are yet undiscovered. For example, there are many proposed extensions of the SM with an entire hidden sector of new particles that connect to the particles of the SM via the Higgs boson. In such 'Higgs portal' models more decays of the Higgs boson into invisible particles than just the neutrinos of the SM may appear.

Another possibility is that the Higgs boson interacts with unseen massive particles, too heavy to be seen directly, that in turn, generates supplementary interactions between the Higgs boson and other SM particles. Measurements of any such interactions can guide us towards understanding the properties of these massive particles, just as studies of the weak interactions in the 1930s, 1940s, and 1950s guided us towards the massive W and Z particles in the SM. The LHC high-luminosity upgrade (HL-LHC) will provide insights into the Higgs boson couplings to the SM gauge bosons and to the heaviest SM fermions (t, b, τ, μ). Together, FCC-ee and FCC-hh will provide the most sensitive probes of such supplementary interactions and whatever massive particles cause them (FCC Collaboration, 2019b, de Blas, 2019), as illustrated in Figure 5.3.

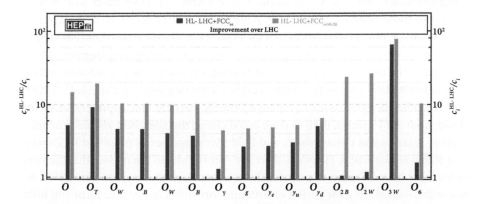

**Figure 5.3** Achievable precisions for modified Higgs and electroweak couplings at proposed next-generation e+e− colliders including FCC-ee

*Source:* de Blas, J., Durieux, G., Grojean, C. et al. (2019). On the future of Higgs, Electroweak and Diboson Measurements at Lepton Colliders. *Journal of High Energy Physics*. 2019, 117. https://doi.org/10.1007/JHEP12(2019)117

## 5.6 Dark Secrets

The shortcomings discussed above do not detract from the success of the SM in describing all the visible matter in the Universe, from the stars to human beings. However, this visible matter provides only about 4% of the overall density of matter and energy in the Universe. Astrophysicists and cosmologists have discovered that there is a much larger percentage of invisible dark matter, and that an even larger percentage of the density of the Universe is not material at all but is spread uniformly throughout the Universe in the form of dark energy.

An astronomer, Fritz Zwicky, was the first to predict the existence of dark matter in the 1930s (Zwicky, 1933, 1937). Zwicky's observations of the Coma cluster of galaxies showed that the galaxies were moving much faster than expected. So fast, in fact, that it was impossible to understand how the cluster held together unless there was some additional source of gravity beyond the visible matter. It took several decades for this radical suggestion to become generally accepted. A key additional piece of evidence for dark matter was provided by Vera Rubin and collaborators in the 1970s (Rubin and Ford, 1970; Rubin et al., 1980, 1985, and 1992).

They observed the motions of stars in many galaxies and found that they were also moving too fast to be held together by the gravity generated by the visible galactic matter. Observations of distant supernovae and the cosmological background radiation in the 1990s also indirectly confirmed the existence of dark matter and established the existence of dark energy, which contributes about a quarter and 70% of the density of the Universe, respectively.

The FCC will be able to shed light on the nature of dark matter or dark energy depending on their natures and how they are related to ordinary matter. We know very little about dark matter, apart from the fact that it generates a gravitational field. The possibility that it might consist mainly or partially of black holes has been extensively considered since the detections of black hole mergers by the LIGO and Virgo collaborations (Abbott et al., 2016), but it now seems that black holes with masses similar to those detected so far can provide only a small fraction of the total dark-matter density. For this reason, it is widely expected that dark matter consists mainly of one or more unknown types of particles that are not contained within the SM.

Two general categories of particles have been proposed by the physics community. One is some novel type of fermionic weakly-interacting massive particle (WIMP), and the other is some type of very light bosonic particle that is present in waves throughout the Universe. Both of these could clump together, help visible structures such as galaxies and clusters form, and hold them together as proposed by Zwicky and Rubin in particular. However, there are constraints on the masses that the particles must have in order to perform these tasks. Dark matter particles should be non-relativistic during the period of structure formation in order to form and hold together dwarf galaxies. This implies, in particular, that WIMPs should be heavier than the neutrinos in the SM. Likewise, observations of dwarf galaxies also set (much smaller) lower limits on the possible mass of a boson dark matter particle.

The fact that telescopes do not see dark matter implies that it does not emit much light, though it might consist of particles with a very small electric charge capable of emitting small amounts of light. Many theories suggest that dark matter particles might have some interactions of strength intermediate between the known weak interactions and gravity. These would have played key roles, together with their mass, in fixing the overall cosmological density of dark matter during the expansion of the Universe. Many proposed extensions of the SM, such as supersymmetry and theories with extra dimensions, suggested the existence of stable, neutral WIMPs that could have been produced soon after the Big Bang and would still be present in the Universe today, providing dark matter. Calculations of the present density of such WIMPs could reproduce the density of dark matter indicated by astrophysics and cosmology if the dark matter WIMP weighs about a TeV, possibly within reach of experiments at the LHC and elsewhere and motivating WIMP searches in laboratory experiments.

A generic prediction of WIMP models is the occurrence of events in which energy and momentum are carried away by invisible dark matter particles that do not leave signals in detectors, often called 'missing-energy' events. Some of these events are expected in the SM when neutrinos are produced in the decays of heavier particles, and the missing-energy events detected so far by experiments at both LEP and LHC are quite compatible with these expected SM sources. FCC experiments continuing these searches for additional missing-energy events beyond those predicted in the SM will have unequalled potential for detecting WIMP candidates for dark matter (FCC Collaboration, 2019c). The sensitivity of the FCC-ee and FCC-hh to invisible decays of the Z and Higgs bosons adds a further dimension to the FCC programme of searches for dark sectors, probing regions of parameter space otherwise inaccessible. For example, the very clean experimental conditions at FCC-ee will allow very sensitive searches for invisible decays of the Z and Higgs bosons beyond those predicted in the SM, as in models with additional neutrinos heavier than those currently known. Moreover, FCC-hh will be able to produce much heavier particles than can be detected at previous accelerators including the LHC. In particular, they will be able to look for missing-energy events due to the direct production of heavy WIMPs, and also events in which WIMPs are produced indirectly via the decays of heavier particles, as may occur in models based on supersymmetry or extra dimensions. FCC-hh searches should be able to discover or exclude WIMPs as dark matter.

FCC searches for dark matter will be largely complementary to those by future non-accelerator experiments, but only the combination of these strategies will be able to pin down the nature of whatever dark matter particle may be discovered. For example, missing-energy events at a collider could be due to particles that are relatively long-lived but not long enough to have survived since the Big Bang. On the other hand, if some non-accelerator experiments were to detect a WIMP, it would be unable to provide many clues to the nature of the underlying theory.

## 5.6  Back to the Beginning, and the Future

*D'où venons-nous? Que sommes-nous? Où allons-nous?* is the title of a painting by Paul Gauguin, which may be translated as 'Where do we come from? What are we (made of)? Where are we going?' The questions raised by Gauguin in the painting shown in Figure 5.4 are universal questions that human beings have been asking, perhaps, in their different ways, for hundreds of thousands of years.

They constitute the primary motivation for the research programme of the FCC, though physicists approach these questions from a perspective that is perhaps rather different from that of the people in Gauguin's painting. The sections above have mainly addressed the second of Gauguin's questions, namely 'What are we (made of)?' The search for dark matter is a natural extension of this question to include all the matter in the Universe, invisible as well as visible. However, it is just one of many ways in which FCC experiments will probe the fundamental physics underlying the evolution of the Universe and seek answers to all of Gauguin's questions.

For a physicist or cosmologist, Gauguin's first question, 'Where do we come from?', becomes the question—what physics has governed the evolution of the Universe from its beginning almost 14 billion years ago in the Big Bang? Measurements of the cosmological microwave background (CMB) radiation inform us about the state of the Universe some 380,000 years after the Big Bang, when atoms condensed out of a primordial electromagnetic plasma of photons, electrons, protons, and light nuclei. These CMB observations provide the most accurate measurements of the amounts of conventional matter, dark matter, and neutrinos in the Universe and also constrain the possibilities for other forms of undetected matter. The light nuclei such as deuterium, helium, and lithium were formed out of protons and neutrons by nuclear reactions some three minutes after the Big Bang. Protons and neutrons were themselves formed a few microseconds after the Big Bang, out of quarks and gluons that had previously filled the Universe with a strongly interacting plasma. Experiments measuring heavy-ion collisions at the LHC are studying the properties of this quark-gluon plasma, which is among the most perfect fluids known.

**Figure 5.4** *'D'où venons-nous? Que sommes-nous? Où allons-nous?'* Paul Gauguin's painting exhibited in the Museum of Fine Arts in Boston, Massachusetts, US

*Source:* Paul Gauguin, Public domain, via Wikimedia Commons

A key aspect of the FCC physics programme will be to extend these studies to the conditions that existed earlier in the history of the Universe, addressing Gauguin's first question, 'Where do we come from?' In addition to proton collisions, FCC-hh will be able to collide heavy-ions with each other or with protons. Therefore, FCC offers the opportunity for experiments observing ultra-relativistic heavy-ion collisions to study the behaviour of the quark-gluon plasma at an energy density orders of magnitude higher than those studied so far and will be able to cast light on its evolution towards the near-perfect fluidity measured at the LHC. FCC-ee collisions will measure the fundamental processes that governed the Universe when it was about a picosecond (a millionth of a millionth of a second) old with unequalled precision and may help reveal whether there is an unseen dark sector of matter and radiation existing in parallel to what we know. FCC-hh experiments observing proton–proton collisions will extend these measurements back to the processes that controlled the evolution of the Universe when it was a fraction of a femtosecond (some $10^{-16}$ seconds) old. Figure 5.5 shows different stages in the history of the universe, emphasising that its evolution in the early stages was controlled by fundamental particles and their interactions.

What else may have happened so early in the history of the Universe? According to the SM, at some moment during the time period to be explored by FCC experiments the Higgs mechanism for giving masses to fundamental particles must have switched on. However, we do not know whether this was a gradual process, or whether it occurred suddenly via a phase transition that might have led to observable signatures in the Universe today, such as a background of gravitational waves. Measurements of the interactions of the Higgs boson by the FCC experiments offer our best prospects for exploring the dynamics behind the generation of mass. Also, at some time during this early era probed by the LHC, WIMP particles of dark matter are likely to have disconnected from SM particles, with their subsequent density determined. It is only by recreating early-Universe conditions in the Universe that

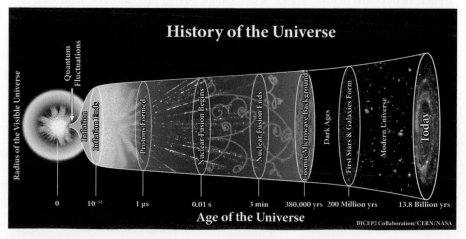

**Figure 5.5**  Different stages in the history of the Universe
*Source:* TheAstronomyBum, CC0, via Wikimedia Commons

we may be able to understand the processes leading to the present density of dark matter.

Another puzzle whose solution may have been found during the FCC era is the origin of matter itself. The Universe today contains over a billion times more radiation than matter, there are no known concentrations of antimatter. Why is there asymmetry between matter and antimatter, and why is there any antimatter at all? As noted in Chapter 2, in 1967 the Russian physicist Andrei Sakharov proposed a possible mechanism based on the microscopic differences observed between the weak interactions of matter and antimatter particles. The differences that have been observed to date in laboratory experiments can be accommodated within the SM, though without a deep explanation. However, Sakharov's mechanism requires some additional source of matter-antimatter differences and posits that the expansion of the early Universe must have deviated from the smooth expansion observed today. FCC-ee and -hh experiments will produce enormous numbers of particle-antiparticle pairs. These will allow detailed explorations of the possible differences between particles and antiparticles, potentially uncovering one element of Sakharov's mechanism that is missing. Another missing element could be identified if FCC experiments can establish whether particle masses were generated suddenly causing a departure from smooth expansion.

What of Gauguin's third question, 'where are we going?' The expansion of the Universe is currently accelerating, driven by an apparently near-constant density of energy in empty space, the dark energy mentioned earlier. If, indeed, it does not change with time, it can be identified with Einstein's cosmological constant. However, according to the SM, although it may be constant nowadays, it would have changed while quarks and gluons morphed into protons and neutrons, and while fundamental particles acquired their masses. These changes would have been many orders of magnitude larger than the density of dark energy today, raising the question of why the cosmological constant is so small today. FCC experiments will cast more light on the processes occurring in the early Universe, and perhaps reveal missing aspects of our current understanding of the dark energy problem. As mentioned earlier, one possibility is that the dark energy density will change in the future, putting an end to the current expansion of the Universe and causing it to terminate in a Big Crunch. This possibility is currently favoured by calculations within the SM based on present-day measurements of the masses of the top quark and the Higgs boson, and the scale of the strong interactions. Measurements by FCC experiments will provide a more accurate basis for these calculations, and possibly also uncover evidence for some extension of the SM that could avert the Big Crunch.

## 5.8 Boldly Going Where Only the Universe Has Gone Before

Every advance in human knowledge raises new, intriguing, and more profound questions. This is true, in particular, in fundamental physics following the establishment of the SM by experiments at the LEP and the LHC. Many questions have been raised in the previous paragraphs, and many possible answers. have been proposed. We do not know which, if any, of these answers are correct. That can only be resolved by experiments. As described above, the FCC experimental programme offers many ways to address the open questions and provide some of the key answers. However, it is also likely that FCC experiments will unearth new puzzles not mentioned above. With apologies to Einstein, we do know what the FCC will be doing, namely reproducing the particles, collisions, and other processes that have formed our Universe. However, we do not know what they are, nor what FCC experiments will discover, and that is the nature of fundamental research.

## 5.9 Marching Together: Brief Lessons from the History of Physics

A brief history of physics suggests that theoretical and experimental physics go hand in hand. Victor Weisskopf, the former director-general of CERN (1961 to 1966), values the dynamics of the experimental processes within the context of particle physics experiments and claims:

> There are three kinds of physicists, namely the machine builders, the experimental physicists, and the theoretical physicists. …. the machine builders are the most important ones, because if they were not there, we would not get into this small-scale region of space. ….. The experimentalists were those fellows on the ships who sailed to the other side of the world and then jumped upon the new islands and wrote down what they saw. The theoretical physicists are those fellows who stayed behind in Madrid and told Columbus that he was going to land in India.
>
> Weisskopf (1977)

The above allegory capturing the dynamic relationship between theory, experiment and instrumentation that defines the pace in particle physics research but also in other fields of fundamental science.

Looking back at the history of physics, one can find numerous relevant examples that led to breakthroughs in areas such as electromagnetism and general relativity. These examples should inform the balance between theory, experiment, and instrumentation, a discussion that is particularly pertinent as we discuss the physics motivation for a post-LHC generation of particle colliders.

Fundamental research that aims to push the boundaries of our knowledge further forward is—by definition—unpredictable. At certain junctures, theory may offer useful guidance, but at various other times in the history of science, experimental results have guided theoretical developments.

Tycho Brahe's main observations of stellar and planetary positions were noteworthy both for their accuracy and quantity. Though a geocentrist himself, his results led to Kepler's laws and the Newtonian revolution in physics. Before Tycho, probably no-one had ever thought to measure the position of Mars with such a degree of accuracy.

Likewise, when Willis Lamb and Robert Rutherford carried out an experiment using microwave techniques to stimulate radio-frequency transitions between the two hydrogen levels, there was no theoretical discrepancy to be solved. Yet the observation of the so-called Lamb shift led to the development of quantum electrodynamics that same year. To quote Freeman Dyson (Cohen et al., 2009): 'Those years, when the Lamb shift was the central theme of physics, were golden years for all the physicists of my generation. You were the first to see that this tiny shift, so elusive and hard to measure, would clarify our thinking about particles and fields.' The minor inconsistencies revealed by the precise measurement of the H-atom spectrum helped to point theorists in the right direction.

Similarly, another observation calling for a theoretical explanation was the $\varphi$ (phi) meson decaying to the theoretically unfavoured kaon-antikaon channel instead of the favoured decay to a $\rho$ (rho) and a $\pi$ (pi) particle. The observed suppression of this decay process by two orders of magnitude, compared to the theoretical prediction, led George Zweig to theorise the existence of quarks[1] (called aces by Zweig): 'if mesons contained aces with the proper quantum numbers, and if the aces in a decaying meson were conserved, that is, became constituents of the decay products' (Zweig, 2013) the decay pattern of the phi meson could be understood. And although Feynman thought that the experiment was flawed, it turned out that quarks do indeed exist and were experimentally observed a few years later. Other instances of experimental leadership include the discoveries of radioactivity and the CMB, which did not come about because of a well-defined theoretical target, but nevertheless opened the way towards a much deeper understanding of Nature.

When Galileo perfected the telescope, he could not predict how many moons would be discovered around Jupiter. Similarly, when studying the feasibility of future colliders, we cannot predict how many new particles we may discover, but only define the questions we wish to address in the spirit of fundamental research. In spite of the exploratory nature of collider projects, future colliders are not merely shots in the dark. Fully exploiting their potential calls for unity between theory, experiment, and instrumentation (Galison and Hevly, 1992; Galison, 1997). FCCs offer a solid, multi-decade-long, research programme with well-defined goals that can greatly contribute to the expansion of our knowledge of particle physics and the Universe.

## 5.10  Shaping a Vision for a New Research Infrastructure for the Twenty-First Century

According to our arguments above, the most efficient and comprehensive approach to thoroughly explore some of the open questions about our Cosmos is a new research infrastructure offering a staged research programme that would combine

---

[1] These were proposed independently by Murray Gell-Mann (who played a preeminent role in the development of the theory of elementary particles) and André Petermann (who pioneered the renormalization group, paving the way for the modern theory of phase transitions), for different reasons.

precision measurements with direct exploration of previously uncharted energies. In December 2018, the Future Circular Collider (FCC) collaboration submitted its Conceptual Design Report (CDR) (FCC Collaboration, 2019; 2019b), exploring the physics opportunities that opened up the next-generation of particle colliders housed in a new 100 km circumference tunnel in the Geneva area. A lepton collider (FCC-ee), as the first step, would push the precision frontier, followed by a 100 TeV hadron collider (FCC-hh) that would allow the direct exploration of previously inaccessible experimental areas. Further opportunities offered by the FCC complex include heavy-ion collisions, lepton-hadron collisions, and fixed-target experiments.

Succeeding in this challenge relies on a number of factors beyond the pure scientific merit of the project, as reflected in the history of previous Big Science projects. Realising an ambitious project like the FCC calls for efficiently building and managing an international collaboration across organisational, sectoral, and national boundaries. Particle physics and CERN are no strangers to this approach. At the heart of this effort lies the development of a global and diverse collaboration; this includes building a large and diverse community of users that seeks to exploit the physics opportunities as well as the means for leveraging resources and mitigating risks during the design, construction, and eventually the operation phase of the proposed colliders. The answers to these questions, together with the scientific opportunities offered by the FCC and results from the technological R&D programme, will inform the final decision on investing in a truly international research infrastructure at the heart of Europe.

The numerical and geographical growth of the FCC collaboration, from the first kick-off meeting in 2014 to the publication of the FCC CDR in 2020, testifies to the attractiveness of the project and the openness of the collaboration-building approach. A number of global R&D efforts were launched during the preparation of the FCC Conceptual Design Report to understand the present technological limitations and identify pathways for reaching the ambitious technical goals of the FCC and to demonstrate the feasibility and sustainability of this project. Adopting a clear long-term vision and a set of target performance parameters for the construction and operation of the FCC has promoted co-operation among diverse groups of researchers from academia and industry within the FCC collaboration, helping to clarify objectives and priorities as well as focus efforts towards them. From a managerial perspective, our goal has been to clearly articulate strategies and sets of goals among all the partners involved, in a transparent and open way, to help align their R&D innovation efforts with their business strategies.

The long timelines involved in this project and the ambitious but tangible technological challenges uniquely position large-scale projects like the FCC to set up an innovation system that maximises the participants' capacity for innovation. This system includes a coherent set of interdependent processes and structures for sharing the desired results with the participants. These processes also assisted in sharing resources and communicating past lessons and technical knowhow, as well as organising regular topical meetings and workshops (including the annual FCC meetings) for companies to exchange their problems and explore solutions. Diverse perspectives are critical to successful innovation. But without a strategy to integrate and align those perspectives around common priorities, the power of diversity is blunted (Massimi, 2019). Clearly defined targets, openness in communication and CERN's previous reputation were catalysers in enabling a culture of trust that allowed this ecosystem to work efficiently and produce results—and the first prototype solutions for many technologies are already being tested and refined. By 2021, the FCC collaboration will count more than 150 institutes including universities, research centres, and industries from 34 countries collaborating to advance the key technologies that will enable the efficient and sustainable realisation of the FCCs.

In addition to the geographical distribution it is perhaps worth discussing the time profile of the FCC project. The implementation of the first stage, the intensity-frontier lepton collider FCC-ee, commences with a preparatory phase of eight years, followed by the construction phase (all civil and technical infrastructure, machines, and detectors, including commissioning) lasting ten years. A duration of 15 years is projected for the subsequent operation of the FCC-ee facility, to complete the currently envisaged physics programme. The total time for construction and operation of FCC-ee is nearly 35 years. The preparatory phase for the second stage, the energy-frontier hadron collider FCC-hh, will begin during the first half of the FCC-ee operation phase. After the end of FCC-ee operation, the FCC-ee machine will be removed followed by the installation and commissioning of the FCC-hh machine and detector, which will take about 10 years in total. The subsequent operation of the FCC-hh facility is expected to last 25 years, resulting in a total of 35 years for the construction and operation of FCC-hh. It is important to note that the proposed staged implementation with FCC-ee as the first step followed by FCC-hh provides a time window of 25–30 years for critical R&D on key technologies that could reduce the cost and further improve the performance for the second-stage energy-frontier collider that will use the same infrastructure. In conclusion, the vision opened by the FCC study offers a solid and credible way to push the energy frontier further within the twenty-first century while advancing novel technologies to do that in a cost-efficient and environmentally friendly way.

Following the recommendations of the last update in 2020 of the European Strategy for Particle Physics (ESPPU, 2020), CERN has launched a feasibility study to understand the environmental and socio-economic impacts of the proposed research infrastructure. The goal is to study in depth the scientific, environmental, social, and economic impact of the project along with the physics opportunities that this research infrastructure could offer. The feasibility study report is expected in 2025 or 2026 as input to the next Strategy update, offering an opportunity to assess the technological challenges of realising the next generation of particle colliders for the twenty-first century. One of the main outcomes expected is the determination of the best placement and layout, balancing the territorial, geological, and physical constraints. The approach that the FCC team has adopted is to mitigate any risks and whenever possible reduce the environmental impact of the project while compensating for any potential impact in line with the principle 'avoid, reduce, compensate' foreseen in the European legal framework and adopted by CERN's Host States. The feasibility study will also serve to optimise the parameters of the two machines and maximise the positive effects of the development of new research infrastructure (RI) in the region.

Currently the FCC project foresees the next steps:

- **2025–2026**: Execution of the FCC feasibility study and production of a report that will inform the next European Strategy Update;
- **2027–2028**: Decision of the CERN Member States to launch the project if the conditions are met, within the framework of the European strategy for particle physics;
- **2030–2031**: Finalisation of the detailed study phase and deliberation in CERN's council for a final decision;
- **after 2033**: Start of civil engineering works, which should last until 2040;
- **mid 2040s**: Commissioning of the first collider (FCC-ee) for operation for around twenty years, alternating periods of operation and maintenance along with the necessary upgrades; and
- **mid 2060s–2070s**: The FCC-ee would then be replaced, in the second phase, by a hadron collider allowing for collisions of both protons and ions (FCC-hh).

As shown during the preparatory phase of the FCC Conceptional Design Report (CDR), the integrated FCC programme minimises the uncertainties that could potentially adversely impact its implementation. An early start of the project's preparatory phase is needed to allow for the timely implementation of the intensity-frontier lepton collider (FCC-ee) that marks the first stage of the project. Residual technical challenges for the subsequent energy-frontier hadron (FCC-hh) collider can be addressed through a well-focused R&D programme during the construction and operation of the FCC-ee.

An eight-year preparatory phase, which includes a feasibility study, is adequate to carry out the relevant administrative processes and develop a funding model for the first stage of the FCC, focusing on a new infrastructure and a high-intensity lepton collider. An immediate and related challenge is the creation of a worldwide consortium of scientific contributors who commit to providing resources for the development and preparation of the scientific part of the project.

## 5.11 Advancing New Technologies for New Discoveries

The proposed FCC will profit from CERN's existing accelerator complex and infrastructure that have developed over time to push the frontiers of knowledge by drawing on the latest technological advances. Today CERN operates several generations of accelerators, in particular: LINAC4 since 2017, the Proton Synchrotron Booster (PSB) since 1972, the Proton Synchrotron (PS) since 1959, the Super Proton Synchrotron (SPS) since 1976, and the LHC, (which was installed in the tunnel that had hosted the LEP between 1989 and 2000) commissioned in 2008 with the first physics results in 2010. The LHC, following the HL-LHC upgrade, will continue its operation until the 2040s, offering more data to tackle some of the open questions in particle physics. It is worth noting that LEP and LHC, like any large infrastructure, went through several phases during their development: in the case of the LHC a design phase (ten years), a construction phase (ten years) and operations (20–30 years).

Looking back at the history of particle colliders, we are reminded that in particle physics, like other scientific fields, scientific advancements are closely coupled with technological breakthroughs. For example, over the past 30 years, the exploration of the infinitely small has gone hand-in-hand with advances in superconducting magnets (Rossi and Bottura, 2012). Specifically, the increasingly powerful hadron colliders, from the Tevatron, commissioned in 1983, to the LHC in 2008, have led to spectacular discoveries thanks to developments in superconducting technologies that were used for building these colliders on an unprecedented scale.

Advances in accelerator technologies must be accompanied by advances in detector technology as larger numbers of more complicated particle collisions are produced. The technological sophistication of the LHC detectors is remarkable, as they include several subdetector systems, contain millions of detecting elements and support a research programme for the international particle physics community. The volume of data that will be produced during the high-luminosity upgrade of the LHC and by future colliders calls for even more sophisticated technologies. Further advances are necessary to enable the processing of larger and more complex data samples that eventually boost performance beyond today's state-of-the-art. For example, at least two areas that need immediate attention for technology development are superconducting materials and gases. Big Science projects such as CERN LHC, and in particular the greenhouse gases (GHG) of the present ATLAS and CMS gas detectors pose a big environmental issue. Resistive Plate Chamber (RPC) detectors are widely used at the CERN LHC experiments as muon trigger due to their excellent time resolution. They are operated with a Freon-based gas mixture of $C2H2F4$ and $SF6$ and these greenhouse gases have a very high global warming potential (GWP). Research is necessary to find environmentally friendly gas mixtures that help reduce GHG emissions and optimise RPC performance at a reasonable cost (Guida et al., 2020).

From an early stage, the FCC collaboration launched a number of R&D programmes bringing together academia with industry while also mixing traditional with new players. In this way combining valuable experience with fresh approaches in a number of technologies is essential to reach the desired performance and exploiting the physics opportunities offered by pushing the energy and intensity frontiers. Tackling the challenges of building and operating a research infrastructure of this scale in a sustainable fashion calls for technological breakthroughs beyond the improvement of existing technologies. From an early stage, and to succeed in preparing a Conceptual Design Report (CDR), the FCC tried to establish an environment characterised by creativity, agility, and openness as the conditions for nurturing research and innovation.

To this end, the FCC collaboration sets thematic priorities and focuses efforts on fields that show particular relevance for the sustainable implementation and operation of next-generation colliders, present great potential for growth and deployment thus maximising the societal impact, and exhibit a high potential for developing innovative solutions that could find applications in tackling other pressing issues of our societies. At the same time, the FCC management has been consistently developing all the competencies in technological skills, training and education that are necessary for the FCC study to offer a progressive research and innovation space, thereby strengthening the viability of this new research infrastructure.

Technology research and development during the FCC CDR preparation phase allowed us to identify the most relevant technical uncertainties and mitigate potential risks while paving the way to evolve the key technologies to the appropriate readiness levels to permit construction and efficient operation. Pushing the boundaries of accelerator and detector technologies for FCC further forward is an important step in the decision-making process for such large-scale scientific projects and is key for ensuring the sustainable and efficient operation of a new research infrastructure that will respect the UN's 2030 agenda for sustainable development.

## 5.12  A Tale of Science and Collaboration

In the following we briefly highlight some of the lessons learned from the global R&D activities launched in the framework of a global Big Science project like the FCC:

a) The FCC collaboration offers a physical and digital space and consequently the spatial and technological proximity among innovators in technology 'hotspots', academia, research centres, industrial parks, and technology incubators that is needed for the accelerating development of technology;

b) The number of different technological domains covered by the FCC study (e.g. beam control, vacuum systems, superconductivity and high-field magnets, radio frequency (RF) cavities, detector technologies, cryogenic and refrigeration, safety, environmental protection, etc.) boost the cross-fertilisation of technologies across various disciplines and result in a broader portfolio of competencies that are fundamental to the competitiveness of technology-based firms;

c) Industry innovation is frequently path-dependent and firms find it costly to break away from existing routines towards radically new or different concepts. The FCC collaborative R&D has encouraged risk-taking and supported different industries to open up to more innovative R&D solutions that they would otherwise not pursue alone. This approach paves the way to more cost-efficient technologies that could be industrialised at large scales, meeting the demand of future large-scale projects and also opening up the potential of using these technologies in market applications beyond HEP, while improving the performance and hence maximising the research potential of future facilities;

d) The ability to build a common vision with the project partners and stakeholders along with a path for turning this vision into reality has been critical for success in the R&D lines. During the first phase of the FCC study that led to the publication of the FCC CDR, it became increasingly apparent that vision can be both conceived in and directly impacted by the context of the times, while it is important from a managerial point of view to possess the ability to oversee that vision's implementation. Vision divorced from context can produce very erratic and unpredictable results;

(e) Alliances like those fostered by the FCC R&D programme are organisationally complex and require considerable resources to maintain collaborative activity compared with more arms-length agreements such as outsourcing. In other words, the collaborative effort that we develop comes at a certain cost and requires the allocation of well-defined resources for setting up a healthy collaboration environment among the different partners;

f) Two important factors that often characterise R&D efforts are risk and uncertainty. This has been the case for the FCC R&D programme. The concept of uncertainty within the innovation process is well-understood, and we will not delve into it in detail here. In general, the newer the sector, the closer it is to 'basic research' in the sense that the outcome of the research can lead to fundamental changes in knowledge, rather than technology. This is the case for many of the technological fields explored within the FCC study, with the domain of superconducting technologies (for RF cavities, high-field magnets, or detector components) being one of the most characteristic examples, given the interplay between instrumentation, theory and experiment that characterises this field. The FCC integrated programme can greatly benefit from such 'blue sky' research and, despite the higher level of uncertainty, the results can have a huge impact on high energy physics and beyond; and

g) Ongoing R&D efforts in the framework of FCC have demonstrated that the rate of technical change is determined not just by the level of uncertainty of technological change, but also by the number of possible directions in which it can develop. Thus, while technological change may not always be perceptible or discrete, it is continuous. It is not, however, determined by one company or concept but by numerous path-dependent solutions being developed independently by several aspiring innovators. A level of optimisation must be integrated into each step during a well-coordinated collaborative R&D effort.

Finally, the FCC study strives to assess the wider socio-economic benefits of collaborative R&D and understand how to maximise them for the FCC study stakeholders involved. To achieve that, from a very early stage the FCC study formed a group of economists, programme managers and policy-makers launching a number of research activities to understand and quantify the wider socio-economic impact

(Florio and Sirtori, 2016). While there is extensive evidence in the literature that innovative R&D leads to considerable economic benefits, there is still little agreement on the methodologies for assessing them. The FCC study invests in creating the space for debating and refining the different methodologies (Beck and Charitos, 2021), profiting from the intense ongoing R&D activities and offering an immediate interaction between the economists, the scientists and the firms working on these R&D programmes.

The discussion above confirms that working hand in hand throughout the entire innovation process is the key to success: from scientists who develop ideas; to innovators who bring ideas into the economy and society; and to people who use the innovations in their everyday lives. To ensure that the FCC research results feed even more effectively into practical application, we are strengthening transfer, supporting open forms of innovation and the development of breakthrough innovations, promoting entrepreneurial spirit and innovative strength in small and medium enterprises, and intensifying our integration into European and international networks and innovation partnerships.

The implementation of the 2020 update of the European Strategy for Particle Physics and the exploration of the feasibility of a post-LHC circular collider like the FCC are adaptive processes. We will therefore tackle its implementation and further technological developments jointly with representatives from science, industry, and society, developing synergies for a participative implementation strategy. At the same time, the success of the FCC feasibility study relies on the involvement and mobilisation of citizens more closely in research and innovation, to inspire the next generation of experts who can join the field and shape the scientific and societal potential offered by a new research infrastructure (RI).

## 5.13  Big Science and Public Investment in Fundamental Science

Ultimately, the value for money to be obtained from such large-scale scientific facilities will depend on the scientific discoveries they help make and the effective exploitation of that science. However, over the past years there has been growing evidence that though the scientific outcomes (and their economic benefits) remain uncertain, RIs bring a number of concrete economic outputs for society extending from industrial procurement and human capital formation to the cultural and educational impact of these facilities.

Given the intangible nature of certain benefits and the long duration of these projects it proves difficult to identify a common methodology for measuring this impact and designing good practices. From its inception, the FCC study together with the HL-LHC worked with a team of economists to develop the right tools.

Cost Benefit Analysis (CBA) represents the most widely used methodological tool to quantify such impacts and its theoretical background and application to large-scale Research Infrastructures (RIs) have been discussed (Florio et al., 2016). Each

RI involves a different set of benefits, costs, and stakeholders that need to be carefully identified and measured at the very beginning of the design of the CBA. Nevertheless, each RI has its own distinguishing features, goals, and time horizons.

Previous studies of the LHC/HL-LHC programme identified six economically relevant benefits: (1) the value of scientific publications; (2) technology spillover; (3) training and education; (4) cultural effects; (5) services for industries and consumers; and (6) the value of knowledge as a public good. The socio-economic impact assessment of the LHC/HL-LHC programme, carried out in the scope of an European Investment Bank (EIB) project by the University of Milano (Italy), has revealed the added value of public investment in research infrastructures. This was the first application of this method and yielded some encouraging results indicating how this impact can be better measured and also on the tools that would allow it to be further maximised. Today, the H2020 EuroCirCol project is a reference case to apply the EU recommended framework for infrastructure CBA to the research community.

The long-time frame of the FCC programme adds complexity to the design of a CBA for a post-LHC collider. However, the CBA of the LHC/HL-LHC serves as a foundation for an evaluation of the societal costs and benefits of different FCC scenarios. The CBA model developed in the frame of the LHC/HL-LHC programme assessment is thus both methodologically appropriate and also necessary for the FCC programme. It could be accompanied by technology forecasting analysis that might help improve the estimation of benefits for firms and other economic agents.

It is assumed that the existing diverse and vibrant set of FCC R&D activities in the field of particle accelerator and detector technologies will continue and will lead to a converging programme for a future research infrastructure, nourished by cross-fertilisation of different particle acceleration technologies, design studies, and the continuous optimisation of facilities in operation. To that end, FCC will continue its unprecedented work with academia and industry and develop an entire ecosystem of innovation and entrepreneurship addressing the sustainable construction and operation of a post-LHC collider as well as societal challenges.

Understanding the socio-economic impact of Big Science demands a large-scale institutional response and is an open challenge for FCC as well as for other large-scale global RIs. There is a rich landscape of potential stemming from public investment in such projects, reaching society long before—and in addition to—the scientific lessons we gain. The methodologies applied and the interpretation of results should be a major subject in public policy, and at grant agencies and universities—reminding us that a project like FCC calls for co-innovations and synergies between multiple disciplines.

## 5.14  An Adventure beyond Particle Physics

Why it's simply impassible!
ALICE: Why, don't you mean impossible?
DOOR: No, I do mean impassible. (chuckles) Nothing's impossible!
Lewis Carroll, *Alice's Adventures in Wonderland*

CERN has always had aspects that reach far beyond those of a particle physics laboratory, since its operation epitomises European unity and its dynamics on a material level. As far back as its establishment in 1954, it has played an important part in the attempt to coalesce the ruined and fragmented European space into a vigorous and unified scientific, technological, financial, political, diplomatic, and social sphere.[2] At present, when the vision of European integration is challenged once more through increasingly intensifying nationalist and populist tendencies, CERN's mission as a unifying mechanism becomes exceptionally relevant again. Thus, a new dynamic project, such as the FCC, would allow CERN to place the heart of global science on European soil once again: a heart that will be able to 'pump blood' around the entire globe, acting as a circulatory network for workforces, research methods and innovations, and presenting a tangible example of scientific, political, financial—and even social—relationships.

Large-scale research infrastructures like the proposed FCC have the potential to catalytically reshape the world around us also through the technological spin-offs that accompany them. We will not delve into the famous technological applications that emerged via CERN (the World Wide Web, PET scans, touch screens, etc.), but we will focus instead on one decisive historical event for post-war science. Shortly before the flames of the war were extinguished, in 1945, the President of the USA, Franklin Roosevelt, tasked the acclaimed Vannevar Bush with proposing guidelines on how science should be supported so that it would meet the practical demands that lay ahead in the peacetime era to come. The issue at hand lay in outlining a funding policy for science that could be expected to stimulate progress in practical matters. Bush suggested that basic research is pivotal in making practical progress. As he argued, technological innovation is not likely to be brought about by research narrowly targeted at the problem at hand.

A superior strategy would be to perform broad fundamental research. The chief argument given was that the theoretical resources suitable for resolving a practical difficulty cannot be identified in advance. Rather, practical success may be made possible by findings that are prima facie unrelated to the problem at hand (Massimi, 2021). Post-war science policy was structured upon this idea, developing not only our scientific but also our technological culture. The same spirit seems to still inspire the scientists of our time.

So, some decades after Vannevar Bush, CERN's former director Rolf-Dieter Heuer claimed: 'If you only do targeted research, you lose the side-routes. You lose the way to use different routes, to go into a completely different domain, and to go into a completely different way of making breakthroughs. If you do not invest in basic research at some stage, you start losing the basis of applied research. The two are intimately interconnected' (Jung, 2012). In the same interview Heuer gives a pertinent example: 'If you look back some eighty years, then basic research completely revolved around trying to introduce the concept of antimatter. Nobody would have dreamt at the time of the introduction of antimatter, as a theoretical concept, that it would be used 40 years

---

[2] For instance, one of the main factors which led to the establishment of CERN was to minimise the effect of 'brain drain' out of Europe and into the US.

later in the hospital. Hospitals that combine the PET with the MRI are using detectors that were developed from our experiments.'

Particle physics finds itself today at a critical juncture, mirroring that of the societies around us, which find themselves in a unique historical period: grand social visions are disfavoured, financial and ideological challenges test the limits of the social fabric, and faith in scientific knowledge is frequently called into question while unscientific narratives swirl within public discourse. In this context, scientific projects such as the FCC could potentially contribute more expansive visions for our societies, operating akin to road signs at crossroads like these.

This is not a guaranteed result, of course, but rather a challenge both for science policy makers to provide opportunities for engagement, as well as for the broader public to debate issues relating to inclusivity, diversity, and sustainability. Let us not forget, moreover, that CERN's own establishment, at another critical historical juncture over 65 years ago, inspired a world that was finding its way out of the darkness of two world wars and the atom bomb.

At present then, when contemporary particle physics is characterised more by an open-ended explorative kind of research rather than research that has been tailored to test any particular theoretical prediction, the situation should not be regarded as unprecedented (ESPPU, 2019). The fact that this particular situation is not terra incognita does not of course mean that there exist ready-made patterns for us to follow. The path towards discovering New Physics will be long and arduous, something that becomes apparent when looking at the numerous unsuccessful attempts through the years.

Our efforts to discover the underlying laws and the fundamental building blocks of the Universe are a universal and enduring endeavour that dates from Leucippus and Democritus to the discovery of the electron and the rise of modern high energy particle physics. The FCC study, designing the next generation of post-LHC particle colliders, continues this extraordinary story of exploration. Discovering the global character of the physical laws allows us to understand both the micro- and macro-structures of the Universe, while curiosity and the ability to learn and pose new questions are part of our shared human experience.

## 5.15 Conclusions

We have discussed some of the open questions scientists face in the current landscape of particle physics, along with the theoretical and experimental evidence for the existence of new physics beyond the Standard Model. Answering the big open questions about our Universe calls for synergies with other fields beyond particle physics, including astrophysics and cosmology. It was highlighted how collider physics, astrophysics and observational cosmology can help to shed light on the questions of dark matter and dark energy. Progress in particle physics could have a tremendous impact on other fields, contributing to our understanding of the origin as well as the future of our Universe.

Furthermore, it is important when debating Big Science projects to recognise the essential contributions made by different communities—not just theorists and experimentalists, but also engineers, technicians and postgraduate students who collaborate to develop new and more efficient, scientific tools that could advance us along the path of discovery. Progress in science calls for unity among the different communities. Rapid scientific development also requires the cooperation of various other stakeholders besides particle physicists, including information technologists and other specialists, as well as various industrial stakeholders and government research laboratories. A project like the FCC requires international cooperation across organisational, sectoral, and national boundaries, which is a basic feature of large research programmes.

We have focused on the FCC as the facility that offers the most diverse particle physics research programme for the twenty-first century. However, we believe that similar lessons apply when thinking about other proposed frontier colliders as well as instruments in astronomy that will help us to explore the twenty-first-century landscape of physics and astrophysics.

What key lessons can we draw from this chapter? The following are some important messages that we wish to share:

1. Answering the grand questions 'How did the universe evolve after the Big Bang? What are we [made of]? What is the fate of our universe?' are universal questions that people have asked throughout human history and they are the main motivations behind the scientific research programme of the Future Circular Collider (FCC);

2. The LHC has shown how Big Science experiments can not only probe fundamental theories such as the Standard Model but also look beyond it to explore how the majority of the mass and energy in our universe could originate from physics that is currently unknown;

3. FCC experiments will cast more light on the processes that occurred in the early Universe, offering unprecedented precision measurements and direct access to new energy regimes;

4. Theory is important, but the history of science reminds us that scientific progress is dependent on a continuous dialogue between theory and experiment—a healthy balance between theory, experiment, and instrumentation is essential;

5. Progress results from asking the right questions and addressing them experimentally—how else do we know what we know to be true?;

6. To answer key questions about the origins, structure, and behaviour of the Universe, international research infrastructures offering staged research programmes are necessary—Big Science research infrastructures such as CERN, ESO, and other scientific facilities unite the global community of researchers and combine their wisdom and intense scientific curiosity;

7. International collaboration across organisational, sectoral, and national boundaries is crucial for a new programme like the FCC and effective

international collaboration is a fundamental tenet for the success of Big Science programmes;

8. Proper management strategies by partners and stakeholders are the key success factors and are necessary to ensure smooth and cost-effective operations—'*short cuts make long delays*';

9. A new programme like the FCC would enable CERN to continue to make possible world-leading scientific research and help CERN to continue to provide leadership for research into new physics, phenomena, and industrial applications. Such knowledge has the potential to spin off many technological and social innovations in medicine, new materials, energy, complex climate change phenomena, and industry applications; and

10. The FCC has the potential to expand our understanding of the fundamental laws of physics, matter, and the universe and open up new frontiers in high energy physics.

The open and diverse FCC collaboration will require a balance between traditional and innovative players with strong industry involvement from the early stages of the life cycle of such a long-term project. Furthermore, in designing any of the next generation of Big Science projects, the study of their broader socio-economic impacts should be considered from an early phase, as this can also maximise the social returns from such a large public investment, by attracting broader engagement and support from the various stakeholders.

# PART 2

# INNOVATION THAT WORKS

Big Science generates valuable fundamental knowledge. Undoubtedly such knowledge is essential for human progress and solving existential threats and challenges. To make this fundamental knowledge useful for humanity, at least a good part of it needs to be converted into useful products, processes, and services. Part 2 explains how knowledge gets translated into valuable scientific, technological, and organisational innovations in Big Science settings.

# 6

# Knowledge Diffusion by Design

## Transforming Big Science Applications

*Christine Thong, Anita Kocsis, Agustí Canals, and Shantha Liyanage*

## 6.1 Introduction

Big Science creates a knowledge base to assist in tackling complex 'wicked' global challenges, such as climate change, medical diagnosis and treatment, public health surveillance, and GPS essential for many emergency services. Big Science refers to complex processes and systems, hence the complexity theory is closely associated with Big Science. Both deal with complex systems and multiple phenomena on a large scale. As discussed in Chapter 4 of this book, Big Science has the potential to lead to big innovation. In 2018, the European Commission communication, a Renewed European Agenda for Research and Innovation, emphasised that research and innovation are crucial for our future and asserted that it is the only way to tackle low economic growth simultaneously and sustainably, limited job creation and global challenges such as health and security, food and oceans, climate change, and energy (European Commission, 2018).

Taking fundamental research carried out in Big Science organisations, such as CERN, ESO, LIGO into the public domain is a challenging task. It requires a combination of outreach educational programmes, technology transfer initiatives, and industry procurement programmes. These programmes are designed collectively to promote public understanding and appreciation of Big Science research initiatives and to provide opportunities for the public and industry to interact with scientists to generate break through innovation.

Design theory and practice can offer a bridge to tackle complex global challenges in climate change, medical diagnosis and treatment, public health surveillance, and Global Positioning Systems (GPS)essential for emergency services. Design practices generally explore, conceptualise, and demonstrate new ways to understand complex thoughts in their most simple forms. This can be done through untangling fundamental and applied sciences through design artefacts. These artefacts may be physical or digital, products, built environments, services, or experiences. Design artefacts

Christine Thong et al., *Knowledge Diffusion by Design*. In: *Big Science, Innovation, and Societal Contributions*. Edited by: Shantha Liyanage, Markus Nordberg, and Marilena Streit-Bianchi, Oxford University Press. © Christine Thong et al., (2024). DOI: 10.1093/oso/9780198881193.003.0007 facilitate innovative applications of Big Science knowledge through codification and abstraction of scientific knowledge.

This chapter discusses how design artefacts and practices explore, conceptualise, and demonstrate new ways not only to just utilise, but also to understand how Big Science knowledge can be diffused through design artefacts. These design artefacts can be physical or digital, products, built environments, services or experiences, and more. Further, the processes used by design practices consider the end-users and implications for citizens from the outset, synthesising with other technical, economic, and societal considerations into tangible outcomes.

Design artefacts and practices facilitate innovative applications of Big Science knowledge. They also offer new possibilities for the codification and abstraction of scientific knowledge. The chapter explores examples of design practices transforming Big Science knowledge to be applied to new societal contexts. Designing to simplify complex systems is possible with careful planning and execution of design practices. Boisot's I-Space Framework and Social Learning Cycles (SLC) (Boisot et al., 2011) are used in a new way to frame how design practice may simplify fundamental scientific knowledge residing in Big Science organisations and facilities. Further, the chapter introduces the SLC as a metaphor or archetypes for translating tacit and explicit fundamental knowledge for innovative activities. The I-Space framework is used to explore how examples of design practices may influence knowledge diffusion across science, technology, and innovation.

Through design theories and practices, Big Science conveys new meaning to knowledge from different actors (users or citizens) in various innovation contexts, and simultaneously diffusing and expanding knowledge. Design practices can mediate complexity, facilitate innovation, and thus unlock new opportunities for the societal impacts of Big Science. Design practices can also help to understand user needs, goals, and behaviours. They can assist in breaking down complex knowledge into more manageable modules that can show the interdependence of parts of the complex system.

Indeed, design knowledge and practice, like complexity theory, may provide a common language and framework for scientists, industry, and government agencies to work together to solve complex problems.

Examples from particle physics (CERN), astrophysics (Melbourne Museum), and dark matter particle physics (Australian Research Council Centre of Excellence) are discussed in this chapter. Authors suggest how design practices may influence the nature of SLCs in the fundamental and applied sciences. Design practices may act as a 'generator' and 'connector' of knowledge, thus shaping scientific knowledge flows to influence reach and impact on society. The examples explore the conception of new objects and machines used in daily life, visualisation and citizen science to engage the general public in scientific concepts and fostering the capability of future innovators to use design practices to leverage the potential of Big Science for sustainable growth and development of society.

## 6.2  Approaches and Practices of Design Disciplines

Design is a broad term and means many different things depending on the context. The word 'design' entered the English language in the 1500s as a verb, with the first written citation of the verb dated to the year 1548. Merriam-Webster's Collegiate Dictionary defines the verb 'design' as 'to conceive and plan out in the mind; to have as a specific purpose; to devise for a specific function or end' (Manzini, 2015). Design comes with a set of cognitive models of purpose, vocabulary, practices, communities of practice, and research set up (Kimbell, 2011). Design is a profession; a research discipline, and praxis invested in changing and challenging the current state, problem context, or status quo into an outcome or improved state. According to Michlewski, design-inspired constructs have to be adopted, or at least discussed, in order to be effective or to have an impact (Michlewski, 2015). Design thinking and concepts can act as a mindset and a process for creative complex problem solving.

Design as a process (Archer, 1979) can be organised into generic stages of discover, define, develop, and deliver organised as a 'double diamond' according to the UK Design Council (2019). Starting with a challenge, and ending up with a tangible outcome, there are always contingent contexts, cultures, and human factors influencing the double diamond, expressed as engagement, leadership, design principles, and a method bank. Iteration underpins the design process, with cycles of iteration within and between stages varying according to the project at hand. The UK Design Council (2021) has identified that design builds a bridge between technological research and innovation and their application to social practice. This is further supported by Verganti (2009) who explains design-driven innovation, where radical change in meaning (breakthrough innovation) peaks at the intersection of technology-push and design-driven practices. However, if design only considers current market demands, incremental change is the likely outcome (Verganti, 2009) and may fall short in trying to address complex societal challenges. This highlights the importance of integrating design practices with Big Science knowledge.

Design thinking is currently a common creative problem-solving approach, used by design and non-design professionals alike for its applicability across different industry contexts and community settings. Design thinking is both a process and a mindset, seeing outcomes that balance human or market desirability, economic viability, and technical feasibility (Brown, 2005, 2009). As a process, like the double diamond, design thinking can be expressed in generic stages; empathise, define, ideate, prototype, and test. These stages can be applied iteratively to an end user-challenge to deliver a useful or improved outcome. As a mindset, design thinking is open, the curious, experimental, and adventure-seeking people with diverse perspectives, will be able to explore innovative challenges and also learning by doing. In other words, design thinking uses abductive thinking and iterates between converging and diverging to realise ambidextrous learning, thus mediating human dimensions and experimentation (Zheng, 2018).

General design processes like design thinking and double diamond frameworks[1] vary in action according to the purpose and aims of the challenge at hand. Design thinking among other design frameworks in human-centred design demonstrates a history of prioritising a human first approach such as user-centred design (Van der Bijl-Brouwer and Dorst, 2017).

Subsets of user-experience design can assist service and digital interactions. Co-design integrates the users and/or stakeholders in the design processes. Human-centred design draws on a range of tools including: observations, interviews, cultural probes, rapid prototype testing, and heuristics to source and integrate user insights.

Further two other design processes include: speculative design (Dunne and Raby, 2013), where design artefacts depict and provoke discourse on the future; and circular design (MacArthur and IDEO, 2017). Design thinking can also have an environmental sustainability focus. In radical innovations, the role of design can be to disrupt the status quo and induce disruptive change (Verganti, 2009).

The types of design artefacts and associated skills are broad. These broad perspectives can be undertaken by design professionals, where design processes are engaged with more readily by other professions including business, science, engineering, and health fields.

Design practices are both thinking and doing, and also include an agent of change. Design has changed in the last 50 years as a discourse and as a problem-solving perspective and tool with cross-disciplinary boundaries. It is a distinct craft, a broker of change, and a precursor of innovation (Cooper, 2019). Design practices feature mindsets, activities, and processes that overlap with other disciplines or can be applied in other disciplines. Designs have played a role in Big Science in the visualisation and image reconstruction of events in LHC experiments and astrophysics. Engineering prototypes and artefacts have been prepared for various components and devices. Design is an integral part of many scientific instrumentation and components in LHC detectors and Hubble telescopes. These design functions contributed to both functional and human interactions with the instrumentation. In most popular innovation such as medical devices and equipment, detector technology from high energy particle physics is extensively used.

However, the approach of design as a professional discipline emphasises even more the importance of people with objects and experiences—finding a balance between people's needs, desires, passions, and emotions with scientific, technological, and economic considerations. Technical, economic, ethical, and desirable requirements are synthesised and socially mediated into concrete artefacts.

These artefacts represent design processes performed by expert design disciplines. Such design approaches can uniquely explore ways to simplify complex large scientific data to be used, understood, evaluated, and experienced by non-scientific audiences.

---

[1] The double diamond framework was designed by the UK Design Council and it consists of four phases: Discover, Define, Develop, and Deliver.

## 6.3  Big Science, Knowledge Diffusion and Social Learning Cycle (SLC) Archetypes

Big Science organisations and experiments have different approaches to the diffusion of knowledge. With a large number of international participants in Big Science experiments, both tacit and explicit types of knowledge diffusion take place. Tacit knowledge is difficult to codify, articulate, or transfer and it is deeply embedded in individual experience, skills, and beliefs. Explicit knowledge is relatively easy to codify and communicate through language or other symbols.[2] Explicit knowledge includes scientific data, methodologies, and scientific communications. Diffusion of both tacit and explicit knowledge is essential for success in Big Science and its translation of knowledge to end users. Different communication processes need to be put in place to establish a culture of knowledge sharing, translating, and dissemination.

The translation of knowledge into useful forms of artefacts that can be disseminated to society is an important function. Various methods of diffusion of knowledge via design practices are employed in the diffusion of knowledge to society. Designs can be part of symbolic language that facilitates to conveying abstract concepts and mental experiences as discussed in theories of embodied cognition and situated learning.

In this section, a conceptual framework of the Information-Space (I-space) is used to understand the dynamics of knowledge flows in social systems (Boisot et al., 2011). The interplay of codification and abstraction underpins the creation of all reliable knowledge.

I-Space was built on the premise that the codification process allows information to be extracted from data (for example LHC data) and subsequently structured and shared within a community. Using codification processes, structured knowledge flows more easily than unstructured knowledge (Boisot and Nordberg, 2011: 32; Ihrig and Child, 2013).

I-Space takes information structuring as being achieved through two cognitive activities: codification and abstraction. Codification articulates and helps to distinguish from each other the different categories of knowledge we use (Boisot, 1998: 42). The higher the degree of codification, less data processing that will be needed to categorise particular phenomena (Boisot, 1998: 44). When a phenomenon is complex or vague in expression, or when the categories we use to apprehend it are not clear-cut, the amount of processing efforts tend to be high.[3]

---

[2] A detailed discussion on tacit and explicit knowledge diffusion can be found in I. Nonaka and H. Takeuchi (1975), *The Knowledge Creating Company, How Japanese Companies Create Dynamics of Innovation*, Oxford University Press, Oxford.

[3] Codification facilitates the associations required to achieve abstraction, and abstraction, in turn, by keeping the number of categories needed down to a minimum, reduces the data processing load associated with the act of categorisation. Taken together, they constitute joint cognitive strategies for economising on data processing. The result is more and usually better structured data. Better-structured data, in turn, by reducing encoding, transmission, and decoding efforts, facilitates and speeds up the diffusion of knowledge within a given population of agents while economising on communicative resources (for details see Martin and Child, 2013).

Graphically the diffusion curve in the I-Space can be presented in a three-dimensional space with the degrees of codification, abstraction, and diffusion on each of the three axes (Figure 6.1). The left part of this curve illustrates how after increasing the degree of codification and abstraction of a given message, it will be possible to diffuse this knowledge to a larger number of actors in the population. Once knowledge has been diffused within a target population, it may be absorbed by that population (a movement down the codification dimension called absorption) and then applied in new specific situations (a movement from the abstract to the concrete end called impacting). The right part of the curve in Figure 6.1 illustrates these movements.

However, the absorptive capacity[4] of each individual determines the level of knowledge absorption, integration and combination (Van den Bosch et al., 1999). As a consequence, knowledge reservoirs may decrease the degree diffusion. The process known as scanning may identify new knowledge that has been created or provide an opportunity for new knowledge creation and also catalyse a new set of knowledge to be diffused.

Such knowledge creation and diffusion as well as reconstituting knowledge happen continuously in Big Science environment. In large collaborations, the CERN and ESO have very skilled experts who are able to solve complex problems considerably faster. Larkin et al., (1980) labelled those experts as having 'physical intuition.'[5]

Taken together, the different processes mentioned above configure a cycle in the I-Space, known as the Social Learning Cycle (SLC) that is continuous in creating new meaning (Boisot, 1998: 58; Boisot and Nordberg, 2011: 36).

As described, this cycle is made up of six steps: scanning, codification, abstraction, diffusion, absorption, and impacting (Figure 6.1). Many different shapes of the SLC are possible, reflecting both the obstacles and the incentives to the learning process. However, one of the tenets of the I-Space framework is that when learning leads to the creation of new knowledge, the cycle will move broadly in the direction of the stages indicated in Figure 6.1 (Boisot et al., 2011) and repeatedly move through the stages, thus creating a continuous cycle. In the case of Big Science, the SLC was largely applicable to the learning process within scientific experiments.

As knowledge flows through the SLC, new meaning is assigned to information by actors according to their social reality thus providing a visual and descriptive model that may represent the continuous co-creation of knowledge. This makes the SLC useful as a framework to discuss, where and how design practice can influence scientific knowledge while simultaneously diffusing and creating knowledge.

---

[4] Absorptive capacity is defined as an organisation's ability to identify, assimilate, transform, and use external knowledge, research, and practice (see for details Cohen and Levinthal, 1990), 'Absorptive capacity: A new perspective on learning and innovation', *Administrative Science Quarterly*, 35(1), 128–152.

[5] A sizable body of knowledge is a prerequisite for expert knowledge and this knowledge must be indexed by large number of patterns, that on recognition, guide the expert in a fraction of time to relevant parts of knowledge store. The knowledge form complex schemata that can guide a problem's interpretation and solution contributing to what is called physical intuition (see J. Larkin, J. McDermott, D.P. Simon, and H.A. Simon (1980), Expert and novice performance in solving physics problems, *Science*, 208, 1335–1342.

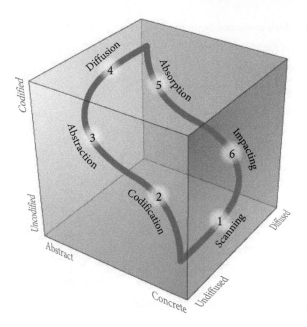

**Figure 6.1** The social learning cycle (SLC) and flow direction of stages

*Source:* Boisot, Max, and Nordberg, Markus, (2011). 'A Conceptual Framework: The I-Space', in Max Boisot and others (eds), *Collisions and Collaboration: The Organization of Learning in the ATLAS Experiment at the LHC* (Oxford; online edn, Oxford Academic, 22 Sept. 2011)

As discussed in Chapter 4, fundamental or basic knowledge needs to co-exist or be co-created with the application of knowledge. Knowledge application and diffusion translate into useful products, processes, and services. In some Big Science knowledge, high levels of diffusion can be achieved by targeting specific or multiple communities and/or industry sectors that are receptive to taking up resultant innovation. For example, in the case of accelerator research for medical applications, maximising the diffusion of technology for public consumption. SLCs that represent knowledge flows in both fundamental and applied science are presented in Figure 6.2.

Fundamental science follows a narrow cycle that goes up and down the codification and abstraction dimensions, representing maximum diffusion as when knowledge is accessible across a specific fundamental science community (e.g. high energy physics or astrophysics). Most fundamental knowledge resides in the public domain and is accessible to a wider community. In Big Science, knowledge is quickly published and disseminated, so not much time is spent at the point of minimum diffusion. The degree to which fundamental science is codified and abstracted remains confined to the specific science community.

Applied science follows a rounder and more compact shape that starts with higher levels of abstraction and codification of knowledge so that the fundamental or applied science cycle establishes the adjacent possible (Kauffman, 2000) for applied science. This means translating fundamental to applied science is closely

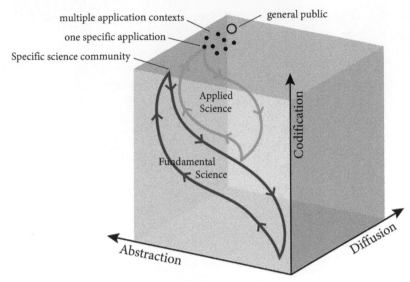

**Figure 6.2**  Proposed fundamental science and applied science SLCs
*Source:* Author — C Thong

associated. This notion assumes that a certain level of codification and abstraction is required for knowledge from fundamental science to inspire the experimentation or implementation of applied science in a specific real-world context.

Figure 6.2 shows the scope to further codify and diffuse fundamental and applied knowledge for additional applications and reach further actors including the general public. We will discuss how this can be done using design concepts.

## 6.4  Knowledge Transformation through Design

The challenges or parallels in both designing for and understanding the social construction of meaning in the diffusion of scientific knowledge are 'bridging the micro to the macro features of society', as communication is between individuals, and across groups, sub-groups and sectors (Spitzberg, 2014: 317). Diffusion of innovations in large-scale, transdisciplinary societal good projects, requires 'design attitude'—design's sense making and problem-solving skill (Rawsthorn, 2018) as part of the equation. Problem solving and sense making are interconnected and design is intrinsic in materialising knowledge in ways that are 'explicit, discussable, transferable and compoundable' (Manzini, 2015: 39).

Designers involved in large scale transdisciplinary projects found in Big Science for example contribute to, architectures of adaptive integration (AAI), defined as the explicit and implicit dynamic structures and processes that characterise collaborations among heterogeneous groups of scientists and stakeholders working to address complex societal problems (Morton et al., 2015). Design methods help integrate non experts in AAI projects to address the context and bridge the gap between research and practical application. Integrative research helps to extend scientific knowledge and innovation applied beyond the problem at hand. This is because, it was intended 'only because it is adapted, extended, or modified by subsequent scholarly contributions that take up different roles in the diffusion process' (Herfeld and Doehne, 2019: 4). Design practices are useful in transferring Big Science knowledge to usable equipment, products, and utility services.

Design practices produce the end user scenarios. User studies are relevant to the success of any campaign or societal uptake. Touchscreens for example had a quick uptake once products needed them as explained earlier, whereas the adoption of seat belts as a lifesaving preventative device (Rogers and Kincaid, 1981) had varying levels of adoption according to socio-economic factors and market laws.

Design practices also integrate a high level of coding in order for users to easily access or understand the intention of the product service and system. As discussed in Boisot (2013), the more abstract and codified nature of messages become, the more likely they are to travel beyond the environment in which they are originated and the greater power they have in establishing remote references. However, they draw on a range of similar principles, for example complex concepts as picture language—Neurath's International system of typographic picture education (ISO-TYPE) (Lupton, 1986; Pietarinen, 2011) and how graphics reveal data (Tufte, 1983) or skeuomorph aeroplane a graphic digital interface that mimics a physical object.

When considering design in relation to SLCs, there are three key points in the cycle where the authors propose design may progress knowledge moving along the cycle, as described in Figure 6.1. First, at step 2 codification a designer can respond to scientific knowledge (which must have some level of codification for the transmission of this knowledge), exploring, generating, and conceptualising new ideas. If the scientific knowledge was touchscreen technology, codification may manifest as sketches, prototypes, and visualisations proposing touchscreens integrated into the dashboard controls of an aeroplane, used on a mobile phone, or how people select which floor they want to exit a lift. While these ideas were not realised, a broader audience can interpret the tangible demonstration of concepts. This may further codify scientific knowledge to assist in 'accelerating' knowledge towards the SLC step 3 abstraction.

Second, between step 3 abstraction and step 4 diffusion, designing artefacts represents scientific knowledge and transmits it through physical and/or digital forms (such as products, graphic visualisation, built environment, and services). For example, industrial design, interaction design, and design engineering practices would consider the various functions the smartphone provides to its user. Advance functions include how to realise these customer centric design requirements via the assembly of various components, external casing, ergonomics that shapes, how the user holds and operates the phone, digital interfaces, aesthetics, etc.

Design practices may also be considered as a connector across SLC typologies for fundamental science and applied science. An example of design practices as a connector may be demonstrated through design artefacts of touchscreen technology. Capacitive touchscreen technology developed specifically for computer control system needs at CERN is shown in Figure 6.3. It was adapted to be fit for purpose in smartphone and tablet applications that are ubiquitous in day-to-day life many decades after the original CERN application. This technology is one example of the long technology gestation period and the slow rate of adoption of technology emanating from Big Science.

These applications further abstract, codify and diffuses the capacity of technology, by connecting the top point of the SLC to higher levels of diffusion. Here, design practices integrate the end user's needs and desires (tactility, ergonomics, aesthetics, features, functions, interactions, and experiences) of smart phones and tablets, and demonstrate what is technically feasible and economically viable when detailing designs for production. Additionally, design practices can challenge what might be technologically feasible by showcasing novel user experiences. The functionalities in design concepts, thereby influence further R&D in technology development to make such performance feasible.

The SLC of new applied science work for sensing technology is made possible by the established knowledge of prior applied science sensing projects, prompted by the needs for a socially mediated design products.

The notion of design practice acting as a generator and connector in Big Science knowledge is interconnected with the diffusion via design artefacts. These artefacts evolve interdependently based on societal needs and readiness. The following sections explore three examples of design practices integrating with fundamental and applied knowledge from Big Science, and the authors propose how different design practices influence such knowledge flow.

**Figure 6.3** Original application area of capacitive touch screen technology; CERN SPS control room in 1977

*Source:* © CERN

## 6.5  Example 1. IdeaSquare, CERN: An Experimental Innovation Platform within the Organisation

IdeaSquare was founded in 2014 at CERN as an interdisciplinary, early-stage innovation platform located within the organisation to experiment with new ways particle physics might benefit society at large. It's an open learning facility designed to generate new ideas in a collaborative environment and to promote experimental innovation and rapid prototyping for innovation-related projects. Integral to all IdeaSquare activities is the culture that supports a positive mindset for innovation that is closely associated with curious and innovative mindsets in Big Science communities (CERN, 2019).

The ATLAS experiment is located right next to IdeaSquare, which offers facilities and tools to support prototyping in order to quickly explore, communicate, and test new ideas. The interiors of IdeaSquare are designed to promote serendipitous conversations, knowledge sharing, and interdisciplinary teams to collaborate and workshop ideas.

Innovation processes take place in IdeaSquare draws on both entrepreneurial and design disciplines. Therefore, IdeaSquare could be viewed as a sort of innovation lab, with expertise in facilitating an ecosystem of activities geared towards design-inspired innovation that makes use of science and technology for the benefit of society.

IdeaSquare facilitates an ecosystem of activities aligned in pursuit of design-inspired innovation that leverages science and technology for societal good. In reality, IdeaSquare works with other Innovation Labs (Design Factory Global Network), businesses, and universities all over the world as part of an open innovation system. This allows for a greater diversity of disciplinary perspectives, including design, that may combine to interpret, explore, and synthesise Big Science knowledge in new ways.

IdeaSquare organises novel forms of expertise interaction and collaborative workshops, three-day hackathon, and short and long R&D projects. Mentors and staff at IdeaSquare facilitate and champion various learning and knowledge exchange initiatives. Three key initiatives driven by IdeaSquare are ATTRACT,[6] Challenge Based Innovation (CBI) courses, and Crowd4SDG. For example, the Challenge Based Innovation (CBI) programmes launched at IdeaSquare at CERN are challenge-driven and student-centred programmes. They allow students with multidisciplinary back grounds (not only science but also business and other disciplines) to work collectively with academics, staff at research institutes and industry around the world. By doing so they could also collaborate with CERN researchers to discover novel solutions that may assist to solve major social, environmental, and economic issues. The concepts of design and simplifying complex knowledge are central to these initiatives.

Table 6.1 summarises the project parameters with attention to various types of design practices and artefacts used to extend the diffusion of knowledge used at

---

[6] For details of ATTRACT see Chapter 12.

Table 6.1  Summary of three IdeaSquare design practice initiatives

|  | ATTRACT 1 & 2 | CBI courses | CROWD4SDG |
|---|---|---|---|
| Project Purpose | To explore commercial applications of imaging and detecting technology, with attention to societal benefit | To explore opportunities for the application of CERN technology to address UN SDG's | To find opportunities for citizen science to develop programmes to monitor UN SDG's |
| Type of Science | Fundamental and applied | Applied | Fundamental and Applied |
| Design Practices | Prototyping, sketching, CAD visualisations, user-centred design, design thinking, circular design | Prototyping, sketching, CAD visualisations, speculative design, user-centred design, design thinking, circular design | Prototyping, sketching user-centred design, design thinking |
| Design artefacts | Physical and/or digital proof of concept prototypes for different types of applications relevant to specific projects | Physical and/or digital demonstration and proof of concept prototypes for different types of applications relevant to specific projects | Design demonstrators and visualisations relevant to specific projects |
| Audience | Primary: stakeholders from industry sectors related to the technology and application secondary: tertiary students, educators | Primary: tertiary students and educators Secondary: Stakeholders from industry sectors identified in specific projects. | Primary: general public, particularly those aged 16–26 in Cycle 1 of the programme secondary: stakeholders from industry sectors identified in specific projects, NGO ecosystem in Geneva |

IdeaSquare. These initiatives demonstrate how design may act as a generator and connector; applied to Big Science through ATTRACT and Crowd4SDG initiatives, and also through CBI courses.

To address the United Nations Sustainable Development Goals (UN SDGs), CBI courses are offered by various universities around the world each using their own design and innovation courses and practices (CBI, 2020).

Tertiary student teams come from different disciplines including design, engineering, business, and social sciences. They collaborate to explore, conceptualise, and develop new ideas, communicating using physical and digital prototypes that demonstrate design intent and function.

As demonstrated in Figure 6.4b, ACTIWIZ 3 software and ROOT software technologies from CERN are proposed to be adapted for 2030 to realise a smart, closed-loop drone farming system informed by real-time data and the proposed new idea of prototyping, codifying, and abstracting Big Science knowledge to act as generators along the SCL cycles.

(a)

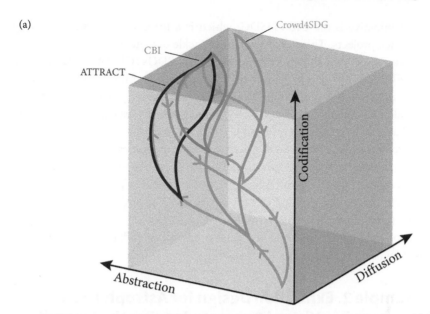

(b)

**Figure 6.4** (a) Conceptual SLCs in relation to Big Science SLCs and (b) proposed digital prototype of future design metaflora

*Source:* Author — C Thong

The idea is to involve tertiary education students and academics in using concept testing with potential end-users in mind. These programmes connect communities, CERN's stakeholders and industry leaders for the diffusion of knowledge on some Big Science issues.

Another initiative is the Crowd4SDG, which is a three-year initiative fostering citizen science projects. These projects aim to tackle climate action through innovations focused on AI to effectively monitor the UN SDGs (CROWD4SDG, 2020). This is an EU funded project supported by the European Commission's Science with and for Society (SwafS) programme. Crowd4SDG uses design practices as part of its Gather Evaluate Accelerate Refine (GEAR) method, where idea exploration leads to functional and demonstration prototypes.

These projects and initiatives act as generators of various types of knowledge reside in CERN to initiate new projects for citizen science. Various programmes act as connectors or conduits of useful knowledge to diverse new actors. The programmes also connect local Non-Governmental Organisations (NGOs) and the participation of the general public from all over the world.

The proposed SLC for these initiatives and programmes differentiates knowledge flows as shown in Figure 6.4a and Figure 6.4b.[7]

## 6.6  Example 2. Exhibition Design for Astrophysics. The Universe in a Virtual Room: Balancing the Integrity of Scientific Truth with a Meaningful Visitor Experience

Visualising the universe in museum spaces has an established history. This example discusses the challenges of co-design between designers, scientists, and interdisciplinary teams in the pursuit of making complex scientific theories, comprehensible, and meaningful for visitors in a museum. The setting of the following project is one of the first Virtual Room visualisation platforms (Figure 6.5a) developed by Swinburne Astronomy Productions (2021) at the Centre for Astrophysics and Supercomputing, Swinburne University for exhibition at the Melbourne Museum.

As part of the exhibition brief, astrophysicists wanted to share observations regarding the 'Einsteinian' universe with the museum audience. In practice this meant communicating content that incorporated supercomputer simulations, separate observations from astronomical observatories around the world, and the explanation of difficult scientific theories.

The role of the designer revealed a triadic negotiation between the visitor, designer, and scientific team to: (a) explore the communication of complex scientific content through visualisation and prototypes; (b) to consider closely the differences in knowledge systems and language while confronting the challenge of highly specialised and unfamiliar scientific content; and (c) to codesign with the scientific team from the perspective of the visitor.

The activity of designing and co-designing together revealed the scope and challenges of deconstructing complex content, codified through design semantics, media and interaction utilising user-centred design as neither the designer nor the audience has sufficient grounding to master scientific knowledge. The tension between

---

[7] A conceptual idea for a drone operated smart-farm to reduce food waste in hospitals. Student work by Lachlan Mackay, Paris Triantis, and Justin Yuan.

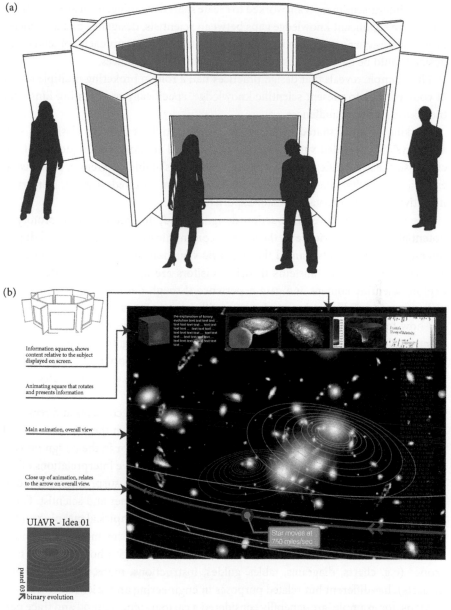

**Figure 6.5** (a) Digital sketch of a virtual room and (b) prototype of an interactive exhibition

*Source:* Author – Kocsis, A. (2010). Co-designing new media spaces (Doctoral dissertation, UNSW Sydney).

how scientific integrity be kept intact versus abstraction and codification for the diffusion of concepts for wider audience experience and learning became apparent as designers, astrophysicists, and exhibition team worked together. The challenge for the designers along with the scientists was how to receive, interpret, convert, transmit, and communicate complex data into visual form for museum visitors.

As an interdisciplinary team worked to create shared meaning, it became clear that there were significant knowledge gaps between scientists, designers, and audiences. From the designer's point of view, the user-centred design approach contends that design should be based on messages that resonate with audiences.

The example reveals that design practices had a role in brokering multiple layers of codified data to present scientific knowledge, specifically representing Einstein's theories to museum audiences.

Working towards shared meaning, co-design activities between designers and astrophysicists revealed various mental models and semiotic techniques. In some circumstances scientific content was not easily translatable. Therefore, prioritising empirical integrity, versus scientific communication required abstract, non-truth narrativisation of the data for visitor experiences.

The designer employed interaction design, sound, and animation to prototype potential visitors experience in the science communication. Prototypes and digital interaction experiments framed the scientific visualisations to explore some complex concepts such as Einstein's theories. Visitors are offered different designs to explore scientific content and gain experience through design. The challenges for visitors are two-fold: (a) how to interact with human interfaces with technology; (b) how to communicate science concepts with technology. The design practices position the visitor with both user and audience guiding the scientist's priorities in structuring the content through the activities of prototyping—screen grabs, motion graphics, diagrammatic expressions of formulae and associated voice/audio (see Figure 6.5b).

Co-designing also helped to break down the key scientific concepts and reassemble components to convey their potential meanings. Although the scientists valued the definitive scientific truth-effect, the inclusion of the visitor in the design process helped the astrophysicists and wider team consider alternative interpretations of the data. Data visualisation, a key concept in the design strategy, surprisingly proved to be contested ground in the semiotic strategy between designer and scientist. Complex scientific information is visualised to provide distinct empirical meaning and to be reinterpreted by visitors using design semantics, colour, form, motion, and audio.

Visualisation, rather than direct collection of data through photographs or telescopes (e.g. charts, diagrams, tables, guides, instructions, maps, 3D images, and datasets), has different but related purposes in engineering and science.

Maps, for example, are generally considered a cartographic method and trace perceptions of the cosmos, as both document and creative expression and these maps offer compelling indirect evidence for the existence of the elusive dominant matter component that shapes our universe (Natarajan, 2021). Another example is interactive visualisation for analysing the results of numerical simulations in computer physics (Berry et al., 2011: 2301). For the information of designers, the diagrams and maps become design strategies for enhancing the dimensionality and density of portrayals of information (Tufte, 1983). All three of the aforementioned visualisation techniques instantiate various modes of communication, exposing the scientists to different modes of scientific communication.

**Table 6.2** Summary of the Universe in virtual room design practices

|  | Museum visitor experience—Universe in a virtual room |
|---|---|
| Project purpose | Codesign to introduce visitors as determinants in science communication. Interaction design methods to present fundamental science content as experiences for visitors |
| Type of science | Fundamental |
| Design practices | Sketching, prototypes, interaction design, exhibition design, 2D and 3D visualisation, co-design, user experience, experimental design |
| Design artefacts | Exhibition design: interpretation of science content via media for the virtual room |
| Audience | Stakeholders: general museum audience; visualisation experts; technologists experimenting with VROOM tech |

Co-design methodology gives scientists the parameters of a semi-immersive interactive exhibition environment to move beyond didactic, empirical, and truncated meaning. Interaction and information design allow the codification of scientific information to provide multiple meanings. These meanings provide new visual relationships to data for visitor experiences.

Co-design played a role of translating cultural models and accompanying semiotic strategies for diverse teams, in order to deconstruct or shift the cycle of codification to abstraction. The learnings reveal cultural model disparities that cast doubt on the designer's ability to serve as a mediator of meaning between audiences and scientists as well as between scientists and designers. Table 6.2 summarises the project's specifications with an emphasis on the various design features that were used to increase the spread of scientific knowledge.

The Virtual Room Design demonstrates how design acts as both a generator and a connector for fundamental science. It creates a new museum experience that engages humans through abstracting and codifying knowledge so it may be communicated to a wider audience.

It can be argued that the design strategies used in these examples are capable of shifting the SLC curve of fundamental science to reach higher levels of diffusion (see Figure 6.6).

## 6.7  Example 3. Design for Dark Matter: Applied Innovation Lab

The Australian Research Council (ARC) Centre of Excellence for Dark Matter Particle Physics (CDM) was established in 2020, as a large-scale, research initiative to explore and understand the nature of dark matter (DM) over a period of seven years. CDM has many partner organisations and institutions across Australia (and internationally), and utilises state-of-the-art detection facilities such as the Stawell Underground Physics Laboratory (SUPL) located in rural Victoria, Australia.

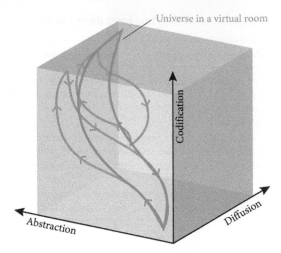

**Figure 6.6** Proposed mapping of the universe in a virtual room SLC
*Source:* Author — C Thong

Big Science research undertaken by the Centre aims to extend the understanding of the Standard Model of particle physics (see Chapter 5 for more details on DM research). ARC funding is highly competitive, and in line with global trends, has increased its emphasis on awarding criteria based on research creating positive impact via societal and/or commercial avenues.

Inspired by IdeaSquare at CERN, CDM integrated an Applied Innovation Laboratory into their educational activities. This is not a physical space but rather a virtual laboratory as CDM members are dispersed across nodes at multiple universities in different states across Australia. With the help of its members, CDM will hold innovation workshops, foster an innovative culture, fund the development of good ideas, train the next generation of workers to apply design thinking to Big Science applications, and enable design artefacts that serve as proof-of-concept prototypes for new ideas that can be developed further through the incubation or accelerator programmes that are now offered by the majority of universities.

The Innovation Lab features two core initiatives:

- Developing '*Kreative Kits*' to give secondary school students, especially those in remote areas, opportunities to learn about the intersection of science and design practices:
- Joining forces with IdeaSquare, CERN's CBI initiative through the Australian hosted programme CBI $A^3$, to include technology related to DM science as well as CERN and ANSTO science. This combines local industry and community considerations with global perspectives.

CDM's core initiatives demonstrate how design may act as both a generator and connector for Big Science projects, propelling SLC's with higher levels of diffusion as demonstrated in Figure 6.7a. Table 6.3 summarises the project parameters of these

**Table 6.3** Summary of two CDM design practice initiatives

|  | CBI A$^3$ course | Kreative Kits |
|---|---|---|
| Project Purpose | To explore opportunities for the application of DM related sensing and detecting technology to address UN SDGs | To use design practices to engage a broader, more diverse group of secondary school students in STEMM topics, and foster the next generation of innovators with the capability to conceive socially responsible applications of Big Science. |
| Type of Science | Applied | Fundamental and Applied |
| Design Practices | Prototyping, sketching, CAD visualisations, speculative design, user-centred design, circular design, design thinking | Prototyping, sketching, CAD visualisations, user-centred design, design thinking |
| Design Artefacts | Digital and/or physical design demonstration prototypes for different future application relevant to specific projects | Conceptual designs, visualisations, demonstration prototypes, digital and/or physical exhibitions of design artefacts |
| Audience | Primary: Tertiary students and educators Secondary: Stakeholders from industry sectors related to specific projects | Primary: Range of secondary school student, their families and educators secondary: those interested in the societal topics addressed in specific projects, potential users, and general public |

two initiatives with attention to various components of design practices used to extend the diffusion of DM knowledge.

*Kreative Kits* provide a physical resource package to support 2–5 day learning programmes consisting of learning materials, plain English technology resources, exercises and guidance on design practices to undertake innovation projects that use DM science and Science, Technology, Engineering, Mathematics and Medicine (STEMM) concepts to address societal challenges, like sustainable food production. The Kreative Kits themselves could be viewed as a design artefact that will act as a generator by codifying and abstracting DM science so that secondary school teachers, students, and their families can understand and interpret Big Science knowledge in new ways increasing diffusion by reaching new audiences. Further codification and abstraction will occur when students use the *Kreative Kits* to explore and propose new ideas that demonstrate the application of Big Science. This alters the position of the SLC in the iSpace, creating a shape that is a connector across fundamental and applied SLC typologies.

Similar to *Kreative Kits*, CBI A$^3$ is a learning programme where design practices guide the exploration of new utilitarian applications for 2030 to address UN SDG's using CERN, ANSTO, and CDM related technology (CBI A$^3$, 2018). The CBI A3 programme, which serves as a generator, will codify and abstract knowledge derived from various knowledge bases. It does this by using a technology card tool designed to explain such technologies in straightforward terms so that non-scientific audiences can suggest innovative ways to use them for societal benefit (via UN SDGs).

Further codification and abstraction occur when students propose new ideas, creating new meaning as Big Science applications are depicted in new contexts.

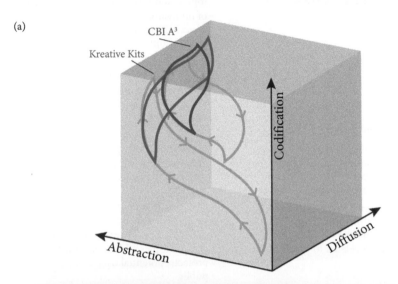

**Figure 6.7** (a) Proposed mapping of Kreative Kits and CBI A$^3$ SLCs in relation to fundamental and applied science SLCs; (b) digital prototype demonstration of a proposed 2030 design halo, an airport screening system to detect illicit and biohazardous substances

*Source:* Author — C Thong

For example, past CBI A³ students have proposed that CERN's Medipix3 chip technology might be adapted for uses other than Big Science applications. These students visualised the future use (by 2030) of image and detector technologies for a new conceptual idea for airport scanning device to further reduce illicit and/or biohazardous substances crossing security borders (see Figure 6.7b).

As a conceptual idea, much further Research and Development would be required to realise technical validation and functional prototyping as part of the process towards commercial implementation, however the explicit communication of conceptual ideas still demonstrates the abstraction and codification of technology into a proposed socially mediated artefact.

Demonstrating the outcomes through physical and digital prototypes allows audiences external to Big Science organisations to engage with knowledge from applied science. CBI A³ reaches new audiences of tertiary students, educators, their interested family and friends, and stakeholders from the societal problem space of each student project. This increases the diffusion of knowledge, and creates an SLC shape that demonstrates CBI A³ as a *connector* of applied science to new and further applications.

## 6.8 Conclusions

The chapter outlines design theory and practices that provide a systematic approach to scientific knowledge diffusion through creativity, iteration, and innovative solutions. Through collaborating with design practices and artefacts, it is possible to unravel complex problems and challenges associated with Big Science organisations and experiments. Further, the design artefacts generated are innovative in their own right, offering new products, experiences, services, and built environments that can address complex or 'wicked' problems and socio-cultural challenges. Overall, design artefacts, such as sketches, drawings, models, prototypes, diagrams, flowcharts, specification and design guidelines and concepts, simplify how complex knowledge can be translated into social and economic practice. They are essential parts that assist to understand complex systems and concepts and provide ways to communicate, test, refine, and build useful solutions.

Designs have played a major role in Big Science in the visualisation and image reconstruction of events in LHC experiments and astrophysics. Using design concepts, engineering prototypes and artefacts have been prepared for various components and devices. Design is an integral part of many scientific instrumentation and components, including LHC detectors and large instruments like LIGO interferometers.

Design practices can codify, abstract, and generate new meaning for Big Science knowledge by synthesising human, technical and economic considerations into tangible design artefacts. These artefacts in turn can create niches and complex scientific knowledge, that may otherwise appear obscure, accessible to anyone outside the specific field related to Big Science.

The I-Space framework and Social Learning Cycles demonstrate how design practice influences knowledge flows across fundamental and applied sciences in order to diffuse knowledge for societal application. The knowledge translation processes, mindset, and skills of design practices act as a generator and connector to Big Science knowledge. Design artefacts facilitate the accessibility of complex scientific knowledge, that may otherwise appear obscure, to anyone outside the specific field related to Big Science. Such transfer can happen in many forms of informal and formal training, education, and knowledge exchange programmes.

Initiatives like IdeaSquare, CERN's Universe in a Virtual Room and Applied Innovation Laboratory, or the Centre for Dark Matter (CDM) in Australia, show international examples of how Big Science can integrate design practices to conceptualise, communicate, and develop human-oriented applications. This serves as a pragmatic model for future Big Science initiatives that wish to increase the range of societal benefits derived from their research.

Design theory and practices can contribute to the development of scientific tools and technologies that facilitate Big Science research initiatives. Enabling design artefacts can help the expression of socio-cultural connections that enhance greater understanding, increase impact, and influence Big Science in society.

# 7

# Big Science, Leadership and Collaboration

*Grace McCarthy, David Manset, Marilena Streit-Bianchi, Viktorija Skvarciany, and Shantha Liyanage*

## 7.1 Introduction

Fabiola Gianotti, CERN's Director General since 2016 and the first female to hold two consecutive appointments in the history of the European Organisation for Nuclear Research (CERN), said that the value of science is central to Big Science initiatives and Big Science research demands high creativity and complex work and interaction. As a result, leading Big Science organisations and experiments demand special kinds of leadership traits and behaviours.

Leadership issues in Big Science can be examined from two different perspectives: (a) leadership roles to steer strategic outcomes for organisations; and (b) leadership styles necessary to achieve strategies, and the goals and outcomes of individual experiments. The authors focus here more on the latter.

Big Science organisations such as CERN in particle physics in Geneva, Switzerland or the European Southern Observatory (ESO) in astrophysics whose headquarters are located in Germany, derive their research leadership from their successful operations in the past which led to impressive contributions to science and to the international physics community.

This chapter outlines the role of leadership in managing complex research organisations and what it takes to deal with advanced technologies and a highly competent scientific community that collaborates to achieve common scientific goals.

The research for this chapter was drawn from interviews with leading researchers and directors who were responsible for creating and steering several Big Science organisations such as CERN and ESO, and large experiments such as ATLAS, CMS, or at LIGO and smaller LHC experiments such as LHCb and ALICE. The authors conducted extensive interviews, ethnographic studies, and case observations to elicit the key leadership traits and behaviours applicable to Big Science projects. From conception to realisation, leading Big Science projects demand well beyond scientific expertise and draw on organisational financial, diplomatic, and other administrative knowledge and skills.

Grace McCarthy et al., *Big Science, Leadership and Collaboration*. In: *Big Science, Innovation, and Societal Contributions*. Edited by: Shantha Liyanage, Markus Nordberg, and Marilena Streit-Bianchi, Oxford University Press.
© Grace McCarthy et al., (2024). DOI: 10.1093/oso/9780198881193.003.0008

## 7.2  Leading Scientists in Big Science Experiments

Big Science leadership is a combination of individual vision and collective efforts based on prior knowledge. A former Director General of CERN and Nobel laureate Carlo Rubbia aptly described this: 'In Big Science, the role of the individual scientist must be carefully preserved. So is that of original ideas and contributions. Our collaborators are as proud and honoured as we are in receiving this Prize' (Carlo Rubbia, Speech at the Nobel Banquet, 10 December 1984).

Big Science is, in fact, about national and international collaborations and commitments to do 'big things' that simply cannot be achieved by individuals alone. It is about collective research endeavours that put the team first rather than an individual first. Big Science thrives due to the collective constructivism of the scientific community.

In the context of a Big Science experiment, leaders are commonly referred to as 'Spokespersons'. At the top level, Big Science organisations such as CERN and ESO, like most enterprises, have an organisational structure governed by a Council. Council is a decision-making authority with delegates from all member and associate member states determining the key policies and taking all important operational decisions concerning scientific, technical, and administrative matters.

The Council is assisted by several committees specialising in the domains of scientific policy and finance. The Director General is the Chief Executive who reports to the Council and has a directorate running departments. Depending on the nature of operations, the departments of Big Science organisations such as CERN are structured according to their major areas of work, e.g. detectors, beams, technology development, information technology, physics, administration, health and safety, etc. The organisational chart of CERN is given in Figure 7.1.

Big experiments such as ATLAS and CMS are managed as independent entities, hosted by CERN and subjected to CERN organisational rules and procedures. Collaboration partners for these experiments are drawn from all around the globe.

Similar decision-making structures are used in other Big Science organisations such as European Southern Observatory (ESO). The Director General leads the organisation and is appointed by the ESO Council, ESO's main organisational and operational units are the Directorates, each led by a Director.

As illustrated above, the flow of decisions in Big Science host organisations is top-down, i.e. decisions flow from the Council to the Director General to the Sectors, Departments, and down to Groups and Sections. However, this is not a linear process but rather a consultative process where decisions are taken in consultation with leading scientists both in-house and outside. When it comes to Big Science projects or experiments, decision processes are more bottom-up processes, led by collaborative efforts.

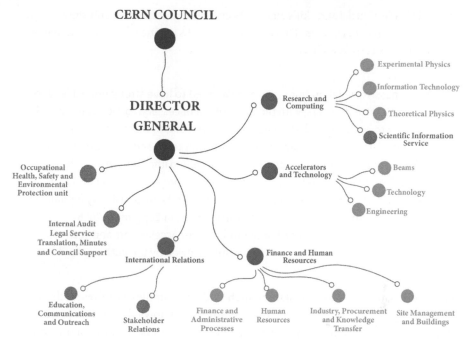

**Figure 7.1**  CERN governance and organisational chart, 2019–2020
*Source:* © CERN

Experiments are staffed by collaborating partners with only a small number of core staff affiliated with the host organisations such as CERN and ESO. In the case of CERN, total staff numbers were 3459 and in ESO, 750 in 2021. Some of these staff, fellows, and users come from different countries and organisations. Usually core permanent staff and students work at the operational sites where instruments are located (detectors, telescopes, gravitational waves interferometers, etc.) whereas some affiliated research partners may work from their home institutions.

Given the level of knowledge complexity and how Big Science experiments are conducted, leaders fulfil specific high-level leadership tasks, including:

- Creating a scientific/technical collaboration to build and maintain major scientific infrastructure;
- Negotiating high-priority large-scale scientific research projects in particular for funding;
- Collecting, organising, and analysing research data and undertaking large and small experiments;
- Managing a highly motivated scientific community; and
- Mentoring and negotiating high-quality research at the cutting-edge of knowledge.

Scientific leaders' tasks and skills emerge as critical to the effective and smooth operation of research experiments. The following remarks by one of the interviewees give a flavour of this responsibility:

> We have massive collaborations now and we need to have strong project leaders, strong accelerator directors, with the authority, if necessary, to move people from one activity to another, and do what needs to be done to restore functioning and performance. It is very important to be sure of what to do and therefore to have all the different committees necessary to get documented information from the different specialists, but ultimately the essential role of a project manager or director is to make decisions, follow their implementation and take responsibility for difficult decisions like in the case of the LHC start-up incident in September 2008.
>
> (Frédérick Bordry, Former Director for Accelerators and Technology, Interviewed by the authors, April 2021)

Due to the nature of fundamental research and the complexity of research infrastructure, Big Science projects often require extensive knowledge and multiple skills developed over many years. A leader's ability to attract and inspire highly accomplished individual scientists and to build effective teams are essential leadership traits. Fellow scientists must be able to respect, trust, communicate effectively, exercise empathy, and appreciate different points of view. Therefore, Big Science leadership entails effective and human-centred approaches as well as a highly pragmatic, responsible, and diplomatic approach. A leader's willingness to engage in dialogue with colleagues and senior management is critical for making decisions that are based on scientific strategy, institutional governance, and the best interests of the scientific community.

In addition, leaders have to respond to economic and political realities. In a knowledge-intensive environment, almost all are equal players in their chosen fields. Leaders need to inspire not only scientists but also the international scientific community, funders and inter-governmental organisations for collaboration and connectivity on the fundamental questions about nature (Robinson, 2021; Smart et al., 2012).

Large scientific infrastructure projects, for example LHC and its experiments, or large interferometers like LIGO and Virgo, require a special leadership style that focuses on clear scientific goals and outcomes with maximum efficiency. Building a strong culture of collaboration is never easy with unequal international partners who have various scientific and political agendas. CERN for example hosts several LHC and non-LHC experiments drawing together various international partners to collaborate. Different leadership approaches are necessary to steer projects of different magnitudes.

CERN and ESO experiments are extremely expensive and require thousands of scientists working collaboratively in each experiment. The funding for such experiments comes largely from the government and can be difficult and competitive to obtain. The persons who lead experiments should have the ability to think systematically, meaning the ability to breakdown problems into the steps that need to be followed.

One of the CERN leaders interviewed said: 'You need to be challenge-driven. I think, a leader that just wants to keep the status quo and avoid challenges is not the right personality for leadership. Leading groups in a complex organisation require complex leadership interactions' (Doris Forkel-Wirth, Former Head of CERN Health, Safety and Environmental Protection Unit, Interviewed by the authors, June 2021).

Big Science is also context-specific and follows specific investigative trajectories based on a robust scientific case. These scientific trajectories in high energy physics, astrophysics, and medicine lead to independent groups who are able to bring about scientific and technological success. The unanimous support of the international particle physics or astrophysics community is essential to continue such research.

Consensus building and democratic processes are indeed the pillars of international collaboration of which leaders need to be cognisant. Leading Big Science requires looking beyond national interests and exigencies. As a result, Big Science leadership is far from an authoritarian or hierarchical process. Rather, it has to be a collegial and cooperative process based on mutual respect for all required expertise. Given the uncertainty of outcomes, a leader must promote innovativeness whilst keeping focused on physics requirements.

## 7.3  Big Science Vision and Political Support

Effective leadership is derived from a collective vision to solve complex scientific problems through collaboration among international scientific communities. Big Science leads to exploring uncharted waters and the leaders have to take firm decisions to create in-house research capabilities and procure external capabilities to build infrastructure that was neither tested nor built before. In LHC experiments such as CMS, the leaders had to translate the physics guidance into an experiment that would be well placed to answer the physics questions posed.

'At CERN physicists have a vision or a theory they need to prove and the engineers have to make this vision or theory a reality. These two disciplines at CERN have been accomplishing unbelievable results...' (Emir Sirage, Executive Director New Space Portugal Agenda and former Technology Transfer Staff, CERN and CEO of the Atlantic International Research (AIR) Centre, Interviewed by the authors, March 2021).

In other words, leaders need to ignite desire and passion to investigate 'wicked problems' in science—meaning problems that are critical to solve, in order to enable significant advancements in knowledge, theory, and practice. Scientific community together with its leaders needs to find resources from national and international governments to continue such research investigations.

Indeed, research leadership and international collaboration go hand in hand. In the case of Big Science experiments such as ATLAS and CMS, the broad scientific culture encompasses a good understanding of the theoretical as well as the experimental and technological context. Leading scientists must collectively agree in a Big Science research portfolio well placed to answer the physics questions posed. Once such research questions are identified and agreed upon, leaders need to work hard to get the endorsement and support of their research colleagues and the funding agencies to implement those ideas.

Leadership in the field of inquiry and past scientific achievement determine future prospects. For example, the European Strategy for Particle Physics outlines the global context of CERN's leadership in particle physics research:

> CERN should undertake design studies for accelerator projects in a global context, with an emphasis on proton-proton and electron-positron high energy frontier machines. These design studies should be coupled to a vigorous accelerator R&D programme, including high-field magnets and high-gradient accelerating structures, in collaboration with national institutes, laboratories and universities worldwide.
>
> (CERN, 2013)

The senior leader's judgement needs to align with such overall strategies in the research field. Outlining the future priority in Big Science, the Director General of CERN explained, 'Europe's top priority should be the exploitation of the full potential of the LHC, including the high-luminosity upgrade of the machine and detectors with a view to collecting ten times more data than in the initial design, by around 2030' (Gianotti, 2019).

The leadership process involves a structured approach to decision-making, and good communication of the evolving situation concerning the decision to be made. Effective communication is very important for building a consensus, in order for decisions to be made by consensus and not by vote or persuasion.

The construction of Big Science experiments like CMS and ATLAS at LHC or LIGO detectors requires deep knowledge and skills and interdisciplinary knowledge of the field as well as a command of associated knowledge.

One of the lead scientists explained:

> Clearly when founding something like an LHC experiment, the ability to innovate and to identify and adopt new ideas, is quite important, as well as making sure that consensus is created the ideas that are most likely to lead to overall success.
>
> (Tejinder Virdee, Interviewed by the authors, August 2021)

Big Science organisations attract highly skilled expertise in dealing with diverse problems that are not only unique but difficult for a single research laboratory to tackle. It is increasingly recognised that the generation of major or significant fundamental research knowledge requires the use of large-scale science and technology infrastructures, which are complex, international, and available for collective use.

The style of leadership in Big Science that emerges relates more to a paradigm of 'scientific leadership'. In such a paradigm, leaders have to be scientifically and technologically highly accomplished individuals. Due to the knowledge-intensive nature of Big Science, such individuals have to adopt leadership traits and behaviours that align with consensual recognition of scientific ethics and values that are different from the traditional business leadership traits and behaviours in transactional, transformational, charismatic, etc.

In Big Science environments, one has to deal with leading a group of highly motivated intellectuals with diverse capabilities, so that, leadership is closely associated with knowledge creation, synthesis, and utilisation and is driven by scientific values and norms (Liyanage and Boisot, 2011).

The leaders in the high energy physics and astrophysics communities have been remarkably successful in addressing complex problems. They have impressive track records of scientific achievements, multidisciplinary skills, and working in international collaborations (Robinson, 2021). Elaborating further on why Big Science leadership is unique, one of the interviewees emphasised:

> There are two interesting dimensions of leadership in Big Science organisations and experiments. To organise, accomplish and run Big Science projects you need to garnish a huge amount of resources and work with so many countries and people. The leaders also have administrative roles, which are quite different from managing intellectual groups. The classical leadership roles, therefore, do not fit into this mould. Leadership authority comes from collaboration and from your colleagues.
>
> (Peter Jenni, Former Spokesperson, ATLAS, Interviewed by the authors, June 2021)

Big Science research is also about collective decision-making enterprise. These decisions are driven by selected scientific missions that are negotiated and accepted by the international scientific community. The research aims, therefore, must converge on significant problems, and ask the right types of questions (Esparza and Yamada, 2007). As in all research, the ideas need to rekindle the human imagination and curiosity so that the scientific community is convinced about the value of their contributions to human understanding and hopefully, over the long run, to social wellbeing.

Different Big Science collaborations such as ESO, LIGO, and Virgo, radio astronomy organisations, European ground-based telescope projects such as the Very Large Telescope (VLT) and the Extremely Large Telescope (ELT), and also Australia and South Africa's Square Kilometre Array (SKA) appear to approach leadership in novel ways. ESO has developed a set of values that underpin the organisational ethos. ESO outlines that communication, sharing experiences, and developing skills are essential and relevant in all parts of the organisation. Efficient and open communication, transparency, and integrity are ESO's core principles (ESO, 2021).

In addition to building these collaborations and steering projects, leaders have the additional tasks of managing budgets and carrying out prudent project management and controls. The LHCb spokesperson emphasised the significance of clearly outlining the expected community benefits and commented:

> There is no use in having the most amazing idea for a project, if that project is only supported by a community that is not going to be able to finance it. There have to be elements of discussion which are not just focused on the data and the best scientific outcome of the project, but also deep consideration of the relevance and emphasis on: What elements will the community be interested in? What can the community contribute with? What does the community want to finance?
>
> (Christopher Parkes, spokesperson LHCb, Interviewed by the authors, April 2021)

The leaders' task is to demonstrate strong scientific cases and put forth convincing arguments to exemplify the expected discoveries, explain the uniqueness of the problem-solution and garner the unanimous support of the fellow scientific community. While the ability to manage the project within the allocated budget and resources is important, the most difficult hurdle is to assembling the right teams with a commitment to big ideas.

Big Science leadership encounters a further exceptional challenge: that is, to be capable of communicating and persuading politicians, bureaucrats, and financiers to support fundamental science. The outcomes of fundamental science are often uncertain and hard to predict. Convincing financiers and politicians to support such projects is a difficult task. From a political and economic viewpoint, a cost/benefit analysis may assure the funders but that is not the only reason—prestige, future investment, and desire for intellectual leadership come into play. In other words, Big Science leaders have to persuade and equip with different techniques and arguments to convince the funding agencies. This requires mature science diplomacy skills and competencies to make science-informed development happen. In Big Science, science diplomacy is not only managing relations between collaboration international agencies and governments, but also integrating different cultures of scientists and social aspirations.

The factors central to leading Big Science initiatives are outlined in Figure 7.2. The most influential factors revealed in interviews were the need for a clear scientific vision, political support, and strength of collaboration and commitment.

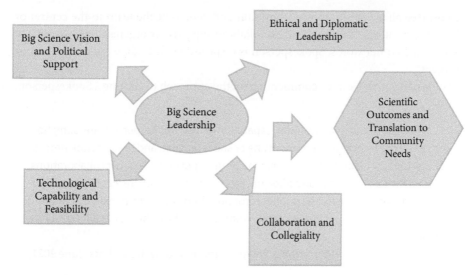

**Figure 7.2**  Big Science leadership processes
*Source:* Created by author S. Liyanage

## 7.4  The Role of the Spokesperson in Big Science Experiments

The Spokesperson is an important concept in Big Science experiments in a collaborative setting. The lead scientist is not even called a leader, but rather a 'spokesperson', a term commonly used in all experiments in CERN, LIGO, Virgo, etc. Unlike a leader in the corporate world, a spokesperson does not have the authority to hire or fire people. This is because in Big Science collaborations, scientists come from different universities and research institutes and are bound by their own organisational human resources policies.

For example, at CERN, a spokesperson is not a leader appointed by the hierarchy. Spokespersons are elected using a ballot within the Collaboration, such as in ATLAS or CMS, for a period of two years, with the possibility of extending that term by a further two years. In LHCb and ALICE the term is for three years.

> Leaders of experiments, such as CMS, labelled Spokespersons (SP), are elected by the Collaborations. I believe what the Collaborators are looking for is a SP who has a good scientific vision, good technical expertise and is considered to be a good manager with good people skills. High importance is attached to scientific credibility that is internationally recognised, along with a broad scientific culture and a significant portfolio of achievement.
>
> (Tejinder Virdee, Former Spokesperson CMS,
> Interviewed by the authors, May 2021)

The Spokesperson has one or two deputy Spokespersons depending on the collaboration. The Spokesperson and Deputy Spokespersons have the responsibility to oversee all aspects of the collaboration and represent the team to the central or host organisation, funding agencies, collaborating partner organisation and outside agencies. In other words, Spokesperson is responsible for leading the experiment or project.

One of the interviewees, commenting on the approach to electing a Spokesperson, said:

> The authority and management capability of the Spokesperson are somehow based on the trust of the community he or she is leading, and the selection process of the Spokesperson and others in management positions. In most collaborations, at least the spokesperson and often the physics coordinator are elected. It is different from most university and public sector positions where the person is appointed rather than elected, which does make a difference to how the managerial structure works.
>
> (Christopher Parkes, Interviewed by the authors, June 2021)

The Spokesperson is central to implementing the collective scientific and technological strategies and ensuring that the collaboration focuses on the key goals and strategies. The spokesperson is nominated based on his or her expertise and his credentials and past achievements. A nomination approach by election ensures the credibility and 'moral authority' of the spokesperson within the physics community.

One of the interviewees explained: 'If you're elected to become a spokesperson of LIGO, it means that you have stellar scientific credentials' (David Reitze, Executive Director, LIGO Laboratory, Caltech, Interviewed by the authors, December 2021).

In fact, this was further emphasised by a former ATLAS Spokesperson: 'Leadership in an experiment is much more collaborative, where you are accepted by your community and by your colleagues, you have credibility and that's where your authority comes from, not from your position' (Peter Jenni, Interviewed by the authors, June 2021).

The Spokesperson needs more than transformational leadership characteristics in order to earn the respect of the community and achieve the expected results.

One of the interviewees, commenting on the style and requisites of ideal leadership explained, 'A considerable transformational style of leadership with convincing dialogue and negotiation skills was necessary to get all on board. Knowledge-led transformational leadership is needed to deal with advanced scientific and technical projects' (Emir Sirage, Interviewed by the authors, June 2021).

Managing the different components of a project within a large experiment can also be challenging, as project components demand a high level of precision and high-quality performance. In achieving complex tasks and high expectations, the Spokesperson naturally needs an empathic leadership style, which helps to lead groups with different intellectual capacities, skills, technical and professional backgrounds, and work as a united team contributing different intellectual inputs.

As one of the interviewees commented on power and control:

> Efficient leadership in Big Science is hence a complex mixture of personalities, with spokespersons having the initial power and then passing the central control

to managerial figures. Large civil works, requiring skills and competencies generally well beyond scientific domains, essentially require engineering knowledge and approach. Hence, no 'democracy' can work, but strict, efficient, managerial decisions are required to realise the complex project outcomes.

(Federico Ferrini, Former EGO Director and CTA Managing Director,<br>Interviewed by the authors, June 2021)

The spokesperson needs to be responsible and carry the baton on behalf of all parties involved in the collaboration. It is essential that acceptable protocols are in place for decision making, operating, and responding to exigencies. Big experiments like ATLAS and CMS may have different management structures but in general the organisation has a lean structure. For example, ATLAS has one resource coordinator and one technical coordinator to support the Spokesperson. In general, the Spokesperson's key responsibilities include: setting clear goals and targets, upholding common values and interests, ensuring political support among partners, and getting new partners to support the cause, ensuring consultative procedures and diplomatic tact in handling disputes and differences, managing external relations and driving with perseverance, and getting the best from everyone in the collaboration.

Although the Spokesperson exerts a significant influence on the way scientific collaborations are led and steered, collective intelligence is required to address complex problems. It is important to note that the CERN accelerator sector is led differently when compared to CERN experiments. The experiments are collaborations, whereas the CERN accelerator sector is led more like arrangements in industry with the creation of a common facility. However, unlike a typical industry, the accelerator sector collaborates with other laboratories and has extensive collective knowledge and technology networks.

As the former Director for Accelerators and Technology at the LHC explained:

No individual is able to understand all the complexities and aspects of running a particle accelerator such as the LHC. Collective intelligence is essential and mandatory. The role of the Director of Accelerators is to listen to and understand the different opinions of the specialists. Then the Director makes the right decisions and takes responsibility for those decisions on behalf of all stakeholders and the committees to which the Director reports.

(Frédérick Bordry, Interviewed by the authors, July 2021)

In Big Science experiments, leadership may be rotated and shared periodically, therefore the success of individual leaders is a relative performance measure. What is more important is the success of the community in achieving the overall goals and objectives. Leadership in Big Science comprises mentoring, foresight, and making the right decisions that are endorsed by the scientific community. Leaders spend time communicating, consulting, and building consensus among the scientific community. The Spokesperson in a scientific experiment needs to build trust and respect and to be transparent and fair when making decisions based on reliable evidence.

One of the interviewees reiterated, 'Leadership decision making is based on scientific and technological arguments that reflect intensive collaborative work. In this vein, consensual leadership qualities are essential' (Ana Maria Henriques Correia, Project Leader Tile Detector and HGTD, ATLAS, Interviewed by the authors, May 2021).

Leadership decisions are often evidence-based and made using good judgement on behalf of the scientific community. These decisions are subject to debate and scrutiny. Disagreements and conflict could arise; however, the leader is ultimately responsible for resolving differences of opinion and making strong decisions on behalf of the community. Although there can be several dissenting views, once a decision has been made, all members of the collaboration accept the decision and proceed to implement it. The leadership process of arriving at a consensus maximises the chances of arriving at the 'right' decision.

The former CMS Spokesperson said, 'Decisions can be made in different ways; sometimes it is the Spokesperson that takes the lead, other times, it is a collection of senior scientists and engineers, or a review body, usually in agreement with the Spokesperson. It then becomes the responsibility of the Spokesperson to make sure that a consensus is built around the proposal' (Tejinder Virdee, Interviewed by the authors, June 2021).

At the operational level, it is critical to take decisions according to scientific community protocols and act accordingly with the minimal amount of bureaucracy or delay.

A former Spokesperson of ALICE, one of the LHC experiments, summarised from his personal experience the main characteristics that guided him in the role in the following terms: 'Be honest, keep a clear view and always keep in mind the main goals. Value people, appreciate and treasure their individual traits, dedicate a lot of time to discussing with them' (Federico Antinori, former ALICE Spokesperson, INFN, Padova, Italy, Interviewed by the authors, June 2021).

The key to effective scientific leadership stems from engendering hope, passion, desire, envy, ambition, resentment, and creative talent of individuals to pursue a common goal of inquiry. Once the project is underway, the lead scientists collaborate to deliver as promised. The leader's success demonstrates their effectiveness and acceptability within a community. In addition, the leader's ability to muster public support to get the project to the next stage of development naturally receives the endorsement of the community. Insights from our interviews affirm that it is the Big Science vision that drives and shapes scientific leadership with the many additional qualities discussed above enabling the vision to be realised.

## 7.5  Ethical Leadership

Ethical leadership is more than having strong values. In Big Science, ethical leadership is crucially important. An ethical leader is someone who respects ethical beliefs has strong values and is motivated by the dignity and rights of others. Big

Science organisations employ diverse groups of knowledge workers drawn from different geographic backgrounds and belief systems. Ethical leadership promotes how basic knowledge can be used effectively for the co-creation of social wellbeing and human progress in an ethical and equitable way for participants (Beck et al., 2020) Ethical leaders recognise and empower individuals to achieve collective scientific achievements and goals irrespective of their personal beliefs.

In Big Science organisations, ethical leadership operates at two levels: (i) at the level of the Director General, Council, and the Member State, where the main decisions to accept or reject proposals are made; and (ii) at the level of the allocation of funds in project operations. At both levels, the overarching force is the drive for fair recognition of individual contributions to overall scientific achievements. Ethical leaders will put processes in place to treat large and small contributors in an equitable manner enabling all partners to access and share research outcomes. Ethical leadership calls for the recognition of individual contributions to scientific excellence in research, education, and innovation and accepting that can happen. This requires looking beyond pure management and project controls.

In Big Science, challenging projects are riddled with uncertainty. It is important for leaders to realise mistakes can happen and all may not go as planned due to unforeseen circumstances. Massimo Tarenghi, an outstanding astronomer and the former director at ESOs, the Very Large Telescope (VLT), Paranal Observatory in Chile, spoke about ethical leadership and the leader's ability to tolerate mistakes and respond accordingly. Tarenghi reiterated that: 'The leadership ability to tolerate mistakes and take responsibility for others' mistakes is an important virtue in collaboration. Track the origin of the mistake; mistakes occur only because the leader has given either the wrong delegation or has not given proper support' (Massimo Tarenghi, Interviewed by the authors, June 2021).

## 7.6  Scientific Capability and Technological Feasibility

Scientific rigour, personal ambition, and influence among colleagues, all have a role to play at different levels of leadership in Big Science. Leaders have to exercise scientific knowledge and technical competencies at various stages of experimentation and contribute to theoretical development. In doing so, scientific leaders and their personalities need to align with the norms and values of science (Sapienza, 2004) generally described as communism, universalism, disinterestedness, and organised scepticism (Merton, 1942).[1]

In leading Big Science initiatives, scientific leaders have to demonstrate that results are unambiguous, and follow the rules of the discipline. Research outcomes need empirical testing and verification.

---

[1] Merton (1942) distinguishes between technical and moral norms and Zuckerman (1988) noted that these moral norms all relate to scientists' attitudes and behaviours in relation to each other and their research.

Science is socially constructed and individual scientists are responsible for maintaining the morality, fairness, and integrity of the scientific enterprise (Ziman, 2000). In Big Science organisations and experiments, knowledge leadership operates as an embedded, path-dependent, and patterned process across networks arising from a powerful culture of collaboration (Mabey et al., 2012). In other words, investigations and experiments evolved over years of investigation and agreed as central problems by the scientific community. Therefore, Big Science research community may uphold the disciplinary values and accepted norms of the physics community as ideals and mostly driven by the reality of how experiments are formulated and carried out.

In diverse and complex scientific settings, leading and managing scientists belonging to different cultures and epistemic groups is never an easy task. One of the interviewees remarked:

> With ALMA, we had a situation in which, three big communities European, American and Asian had to put together the work. I have to say that to being the Director of this kind of cooperation has been the most difficult challenge to overcome because to putting together these three communities, proved to be extremely difficult, as there were three different ways of doing things, [....]. I found myself playing the Magician King.
>
> (Massimo Tarenghi, Interviewed by the authors, June 2021)

Scientific methodology connects and coordinates highly intelligent groups of people. What attracts them to long-term research in particle physics, astrophysics or biomedical research is the excitement of the fundamental questions that the field has to offer. This is particularly the case for young scientists who are attracted to conduct their doctoral work, are looking for guidance and leadership in the field, and for opportunities to engage in both theoretical and experimental work.

Individual scientists in these Big Science experiments have to be conversant with state-of-the-art research approaches in order to choose their own preferred path of research inquiries that are innovative and novel. The role of the leader is to steer the research community to realise communal objectives and values.

The diagram below (Figure 7.3) shows the technological mapping and trajectory development involved in accelerators and experiments from the LHC to the High Luminosity LHC. The accelerator has gone through a significant shift in machine technology from a 7 TeV beam in 2011 (i.e. Run1) to 13 TeV beam in 2018 (i.e. Run 2) (see Chapters 2 and 3). The most recent upgrades (i.e. Run 3) pushed the energy of beam collisions to 13.6 TeV on 5 July 2022. These accelerator developments require considerable advances in luminosity, magnets, and other associated equipment upgrades both in machine and detectors, involving not only advanced technology but also expert human skills. At the experimental level, leadership emanates from physicists and engineers able to deliver outcomes and push research forward. Leadership challenges extend to the future scientific and technological developments and upgrades of the LHC, also known as high-luminosity LHC (HL-LHC) and to the proposed Future Circular Collider (FCC), which will be more powerful than the current LHC as shown in Figure 7.3

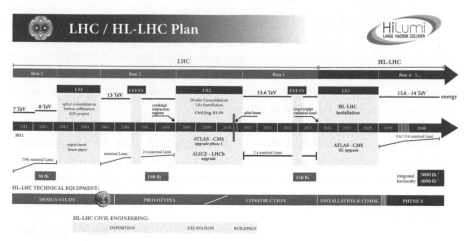

**Figure 7.3** The LHC/HL-LHC plan
*Source:* © CERN

The complexity of Big Science projects also drives the level of formality in leadership decision-making. Differences in control and management arise in big experiments because of the size of the collaboration composed of physicists, engineers, and technicians. The Spokesperson of LHCb commenting on this aspect said:

> In ATLAS and CMS, I think, given the size of the collaborations, they probably decided quite early on to have structures more regulated than the previous collaborations had. [...] My feeling is that there is a strong sense of autonomy and individuality in LHCb which is somewhat smaller than ATLAS/CMS. ... in terms of the approaches, it is probably a little less top-down and a little less administratively managed than in CMS and ATLAS. One explicit example is the way LHCb addresses scientists' contributions outside the physics publications in ATLAS/CMS they are called service contributions and accounted for, in LHCb there is no counting and instead a community expectation to make contributions to the general experimental development.
>
> (Christopher Parkes, Interviewed by the authors, June 2021)

Since research publications are the central mechanism for the dissemination of scientific knowledge, they are subjected to scholarly scrutiny and the contribution of each team member needs to be carefully identified. Authorship is determined by intellectual and technological contributions to scientific discoveries. These discoveries follow scientific norms and publication protocols. The publication policies are determined by the leaders of experiments in consultation with collaboration and follow accepted norms and policy guidelines such as CERN's Open Access Policy.[2]

---

[2] See for details in https://cds.cern.ch/record/1955574/files/CERN-OPEN-2021-009.pdf.

The complexity of Big Science research means that the leader needs an over-all understanding to make informed decisions, to ensure scientific integrity, and to be conversant with research outcomes and procedures that are subjected to inter-national and peer reviewed scientific scrutiny. Leaders need to accommodate the complexity and impact of current research and future research agenda. For example, a leading researcher in dark matter commented:

> We live in a vast sea of dark matter, but its composition at the fundamental level remains an enigma. [...] While the pragmatic goal is to probe the theoreti-cally allowed parameter space until interactions from cosmic neutrinos take over, these experiments might well herald one of the greatest discoveries in twenty-first century physics.
>
> (Baudis, 2021: 24)

Leadership in Big Science needs to accommodate what is feasible now and in the future scientific, technological and social requirements.

## 7.7  Collaboration and Collegiality

Long-term collaboration is a key element in Big Science projects. In the case of CMS and ATLAS experiments, the letters of intent for the construction projects were sub-mitted on 1 October 1992. It took nearly two decades to confirm the existence of the Higgs boson (confirmed on 4 July 2012). This was a considerable commitment for those who stayed committed to the original cause. Since the discovery of the Higgs boson in 2012, ATLAS, and CMS continued their journey to explore Higgs boson's interactions and the search for new and unknown phenomena beyond the Standard Model. Leaders need to convince others that there are many more discoveries yet to come and build on theory and experimental development.

In Big Science, different epistemic cultures are drawn together to achieve desired research outcomes. Experiments at the LHC are planned to continue for over 20 years, which involve substantial involvement of people's careers and commitments. In other words, the clock started for experiments, when these experiments went online (in 2008), but the collaborations in fact can be dated back to 1992.

During these long hauls, one of the most difficult tasks for the Spokesperson, leader, and director of Big Science operations is to extinguish fires that can flare up from time to time due to differences of opinion, varied scientific approaches and per-sonalities of individuals. After all, scientists are human beings with emotions. This diversity of personnel can also spark creativity and serendipity.

Since its inception in 1952, CERN's sustained leadership in the high energy physics community reflects a strong culture of collaborative leadership. In collaborative lead-ership, hierarchy is less prominent, and scientists work together in subtle ways. Collaborative decision-making is part of Big Science leadership which contributes to the success of teams.

The former CMS Spokesperson, commenting on the collaborative leader's role as a facilitator of open debate, explained:

> In this sense, the leader is 'primus inter pares'. Constant tensioning of propositions, in such a (relatively) small circle is quite important to make sure that the decisions or propositions taken forward into the collaboration are most likely to be accurate. Leaders should especially seek out the opinions of people who take an opposing view as something may have been overlooked that could trigger a change of view. To encourage such deliberations, there must be open channels for criticism. This again requires leaders to have inner confidence.
>
> (Tejinder Virdee, Interviewed by the authors, June 2021)

Building a robust epistemic culture among collaborating groups is the resounding success of some Big Science initiatives. Collaborative leadership is entrenched heavily in some epistemic communities.

For instance, the Executive Director of LIGO said:

> One of the things that makes us a little bit unique is that we are the gravitational wave community. We actually work in very close collaboration with other gravitational wave detectors and the one that we've been working with for a long time—almost 15 years now—is the Virgo detector, which is based in Italy. It is a consortium of Italians, French, the Netherlands, Poland, and many others who are involved in it now.
>
> (David Reitze, Interviewed by the authors, June 2021).

Big Science communities thus develop effective modes of collaboration, shifting from personal focus to team focus through informed consent, motivation, and recognition of individual rights to their own thoughts and creativity. Leadership paves the way to nurture such collaborations and extract critical ideas from the scientific community to ensure that the community achieves longer-term outcomes. Effective collaboration requires better communication with key stakeholders including scientists, technologists, and financiers (Voegtlin et al., 2020).

A trend towards collaboration or collectivism has in fact become increasingly visible during the last decade. Collectivism can also nurture and help create social capital. Nahapiet and Ghoshal (1998) advocate a relational dimension that creates a network of trust, reciprocity, obligations, respect, and friendship that facilitates the sharing of tacit and explicit knowledge. Big Science leaders rely on the cognitive dimension—shared understanding, interpretation, and reflection—that among colleagues that allows effective communication, network building, learning communities, and unambiguous interpretation of meaning. Collective leadership dissipates ambiguity and predicts the intentions and actions of fellow members when it comes to controversial, different ways of doing things.

Notwithstanding the emphasis on collaboration in Big Science 'ideas do tend to come from individuals, and so the individuals still play an important role in large teams', according to Tejinder Virdee, former Spokesperson of CMS. It is the role of the leader to optimise the balance between individual and collective decision-making. The most important task of the leader or Spokesperson is to motivate each team member to the best of their ability, an individualised approach to leadership first described by Yammarino and Dansereau (2002).

Collaborative research needs a distributed leadership approach that appreciates and upholds contributions from all involved without hierarchy. The distributed leadership recognises that there are multiple leaders and that leadership activities are widely shared within and between organisations (Harris and Spillane, 2018). Complex experiments such as CMS and ATLAS require a multi-layered matrix of leaders in scientific, technical, resource, and operational matters with a network of teams and collaborators.

A former CERN senior scientist and successful entrepreneur explained, 'Leading means to conduct and drive a team towards common objectives. There are two ways to do it – top-down or bottom-up approaches. Either way leadership needs to be shared' (Stefano Buono, Interviewed by the authors, June 2021).

The LHC and non LHC experiments at CERN have spawned not only classical methodologies but also highly innovative and novel techniques and advanced methodologies such as machine learning techniques. These novel and innovative approaches are making ground-breaking advancements in knowledge and understanding, contributing to substantial improvement in performance and extending the boundaries of research. For example, recent research in Graph Neural Networks (GNNs) is one such area. Leaders need to be open to such new technologies that are always evolving in a dynamic research organisation.

Arriving at the right decision requires a transparent process based on strategy, finance, and reliability of knowledge and judgement, including recognition of noise (Kahneman et al., 2021). Nowadays large companies from digital corporations to insurance companies are trying to counteract shortcomings in their decision-making by aggregating different opinions. Aggregating different opinions and listening have long been widely used by leaders in the HEP environment as discussed in this chapter.

## 7.8 Scientific Leadership and Gender Issues

The role of women in particle physics is a poorly researched subject which deserves more attention. Even among major Big Science organisations such as CERN, female scientists were about 20% of the total staff employed in 2021 (CERN Personnel Statistics, 2021). ESO has female staff representing 26.4% of the overall headcount by the end of 2021 following the implementation of a project called 'Status of women at ESO' in 2005–2006 when the numbers were about 18% of the total staff members. Physics leaders have adopted several steps such as the creation of GENERA (Gender Equality Network in Physics in the European Research Area).

Fabiola Gianotti (DG of CERN) explained, 'Science has no passport, no gender, no race, no culture, no political party. I said that science can play a key role in connecting people and creating a shared future in a fractured world, because science is universal and unifying' (Sciolino, 2018). Attracting women to research careers has been a difficult task as women often face the challenge of balancing personal interests and family responsibilities.

This was further confirmed by Suzie Sheehy (CERN Courier, 2021c), who remarked: 'I think it is important to set up a kind of work/life integration that supports well-being while allowing you to do the work you want to do.' She also commented on the increasing pressure on early and mid-career researchers: 'The hardest thing, and I think a lot of early/mid-career researchers will relate to this, is that academia is an infinite job: you will never do enough for someone to tell you that you have done enough. The pressure always feels like it's increasing, especially when you are a post-doc or on the tenure track or in the process of establishing a new group or lab. You have to learn how to take care of your mental health and well-being so that you do not burn-out' (Sheehy, 2022).

Interviews with women physicists who have had or are currently occupying leadership roles in Big Science organisations indicate that the support of their families has been essential from the start of their careers. It is not a stereotype to outline how much cultural male dominance and lack of financial resources have been and still are a barrier for women, preventing them from getting into higher education and also from obtaining higher-level leadership positions.

While gender may have an influence on leadership style, technical credibility is essential. As one of the CERN scientists said: 'You need to be technically competent and your team must be technically competent, only then you are accepted as a woman leader' (Doris Forkel-Wirth, Interviewed by the authors, April 2021). However, she also pointed out that her personality has changed over the years, she had to become tougher, and, in a way, she had to imitate a 'masculine' style of leadership. She suggested that leading a research team is different from leading other types of teams:

> All types of teams should be led through unrelenting motivation over time; by convincing and creating interest and curiosity. However, the researchers are generally self-motivated, and the management style can be very collegial. Researchers also need to be frequently reminded to take care of their work-life balance. Leading other types of teams might sometimes require a more hierarchical management style.
>
> (Doris Forkel-Wirth, Interviewed by the authors, April 2021)

In earlier times, it was never a problem for a female to reach a certain level of responsibility as a coordinator or a specialist. Female scientists were happy to be recognised just for the good of their work and never thought about becoming leaders, assuming they would be asked if needed to cover important positions. In reality, to get nominated and to remain in a senior position, a woman has to show not only that she is competent and possesses the ability to be trusted for her decisions, but she also has to establish good networks and develop social capital. One of the leading female physicists explained:

> Over the last 5 years I was responsible for a new detector, the High-Granularity Timing Detector (HGTD), to be installed in the ATLAS experiment in 2026, aimed at operating in the high luminosity LHC (HL-LHC) period. HGTD will provide timing information to tracks in the forward region with a time resolution of 30–50 picoseconds/track, a precision that is 6 times better than the spread of the collision time, allowing to distinguish tracks in collisions occurring very close in space but well-separated in time. This will improve the outcome of physics analysis.
>
> A new idea started in 2015 with a few enthusiastic colleagues and it was necessary to create a new team with several ATLAS institutes, leading to a mature detector concept and an approved project in 2020. HGTD is an 11 million Swiss Francs project that is now under construction by approximately 300 collaborators from all-over the world until March 2021. The components of the detector, now under construction, are paid for by 17 funding agencies, with established responsibilities detailed in a Memorandum of Understanding document. The detector components will be finally assembled at CERN, before their installation in the ATLAS experiment in 2026.
>
> I had the privilege to lead all the steps from the initial detector concept, consolidate the collaboration, and bring this proposal up to its approval in ATLAS and CERN external committees. This work was a 'leadership challenge and a very nice experience'.
>
> (Ana Maria Henriques Correia, Interviewed by the authors, May 2021)

Most female scientists are courageous and persevere to drive their project through to fulfil the scientific vision, while always keeping in mind what is best for the group they are leading and the organisation they are working for. Female leaders have proven capacities to deal with others in diplomatic terms, to drive and instigate changes as well as to keep the skills of their collaborators at the highest level needed to cope with the challenges posed by Big Science research. Hence female leaders in Big Science have much to offer society.

Ana Correia, the leader of the Tile Calorimeter Detector, further shared her views on the challenges women face when leading teams:

> Women who choose a scientific career in physics are very motivated, even if conciliating a very intense professional life with their private lives is not easy. Leadership decision making is not arbitrary; it must be based on scientific and technological evidence. The Spokesperson's or project director's technical competence is paramount to leading, convincing, and working collaboratively in teams. Consensual leadership qualities are essential.
>
> (Ana Maria Henriques Correia, Interviewed by the authors, May 2021)

Accomplishments in Big Science receive recognition independent of an individual's gender. Big Science experiments have their own peculiar place with many female physicists having contributed to significant advancements in the field. In recent times, women's contributions have become more visible and more recognised, an example being the attribution in 2021 of the Dirac Medal to Alessandra Buonanno jointly with Thibault Damour and Saul Teukolsky for their theoretical research on gravitational waves (ICTP, 2021).

## 7.9  Scientific Outcomes and Translation to Community Needs

Leadership in Big Science projects is expected to generate important scientific outcomes with a significant impact on society. While CERN was created in 1954, many other large international facilities have joined the Big Science league since the 1960s. They include the European Southern Observatory (ESO), Fermilab, the Institute Laue-Langevin (ILL), the European Synchrotron Radiation Facility (ESRF), the Joint European Torus (JET), the European Molecular Biology Laboratory (EMBL), the International Thermonuclear Experimental Reactor (ITER), the Laser Interferometer Gravitational-Wave Observatory (LIGO), and several other ground and space telescopes. These research infrastructure facilities are making a significant contribution to useful innovations that benefit society and to our understanding of nature and the universe.

Leadership in Big Science is not confined to scientific experiments but also extends to technological innovations that benefit society. The way these organisations assemble and share research data among international partners and other scientists may vary. Accountability and transparency are required to ensure that research outcomes are reliable and accurate and that communities trust the outcomes achieved. However, Rolf-Dieter Heuer, a former Director General of CERN, observing an increasing gap between science and society, wrote: 'we live in an age where curiosity-driven science touches almost every aspect of our lives, yet science has been growing apart from society and culture for decades' (Heuer, 2018).

Science and knowledge are considered pure public goods which are non-rivalrous and non-excludable (Thursby and Thursby, 2007). This means that once knowledge is generated, it is neither depleted nor diminished by use. Big Science organisations such as CERN, LIGO, EGO, and ESO typically produce massive amounts of data, information, and knowledge for public consumption, using specific knowledge development and sharing processes unique to each organisation. As these organisations rely on public investments, leaders are conscious of the need for the dissemination of knowledge and knowledge for the public good; hence, they often do not expect to profit from the fundamental knowledge generated.

Big Science organisations such as CERN and ESO work closely with the industry to help ensure that the outcomes of Big Science generate public good. The process of building relationships with industry is rather complex and long term. Unlike networks with international research partners, relationships with industry can be contractual and designed for specific procurement processes. Leadership and management of technical relationships with industry generally rely on large laboratories which determine the quality, reliability, and specifications of technological products required to assemble Big Science infrastructures.

## 7.10 What Is Unique about the Scientific Leadership Style?

Big Science leadership evolves as a process rather than a specific structure. Unlike in business organisations, most leaders gain their leadership qualities through experiential learning. A leader of CERN experiments said: 'Particle physics is a team effort, quite distinct from the way an army or a corporation is organised. Our leaders are not generals or CEOs, but colleagues called to serve for a time before returning to the rank and file' (Shipsey, 2020).

Big Science projects require leadership not only to initiate a project but also to keep it going until it is successful, possibly many years later. Leadership, therefore, is not vested in an individual but in a team. It is distributed throughout the collaboration and is often negotiated in a harmonious manner (Liyanage and Boisot, 2011). This collaborative approach was emphasised many times by our interviewees. For example, CERN's senior engineer said: 'Leader of a collaboration at CERN has no executive power over other participants and therefore, listening, discussion and compromise are essential elements. Specific outcome-oriented projects require directed leadership styles with democratic leadership approaches' (Michael Campbell, Spokesperson, Medipix 2, 3, 4 Collaborations, CERN, Interviewed by the authors, June 2021).

Leaders commented that their colleagues were coached and mentored to become potential leaders with increased responsibilities, allowing them to develop leadership skills by learning by doing.

It was also necessary to multi-task and be able to lead projects which use extensive data collection, storage, and analysis. In some Big Science areas such as astrophysics, big data has revolutionised the way of conducting research in this field. Big data collection, sharing, analysis, and dissemination have become essential, therefore, how leaders approach sharing and managing access to data.

As a scientific community, they had to learn how to benefit from the enormous amount of data collected around the world. The astronomer is no longer an individual alone with a telescope under the stars—they are a collective force needing dynamic leadership. Astrophysics relies on big international collaborations and leaders are conversant with the nuances of cross-cultural collaboration.

## 7.11   A Leadership Model for Big Science

Effective leaders rely on team effort and collaborative enterprise at all times—not just on a leader's or individual's effort. Our interviews show that Big Science leadership traits uphold value, trust, integrity, openness, perseverance, understanding, account-ability, compliance, learning, and inclusion. Leadership behaviour revolves around epistemic culture and the value of collaborative processes.

Basic differences exist in the leadership models of Big Science. A recent OECD study of large research infrastructures (OECD, 2010) identified that leadership style is traditionally that of collegiality and consensus. The researchers tend to self-select leadership styles based on shared technology preferences, proven expertise, and a strong record of success in the relevant scientific areas. Our interviews confirmed a 'light' management style or consultative approach in leadership processes in most large detectors or telescope experiments. This is because certain subsystems can function suboptimally (or, indeed, not at all) without undermining the entire experiment. Such a 'light' approaches may not always work in the overall functioning of large complex infrastructure such as LHC, which needs a very high standard of perfor-mance and reliability. Therefore, the leadership and organisation must respond to the complexity of the organisation to match the required level of quality and standard.

In Big Science projects, individuals play a key role in driving brilliant ideas. Some charismatic people may lead these experiments. Although charismatic leadership was dominant in the 1970s and 1980s in some Big Science organisations, modern leadership evolved to be more moderate and consensual.

In view of next generation experiments, the European Committee for Future Accelerators (ECFA) has carried out a survey to explore how individual achieve-ments at present get recognition in LHC collaborations with the aim of identifying how leaders and organisations can strengthen individual recognition.

These findings suggest that there are considerable obstacles for young scientists to gain individual recognition and publish their work on time. Building next genera-tion leaders and scientists to devote most of their lives to complex experiments is a challenge. To a large extent, the excitement, curiosity, and challenge of solving future complex projects are the magnet that all leaders need to use.

The focus on common creative endeavours, i.e. in the pursuit of understanding and knowledge in a largely dispersed international community, where scientific goals substitute for market objectives, represents today the distinctiveness not only for accelerator physics and astrophysics, but also in other fields such as molecular biol-ogy and brain research. One of the interviewees differentiating Big Science leadership from conventional leadership, emphasised:

> To be a leader in the CERN model is a service to a community rather than a simple exercise of power: seeking consensus. It implies believing that the collaborator's point of view is important and valuable and that listening is not a waste of time.
>
> (Lucio Rossi, former HL-LHC Project Leader, University of Milan,
> Interviewed by the authors, June 2021)

Big Science leadership in this light puts a strong emphasis on team cohesion and motivation (Robinson, 2021). This is particularly true for some accelerator projects, where most of the leaders involved in promoting, conceiving, and designing these projects may not be present at the stage of generating data and analysis. The continuity of knowledge and skills needs to be passed on to the younger generation and effective communication processes among research peers are therefore central to the future development of the field. In the case of CERN, there is an extensive process of documenting and archiving all forms of communication and regular seminar series allow this communication process to be seamless. It is important to recognise that communication skills follow a two-way process, as one of the interviewees explained:

> Not only is communicating very important, but so is listening. In a sense, there can't ever be enough communication. Good communication lets people know what is expected of them and listening ensures that they are comfortable with what is expected.
>
> (Tejinder Virdee, Interviewed by the authors, July 2021)

Big Science leadership models or theories have to focus more on the scientific community and not so much on the leader's task and performance. This is because knowledge creation and dissemination in Big Science is a complex undertaking as well as a collective process. Research publications in many fields, including biology can in fact have over 1000 authors, signifying the extent of collaboration, the convergence of ideas and collective spirit. In 2015 a joint research paper from ATLAS and CMS published on 14 May in *Physical Review Letters* recorded 5154 authors (Aad et al., 2015).

Therefore, leadership in Big Science calls for a more inclusive leadership style that encompasses knowledge-driven, empathetic, collegial, consensus, and organisational management approaches. Our interviews further confirmed the importance of common visions, consensus, collegial decisions, the lack of imposed formal authority in managing high risks projects as found in IMD (Institute of Management Development, Switzerland) study on the project management of ATLAS collaboration (Marchand and Margery, 2009).

The national, international, and professional cultures of physicists and engineers who have worked on or are working on Big Science projects are the central forces for borderless development, innovation, and education. A combination of traits/criteria, hard skills/soft skills discussed above could materialise as a representation of a leadership model appropriate for Big Science.

What is the optimal leadership approach in Big Science? Our research and interviews showed that scientific or intellectual leadership is preeminent which entails elements of collaborative, visionary, strategic, and ethical leadership. With increasing complexity of experiments and even more complex accelerators like the FCC, future research leaders need to focus on the exact contributions of the above-mentioned factors to scientific leadership processes in Big Science.

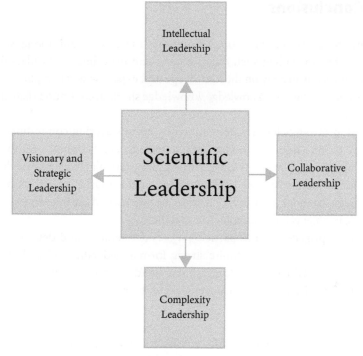

**Figure 7.4**  Emerging model of scientific leadership in Big Science
*Source:* Created by author S. Liyanage

The leadership of Big Science experiments is evolving as projects become complex, interdisciplinary, international, and large scale and the emerging model of scientific leadership has intellectual, visionary, collaborative, and complexity traits as shown in Figure 7.4.

According to our research findings, participative and transformational leadership styles are the most powerful ways to foster and develop complex technological collaborations. Participative leadership helps to solve conflicting views and converge on an agreement that is acceptable to the majority. Even in scientific areas, there are many alternative routes to get to the same outcomes.

To be respected, scientists have to recognise the achievements and capabilities of individuals and agree on the norms of science as guiding principles for knowledge production. Leaders need to exercise moral authority to recognise the intellectual contributions of fellow scientists and give credit where it deserves. To function as ethical and transformational leaders, they have to uphold high morals and values that are respected by collaborating partners. In Big Science experiments, over 1000 scientists can collaborate to produce a single paper. As evident in most CERN and LIGO publications, the authorship has extended to more than 1000 people, which is a significant change in the scientific publication process.

## 7.12 Conclusions

Big Science organisations have complex and multi-faceted organisational structures and approaches to leading such organisations and experiments. Leadership traits and behaviours rely heavily on the knowledge and expertise of its employees and/or collaborators, cutting-edge knowledge, knowledge sharing, continuous learning, and innovation.

Leadership in Big Science is determined by an unwavering commitment to scientific excellence, a quest for fundamental knowledge, and commitment to international collaboration. Hence, it is difficult to assign a particular leadership style to such organisations. Nevertheless, leadership traits and behaviours can be identified based on knowledge, complex knowledge relations, networks, and the international flavour of scientific problems and technological specificity.

Big Science projects such as particle physics accelerators and detectors and the Very Large Telescope (VLT) require strong formal leadership to build, construct, manage, and maintain. Effective leadership to handle Big Science projects requires a combination of epistemic, technical, management, and social skills. Some of the key leadership traits include:

- Leaders possess a high level of scientific, technical, and interpersonal skills to inspire fellow scientists to work together. Big Science projects draw on a multitude of different epistemic cultures, personal interests, and expertise in different disciplines and sub-disciplines;
- Leaders deepen their diversified expertise as they rise through the ranks, often taking on different roles, acquiring experiential skills, that cultivate flexibility and different levels of interaction with knowledge workers, technology developers, managers, and financiers. Management and coordination skills tend to be equally important as scientific credentials and skills;
- Leaders call for an inclusive leadership style that builds a supportive culture, where everyone feels recognised, included, valued, and respected. From traditional leadership theory points of view, this implies combining transactional, transformational and charismatic leadership styles and combining empathic, collegial, consensus building, diplomatic, and strong decision making traits; and
- Leaders should be open minded, unbiased, objective, and pragmatic. After all, projects have to be completed within expected goals and allocated budgets. A high level of organisation and managerial skills are necessary to launch long-term Big Science projects.

The most important factors in Big Science leadership are the need for knowledge intensive drivers with a clear vision, focused strategy, and scientific credentials. These drivers must combine individual commitment, precision, tolerance of failure and intensive collaboration. Our interviews with leading scientists show the importance of listening to their colleagues and seeking their input into decision-making processes. A collective agreement on the right idea is even more important.

Our interviews suggest that leadership is about creating value and recognising the contributions of others. It involves understanding the decision-making processes and making a sound judgement based on facts, evidence, and intuition. The challenge for leaders is to push the boundaries of science and technology to attain successful outcomes. Leaders need to be transparent, fair, and responsible in Big Science projects, and build trust and respect to conduct experiments and reward those who perform well.

Remarkably, the leader in an experiment is called a 'Spokesperson'—not a leader. Spokesperson is 'first among equals' and represents the scientific teams and community to steer scientific, technical, and organisational leadership. Such leaders have considerable power to steer without a rudder. Spokespersons cannot hire and fire other scientists as they are employed by the collaboration. To be elected as a Spokesperson, one must have stellar scientific credibility recognised internationally.

Big Science leaders' skills go beyond the laboratory, requiring leaders' intuition to communicate intentionally and effectively with numerous organisations, industrial partners, and the wider scientific community. With the growing complexity of Big Science experiments, leaders need to appreciate diversity in the scientific community and be more inclusive. The leadership roles of women in Big Science organisations and experiments are not well researched and require greater attention and research.

Female scientists face many challenges. The difficulty of balancing work and family obligations and the general inability to draw women into physics studies were both emphasised by several of the female scientists interviewed. However, these issues are not only confined to Big Science organisation, but are also a widespread issue among many professional communities including science.

Big Science leadership is also about steering complex scientific projects. Skills required are diverse. A leader has to be a visionary, strategic thinker, collaborator, mentor, effective communicator, problem solver, people's person, ethical and integral leader, and be resilient to deal with highly complex and unexpected situations. Trust is a two-way process—a leader has to be trusted by and in return trust the collaborators.

With all that, a leader has to contribute to scientific outcomes and scientific goals as an individual and as a member of a team in order to earn respect within the scientific community.

Clearly leadership lessons in building a complex machine like the LHC experiment and technological complex experiments like ATLAS, CMS, and LIGO demonstrate the ability of humans to identify, generate, innovate, and adopt new ideas and innovative management styles. In doing so, leaders have to work with teams and built around the ideas, evaluate and select those ideas in complex experimental settings. A combination of scientific and technological leadership traits, which collectively can be identified as 'scientific leadership', is useful to enact knowledge leadership in Big Science. Such innovative and scientific leadership style is unique to Big Science where politics, science and diplomacy converge to intertwine with empathy, resilience and humanity that support the culture of collaboration.

## Acknowledgements

The authors sincerely appreciate and are grateful to so many leaders in CERN, LIGO, ESO, and other Big Science organisation for consenting to interviews and numerous email conversations about leadership issues in Big Science. We are particularly grateful to Federico Antinori, Frédérick Bordry, Stefano Buono, Michael Campbell, Ana Maria Henriques Correia, Federico Ferrini, Doris Forkel-Wirth, Peter Jenni, Christopher Parkes, David Reitze, Lucio Rossi, Emir Sirage, Massimo Tarenghi, and Tejinder Virdee.

# 8

# The Evolution of Astrophysics towards Big Science

## Insights from the Innovation Landscape

*David Reitze, Alan R. Duffy, James Gilbert, Mark Casali,*
*Elisabetta Barberio, and Shantha Liyanage*

## 8.1 Introduction

In a recent interview, cosmologist Martin Rees suggested that scientific investigation encompasses the search for understanding of the very small and the very large, and on an even grander scale the very complex (Rees, 2021). The first two belong to the realm of fundamental physics, while the last encompasses virtually every other field of science including biology, atmospheric science, psychology, medical research, much of astronomy, oceanography, and economics to name but a few. It is obvious to say that the advances in these three categories over the last hundred or so years have been revolutionary and form the foundations for the large international scientific collaborations and experimental infrastructures that have come to dominate scientific research in the twenty-first century. Certainly, astrophysics has revolutionised the scientific landscape and is increasingly winning the hearts of non-scientists alike.

Big scientific infrastructures designed to investigate the big ideas certainly generate technologies, spin-offs, and benefits for society, but interestingly the big ideas themselves have also had direct and immense consequences for our lives.

The first two categories, the very large and the very small, are often thought of in the public mind as noble investigations into the nature of the universe, but with few practical consequences for everyday life. These categories include (among others) three different theories—Special Relativity and General Relativity (GR), in which the former is a special case of the latter, and quantum mechanics. Surprisingly, it turns out that far from being esoteric these theories have played a critical role in shaping the modern world.

Indeed, they have been crucial in advancing the frontiers of astrophysics in the past century. This chapter analyses the process of, and benefits from, technological innovation in large-scale collaborations using astronomy and astrophysics as examples. It includes radio astronomy, infrared astronomy, gravitational waves, dark matter, and

David Reitze et al., *The Evolution of Astrophysics towards Big Science.* In: *Big Science, Innovation, and Societal Contributions.* Edited by: Shantha Liyanage, Markus Nordberg, and Marilena Streit-Bianchi, Oxford University Press.
© David Reitze et al., (2024). DOI: 10.1093/oso/9780198881193.003.0009

dark energy. Organisations such as CERN in Switzerland, and ESO in Germany are drawn as examples of 'Big Science Chases Big Ideas' to highlight the need for, and power of, grand-scale collaboration. Based on experience in working in such facilities, the authors highlight and examine key technologies developed in support of scientific breakthroughs achieved in these domains.

## 8.2  Big Science Ideas

Einstein's 1905 Special Theory of Relativity introduced the notion of the relativity of time and the absence of any absolute clock in the universe. When applied to mechanics, the theory leads to the famous equation $E=mc^2$ and the realisation that there is an enormous amount of energy in even a very small amount of mass. As this became better understood in the twentieth century, it led to investigations into nuclear fission and the famous Einstein–Szilard 1939 letter to Roosevelt which warned of the development of a nuclear weapon by the Nazis. The Manhattan Project and the devastation of Hiroshima and Nagasaki followed and the nuclear age was born.

General Relativity (GR) is certainly more remote from everyday affairs. The description of gravity by Einstein in 1915 as a geometric phenomenon of spacetime rather than a mysterious force-at-a-distance enabled a completely new approach to understanding the evolution of the universe. Predictions of the expanding Universe, black holes, and gravitational waves dropped naturally out of his equations. Yet these large-scale phenomena, amazing as they are, seem a long way from the concerns of human beings. This theory, more than others, seems to be of interest only to physicists and mathematicians.

Global Positioning System (GPS) satellites have become an indispensable part of modern life. GPS satellites rely on the extremely precise nanosecond timing of signals (see Figure 8.1). The speed of satellites relative to us on the ground makes their local time about 7 microseconds per day slower than ours (a consequence of Special Relativity), while their different gravitational energies in orbit result in their local time passing 45 microseconds per day faster than ours (due to GR), for a total difference of +38 microseconds per day due to relativistic effects. This may seem small, but if the GPS timing calculations in our phones and navigation systems neglected this correction, there would be incremental positional errors of many kilometres each day. Remarkably, understanding the nature of space and time in our universe has allowed us to drive around cities without getting lost.

At the other extreme, the invention of quantum mechanics as the physical theory of the very small has had an even greater impact on human life and civilisation. Unlike Einstein's work on relativity, the development of quantum mechanics led by researchers in German and other European universities in the first decades of the twentieth century and in the US after the war, was a community effort with many contributors over many years shaping and reshaping the theory. It likely had to be so, because while GR is firmly entrenched in the classical traditions of physics, quantum mechanics is a radical shift in worldview and too great a leap for one person to

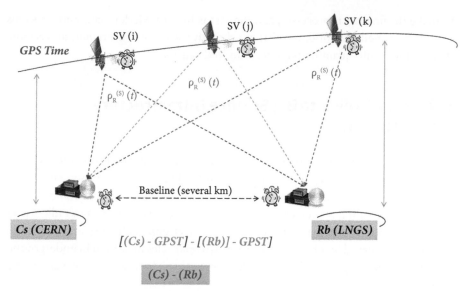

**Figure 8.1** Global Positioning Systems (GPS)
*Source:* © CERN

have been able to make alone. Even today scientists argue as to its actual meaning and interpretation.

Yet the theory, honed over many years, has become an incredibly accurate description of events and interactions at the microscopic scale. Since our bodies and the matter that surrounds us are made of atoms, one might expect such a theory to be extremely useful in understanding and modifying our material world, and it certainly is.

The quantum mechanical theories, rules and approximations make up a kind of toolbox that anyone investigating the realm of nuclei, atoms, and molecules can use to make calculations and predictions. Material science, which has produced exotic substances such as graphene and superconductors, chemistry which fills the world with plastics (for good or ill) and bio-sciences which have transformed our lives with vaccines and better health, all use the quantum toolbox to make advances. But few more dramatic practical applications of quantum mechanics can be found than in the invention of the transistor.

It is important to understand that the transistor is inherently a quantum device, in that the phenomena which underpin its operation and the scientific theory which describes them, are quantum in nature. It is therefore not surprising that the three members of the team at Bell Labs who invented the first point-contact transistor in 1947 (Shockley, Bardeen, Brattain) were all physicists and not engineers. The device they invented was improved and developed, until the invention of the monolithic integrated circuit in 1959 by physicist Robert Noyce (later to co-found Intel). This latter development transformed electronics and allowed large numbers of transistors to be placed on a small piece of silicon at a low cost. In 1970 this technology was used to make the first microprocessor, and the rest is history. Modern computer systems have transformed the world, and all this is directly traceable to the remote and erudite investigations into quantum physics a hundred years ago.

## 8.3   From Galileo to Big Science Infrastructure for Astronomy

Astronomy in the twenty-first century has grown into large world-spanning collaborations, exploiting every technique possible to discover new phenomena in the universe and find evidence to confirm or refute astrophysical theories. The electromagnetic spectrum has been extensively explored from gamma rays to radio waves from the ground and in space. In addition, non-electromagnetic messengers are also commonly observed in different ways from astro-particles measured in underground detectors, high energy particles hitting the upper atmosphere, and meteorites landing in Antarctica.

More recently a completely new technique has proven its worth, namely the detection of gravitational waves as Einstein's long-predicted ripples of spacetime in General Relativity (GR). However, it is the more familiar optical telescope that represents a wonderful example of an astronomical technology that has developed from concept to maturity over many centuries enabling many scientific discoveries along the way.

The origin of the telescope is somewhat unclear, with evidence that some experiments took place as early as the sixteenth century. In 1608 Hans Lippershey attempted to patent the first working device consisting of a convex primary lens and a concave secondary eyepiece. Galileo Galilei heard of the concept in 1609 and improved Lippershey's design, culminating in the famous metre long telescope with which he conducted astronomical observations (see Figure 8.2a). Galileo was to apply this telescope to methodical astronomical observations of the planets, Moon, and Sun, thereby finding evidence in support of the Copernican theory that the Sun lay at the centre of our Solar System.

It is a long way from Galileo's first steps with telescopes to the massive modern observatories situated at the tops of mountains (see Figure 8.2b). What has driven that transformation from small to big? The key principle driving such an increase in scale is that the benefits significantly outweigh the costs and technical challenges. In the case of particle accelerators, size is driven by the goal of achieving high collisional energies. In astronomy, objects such as galaxies exist at very great distances, making them exceedingly faint. The faintest galaxies observed can result in as little as one visible photon arriving per square metre per hour on earth. Useful observations of such objects must therefore collect light from many square metres and/or have very long exposures. It makes little difference whether the photon is at visible wavelengths, radio, X-rays or gamma rays, the gain is the same—the bigger the aperture, the fainter the objects that can be seen.

It was the invention of reflecting optics in the seventeenth century, in which a reflective main mirror replaced the objective lens, which became the great

**Figure 8.2a**  An artist's concept of Galileo Galilei exploring the night sky with his telescope
*Source:* Getty Images

**Figure 8.2b**  The Very Large Telescope on Cerro Paranal in Chile, consisting of four 8.2m aperture telescopes
*Source:* Iztok Boncina/ESO

breakthrough enabling large telescopes to be made. Reflecting optics were developed as early as 1636; however, it was Isaac Newton who built the first functioning reflective telescope in 1668. The 'Newtonian' design he invented is still used in amateur telescopes today. The technology led to the beginning of giant telescopes such as William Herschel's 1.2m diameter telescope in 1789 with which he discovered Enceladus, the moon of Saturn, on the very first night of use, which remained the largest in the world for many years (Sime, 1900). In 1845 William Parsons (Lord Rosse) took another step forward with the construction of a 1.8m telescope using the optical design of Newton.

There has perhaps never been a more famous large telescope 'system' than the 200 inch (5.1m) Hale telescope on Mt Palomar, CA, US, opened in 1949. Not only must the main mirror be supported to avoid distortions under its own weight, but it must also be able to handle changes in shape caused by temperature differences. Furthermore, such large structures invariably flex as they move around the sky which misaligns the optics. The 200-inch primary mirror was made of Corning pyrex low-expansion glass, with the weight reduced two-fold by forming the back of the mirror with a honeycomb structure, steps which enabled the mirror to cool quickly at night and experience low deformation from its correct shape. The mirror's reflecting surface was made of vacuum-deposited aluminium.

Technologically, large research facilities such as the Hale telescope should be thought of as systems. That is, there is no single magical technology that enables their function but rather a range of sub-assemblies and modules that must work together for the whole thing to function correctly. There may often be one or two specific innovative leaps, but looking deeper one will find a myriad of technological advances in everything from metrology, materials, precision machining, electronics, cryogenics, computing, and even large steel structures. In many cases, the key innovations would not even have been possible without the supporting technology.

The Hale telescope remained the world's premier telescope for visible astronomy for many decades. Virtually all large telescopes up to this time were financed by individual wealthy people or foundations. The Hale itself was constructed using a $6M grant from the Rockefeller Foundation. However, after World War Two (WWII), increased public funding for research and universities allowed a different approach. National facilities funded by government grants and university contributions opened up access to 4m class telescopes to astronomers around the world with major new facilities built in the mainland USA, Hawaii, Australia, and Chile.

The large number of the 4m-class telescopes developed around the world in the 1960s and 1970s provided a wonderful infrastructure for deepening and broadening astronomical investigations. At the same time, they highlighted a growing barrier to increasing telescope size, as monolithic 4m aperture mirrors approach several practical limits. Not only do large mirrors weighing many tonnes struggle to keep up with external air temperatures resulting in temperature gradients and optical distortion, but their large size requires ever more specialised (and expensive) casting and recoating facilities. In addition, their large weight requires a correspondingly massive support structure in steel.

Several new approaches enabled monolithic mirror telescopes to make some further progress against the size barrier. A process developed in the 1980s using lightweight spin-cast main mirrors allowed effective monolithic mirrors to reach 8m in size. In addition, adaptive optics, a technique that introduces modern computing to control force-applying actuators on the back of the mirror, allows the shape of the mirror to be actively corrected and distortions removed during observing, an approach tested by the European Southern Observatory (ESO, see next section) at the 3.6m New Technology Telescope (NTT) in Chile, and later used as the basic mirror technology for the 8.2m Very Large Telescope (VLT) shown in Figure 8.2b. However, at 8m in diameter this technology reaches a fundamental limit for further scaling.

Yet there is a way forward. Combining these technologies—a deeply pocketed, hexagonally light weighted main mirror and computer-controlled actuators restoring the correct shape—leads to a natural innovation. By making the pocketing deeper and increasing the number and precision of actuators, the continuous glass surface can instead be broken up into multiple, independently positioned hexagonal mirror 'tiles'. Since each tile and support mechanism may now be a metre or so in size, the manufacturing problem is largely removed and a reflecting surface with a much larger aperture may be made with a sufficient number of these independent assemblies.

The entire telescope would thus work in a giant computerised control loop, constantly measuring the errors in the surface and making tiny (nanometre-scale) adjustments to the relevant tiles. The pioneer in this technology was the W.M. Keck observatory in Hawai'i, US with its twin 10m diameter, 36 tile telescope. Supported by both private and public funding, it was the first telescope to use this technology and became operational in 1993, though reliably positioning the tiles so that the telescope really operated as a single dish limited by diffraction took many years. The 'segmented mirror telescope', pioneered by Keck, opened up a new pathway to ever-larger astronomical mirrors. In practice, there is no fundamental limit to how big a telescope can be constructed from smaller segments, other than the practical engineering limits of large structures. With major engineering companies used to the challenges of building giant bridges and stadiums, the stage was set for the latest step in the development of the optical telescope.

As of 2021, three major projects are underway around the world to build the next generation of segmented giant telescopes. These are now so expensive (multi-billion USD) that only international collaborations can raise the funds required for their construction. Both the Thirty Metre Telescope (a US-led international consortium) planned for Maunakea, HI, USA and the Extremely Large Telescope (ELT, by ESO) under construction on Cerro Armazones, Chile will use many hundreds of 1.4m hexagonal tiles to make mirrors 30m and 39m in diameter respectively. The Giant Magellan Telescope (US-led international consortium) located in the Atacama Desert in Chile will instead use only seven mirrors, but each 8.4 metres in diameter, made using the University of Arizona spin-cast technology. When completed each of these three monsters will have a light-collecting area a million times that of Galileo's refracting telescope.

## 8.4  The European Southern Observatory (ESO)

Of all the global cooperative ventures in ground-based astronomy, perhaps none has been as ambitious as the European Organisation for Astronomical Research in the Southern Hemisphere, commonly referred to as ESO, a 16-nation intergovernmental research organisation for ground-based astronomy. ESO was created in 1962 to give European scientists access to facilities comparable to those in the USA, which were at the time unmatched in the world. The organisation was born out of a new spirit of intra-European cooperation which emerged after WWII and to this day that spirit, rather than national competition, drives its decision-making and operating principles (ESO, 2021).

Among the biggest ESO facilities is the Very Large Telescope array (VLT), which consists of four 8.2m telescopes located in Cerro Paranal in the Atacama Desert of northern Chile and was commissioned in 1998. The VLT unit telescopes generally work separately as four independent telescopes. However, uniquely they are also able to combine their light in a coherent way to form a single giant aperture via a system of mirrors in underground tunnels where the light paths must be kept equal to distances less than 1/1000 mm over 100-metre light paths. In this way, extremely high angular resolution can be achieved on bright astronomical objects. Construction of the VLT was a massive undertaking made possible by pooling the resources and industrial capabilities of all the ESO member states. As with CERN, cooperation in science leads to infrastructure facilities beyond the capabilities of any single European nation.

ESO's experimental sites—including the ALMA observatory and the very large telescope (VLT) in Paranal, Chile are given in Figures 8.2b, 8.3, 8.4 and 8.5.

Given the pre-eminence of US astronomical facilities in the 1950s and 1960s, the young ESO had a lot of catching up to do. With no good astronomical sites on mainland Europe, the new organisation made its most consequential early decision—to base an observatory in the Southern hemisphere at La Silla in Northern Chile. Not only was the southern sky much less explored, but it provided easy views of some very important objects, namely the centre of our own Milky Way galaxy and the two Magellanic clouds (two small galaxies orbiting our own). Chile also had multiple excellent sites in the Andes with dark skies and stable air. North American astronomers reached the same conclusion and established the first Chilean observatory at Cerro Tololo in Northern Chile in 1965. Figure 8.4 illustrates Telescopes hosted by ESO's original observatory, La Silla (Figure 8.4).

The first ESO observatory at La Silla in 1969 paved the way for a long-term association with Chile, followed by steady progress and growth culminating with the construction of the 3.6m telescope in 1976, bringing the organisation into the era of 4m facilities shown in Figure 8.4. At this point, ESO had developed a facility comparable to the best national observatories in the world. However, it was the next major project which took ESO to the real forefront of optical ground-based astronomy. Another important project in the field of astronomy is the ALMA (Atacama Large Millimeter/submillimeter Array) is a powerful radio telescope located in the Chainantor plateau (5000m) of the Atacama Desert, northern Chile. It comprises

**Figure 8.3**  ESO's experimental sites—the ALMA Observatory and the Very Large Telescope in Atacama Desert in northern Chile
*Source:* © CERN

**Figure 8.4**  Telescopes hosted by ESO's original observatory, La Silla
*Source:* Sangku Kim/ESO

66 individual antennae, each with a diameter of 12 metres. A view of ALMA (Atacama Large Millimeter/submillimiter Array) in the Chajnantor plateau (5000m) of the Atacama desert, Chile is shown in Figure 8.5.

**Figure 8.5**  View of ALMA (Atacama Large Millimetre/submillimetre Array) in the Chajnantor plateau (5000 m) of the Atacama Desert in northern Chile
Credits: ALMA, photo Juan Rojas

ESO now employs over 700 staff members and receives annual member state contributions of approximately €162 million guaranteed by international agreements. Most importantly this allows confident planning and scoping of new facilities such as the 39m aperture Extremely Large Telescope (ELT), a 1.3 billion Euro project, without major crises or delays caused by unexpected failures to secure funding (ELT, 2021).

How do astronomical observatories, such as those run by ESO, function? At an infrastructure level, astronomical observatories would seem to be very similar to large particle physics facilities such as CERN. However, the nature of astronomy drives the scientific culture in a quite different direction. While an accelerator facility such as CERN runs high energy physics experiments in a controllable and reproducible setting, astronomy's vast distances restrict the science (except for meteoritic studies) to be observational only. Indeed, astronomy may resemble fields like botany in that an important part of its discoveries come from simply surveying, looking and searching. The universe has been kind in that respect and has filled space with extraordinary and sometimes scarcely believable objects such as neutron stars, vast galaxy clusters, and black holes millions of times the mass of our sun. The plethora of fascinating objects means that to this day telescopes are used by relatively small research teams (compared to large teams of particle physicists working on high energy physics experiments).

The discussion so far may give the impression that telescopes and in particular their increasing size are all that matters for progress in the field. While the light-collecting area is an important factor in our increasing ability to study the universe, it is by no means the only one. Telescopes are designed to bring light to a focused image; no measurement has yet been made. A scientific instrument needs to be placed at that focus if a measurement is to be made and here, the great commercial developments in semiconductor materials, optics, and electronics in the twentieth and twenty-first centuries have extended the sensitivity of facilities dramatically beyond that of the human eye. That aspect is discussed further in the following sections of this chapter.

A state-of-the-art scientific instrument can enable major new discoveries. While modern telescopes are built once in a generation, instruments are designed, built, and exchanged by university or laboratory teams on a more frequent multi-year basis. There is a good reason for this. If the optics and structure are well designed, the telescope itself will be able to operate over a wide wavelength range with high efficiency and near the diffraction limit. Few improvements are necessary, even over decades. The Palomar 200-inch was still making important discoveries in the 1990s, 40+ years after construction. On the other hand, the conversion of photons to electrons and then to data at the telescope focus is a very imperfect process.

Only recently, for example, has it been possible to even fill the focal plane of a large telescope with efficient light detectors so that every accessible object in the field of view is imaged. Yet, even in that case, detected photons cannot have their energies measured, so the entire spectral dimension of the objects is missing. Clever systems called multi-object spectrographs allow spectral information to be derived for many objects simultaneously. But in that case, complete coverage of the field of view must be sacrificed. It is the imperfect nature of instrumentation that allows scope for improvement and hence a lot of research interest. In the case of ESO, teams of scientists and engineers in European universities, institutes and observatories are constantly seeking to improve performance. Large modern instruments can cost 50 million euros or more to construct, often using technologies developed for commercial, industrial, or military use and adapted to astronomy.

One of the greatest outcomes of this new era is the enormity of the data challenge, which is driving a fundamental shift in the nature of astrophysics itself. Unlike other domains of science, where the repeatability of the experiment is a key tenet of the scientific method, astronomy has only the universe itself to observe. We can, however, look to different parts of the universe to see repetitions of the experiment in galaxy formation itself, for example. We can also use visible changes as a way to estimate the intrinsic variance of the measurement itself. Yet with the Square Kilometre Array (SKA) to be built in Australia and South Africa we will soon have surveyed almost all of the observable universe itself, or at least to a certain minimum object size, which presents an inescapable sample variance limit termed *cosmic* variance.

Moreover, we have historically reprocessed the data recorded in previous observations, finding new and exciting phenomena. This has led most recently to the discovery of an entirely new (and still unknown!) class of events, Fast Radio Bursts, in reanalysed Parkes Radio Telescope data (Lorimer et al., 2007). It also facilitates the testing and refinement of models as data is explored in different ways. With the vast sizes of upcoming facilities, we can no longer hope to store the data in its original form, and instead will have to process it in real time and save the 'interesting' events. That means an algorithm based on artificial intelligence (AI) will determine what the cosmos looks like to us in the future. Unless we create an especially clever and flexible algorithm, we risk losing the unknown discoveries that surprise us—discoveries that in this chapter were reliant on human intellect to discern as something worthy of further ultimately revolutionary study, and perhaps a risk that we must consider in the new age of automated searches.

Beyond the search for new science, the era of big data also changes the manner in which we undertake our research. We see the computer facilities brought physically closer to the information generating facilities themselves and even after the filtering by AI we see vast datasets saved locally and not distributed globally. This then brings the queries, and scientific enquiry, by astronomers to the site itself—accessing local resources to sift through and make sense of the data via Structured Query Language (SQL) interfaces with only these final results sent to the end-user at their own local institute, if ever. We have had experience of this with the likes of the Atacama Large Millimeter/submillimeter Array (ALMA) run by ESO and the US National Science Foundation (NSF) in Chile and its global network of several nodes that are the only sites to retain full copies of the refined results. It is science as a service, and in this way, astronomy is following industry's model of cloud computing as closely as the new microelectronics companies led the field's direction in the last century.

Big Science deals with the most exciting and creative discoveries known to humankind. In more recent times with the aid of the Event Horizon Telescope (EHT), astronomers have unveiled the first image of the supermassive black hole at the Centre of the Milky Way Galaxy.

It was only in 2019 that the first-ever image of a blackhole was captured in the Messier 87 galaxy (M87*) 55 million light years away from Earth (The Event Horizon Telescope Collaboration, et al., 2019). On 12 May 2022, the first direct image of Sagittarius A* (SgrA*), a black hole 27,000 light years away from Earth in the Milky Way galaxy, was produced by the Event Horizon Telescope (EHT) collaboration. For the first time in human history, the shadow of the supermassive black hole in the centre of Messier 87 (M87), an elliptical galaxy, was witnessed (shown in Figure 8.6), which is the most powerful source of radio energy among the thousands of galactic systems visible in the constellation of the Virgo cluster.

The EHT's recent main image of Sagittarius A* is an outcome of a combination of thousands of images extracted from EHT observations (Event Horizon Collaboration et al., 2022). The obtained results are indeed quite exciting because the size of the ring from the shadow of the supermassive black hole that is located at the centre of our own Milky Way galaxy about 25,000 light years from Earth exactly matches that predicted by Einstein's General Relativity.

Furthermore, now that clear pictures of two black holes of different sizes are available, researchers can use supercomputers to combine, analyse, and share data that tests the theories of how black holes interact with their surroundings; and how gas and gravity behave around these supermassive black holes (ESO, 2022). EHT is a collaboration of 300 researchers from 89 institutes, and a network of a number of telescopes around the world including ESO telescopes (ALMA, APEX). The EHT collaboration uses a technique known as 'Very Long Baseline Interferometry' to focus multiple telescopes located in different parts of the world to make a world-sized single virtual observatory combining human ingenuity, technological precision, and mathematical excellence.

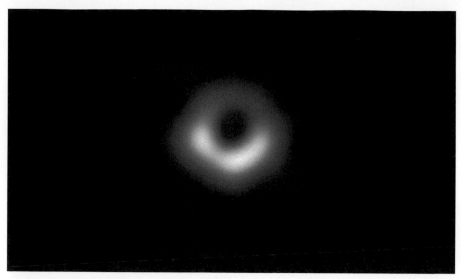

**Figure 8.6**  First image of the M87 black hole
*Source:* EHT Collaboration

## 8.5  Innovation in Global Astronomy Seen from Down Under

As discussed in the previous section, while the fundamental science behind the optics of astronomy has barely changed in centuries, innovations have been driven by novel mirror production methods coupled with adaptive optics, by the detectors and the exponential growth in both microelectronics performance, and by more sophisticated analysis and electronics/noise suppression. These innovations come from a rich and diverse range of expertise, enabling the field of astronomy to leverage advances in fields far beyond its own. In the last decades advances in adaptive optics to fibre optics and precision-crafted coronagraphs have ensured that even optics have been revolutionised.

Each technological advancement has seen leaps in our scientific framework, with the careful measurement of the planetary motions with the first telescopes of Tycho Brahe leading Joannes Kepler to propose his three laws (or mathematical relationships) between the planetary motions that supported Nicolaus Copernicus' heliocentric model and, of course, Galileo's seminal observations already mentioned. Ultimately the scientific model that naturally explained these laws was Isaac Newton's model of gravitation, where the vacuum of space allowed the simple gravitationally influenced motions to be clearly observed.

Such measurements of planetary motions would improve until new discrepancies between Newtonian gravity and what was observed could no longer be ignored. In particular, the anomalous motion of Mercury would provide significant support for GR which could more accurately predict such behaviour. Ultimately, the final prediction of GR, that of gravitational waves, would involve a century-long hunt resulting in a Nobel Prize in 2017 for the work at the Laser Interferometer Gravitational-wave Observatory (LIGO) in the US. It also allowed a new era of multimessenger astronomy to begin, using a range of wavelengths as well as gravitational waves to probe high energy astrophysics events.

## 8.5.1  A Standard Cosmological Model

Paradoxically, in the early decades of the twenty-first century, astronomy has never known so little about the universe. The concordance model (Sollerman et al., 2009) of our cosmos is known as Lambda Cold Dark Matter ($\Lambda$CDM), named after the two dominant (and unknown) components acting within GR. As mentioned in Chapter 5, fully 95% of the universe's (rest-mass energy) composition lies in either dark energy (70%) or cold dark matter (25%) whose exact natures have yet to be conclusively determined (Aghanim et al., 2020).

Dark energy is the term for the driving action behind the measured accelerating expansion of the Universe, with growth rates in agreement with the model of a Cosmological Constant (termed Lambda), an energy associated with the vacuum state of spacetime itself. The cold dark matter is a gravitating particle with little to no interaction with the electromagnetic field, of non-relativistic motion (hence the cold in the thermodynamic sense of the word), and the subject of global particle physics hunts.

Indeed, Australia is at the forefront of such endeavours with the Australian Research Council (ARC) Centre of Excellence for Dark Matter Physics, a seven-year programme that spans multiple detection avenues. One programme is searching for dark matter particle production in the Large Hadron Collider (LHC) at CERN, hinted at by a discrepancy in the mass-energy budget between what goes into the collision and what comes out, with escaping dark matter particles avoiding triggering detectors. The Centre will also explore detection efforts as the dark matter particle travels to us, or rather as the Solar System travels through the dark matter cloud that gravitationally binds the Milky Way together as inferred from astronomical observations.

It is astonishing that just a few measured numbers within the $\Lambda$CDM concordance model can describe so much of the rich diversity of astronomical observations, particularly when the precise nature of the majority terms (dark matter and dark energy) has yet to be conclusively determined. From the afterglow of the Big Bang, a remnant all-pervading electromagnetic signature from across the sky now long since cooled to microwave wavelengths (hence known as the Cosmic Microwave Background) measured by satellites such as WMAP or Planck (Howell, 2018), to the predicted statistical distribution of relatively nearby galaxies as mapped in their millions by the Sloan Digital Sky Survey or the 2dF instrument at the Australian Astronomical Observatory, the concordance model has been able to withstand an astounding variety of experimental probes (Aghanim et al., 2018). One enduring feature of modern astronomy has been this rich diversity of technologies and techniques to test our understanding of the cosmos, a drive that has confronted innumerable challenges and delivered spectacular leaps in technology.

## 8.5.2 Global Teams

These collaborations and leaps in technology and innovation can begin from the smallest of ideas, or chance discussions, a feature of astronomy for the late half of the twentieth century in which outrageous capabilities are advanced with the simplest of statements 'wouldn't it be nice if' and the journey to that advancement is realised in the decades since.

Such guidance can be provided by decadal plans, such as the American Astronomical Society, US National Academy of Science or the Australian Academy of Science, in which the community is consulted and engaged to prioritise astrophysical areas of focus and hence technological development for the decade ahead. That focus on supporting and growing research excellence, with large multi-decadal long institutions to foster and safeguard that institutional memory, provides the learning and testing grounds of internship and training programmes and drives ever onwards and upwards in capability, is shown in agencies from the National Aeronautics and Space Administration (NASA) to the Commonwealth Scientific and Industrial Research Organisation (CSIRO) in the twenty-first century.

A rich diversity of wavelengths, as well as scientific demands, have resulted in enormously challenging and costly undertakings that can only be supported by international partnerships. Astronomy has always been a collegial endeavour, with international (and free) access to radio astronomy data supported through an open skies policy (Nature, 2021) which maximises facilities', rather than individual scientists', research outputs. These collaborations range in scale and complexity from private–university like the W.M. Keck Observatory to the national programmes that resulted in the USA's Very Large Array (VLA) or The Low-Frequency Array (LOFAR) facility in the Netherlands to formalised, international treaties with the likes of the ESO and its VLT/ELT. Utilising such diverse facilities is itself a challenge, necessitating the rise of standardised open-source interface tools such as the Virtual Observatory to ease astronomers in access to such systems.

Arguably the costliest of all such astronomy missions is in space. Space technologies have benefited from astronomy's need to escape the Earth's atmosphere by exploring wavelengths otherwise blocked or distorted by the air. These include Big Bang related science missions such as WMAP and Planck, or more general discovery facilities like Hubble or the James Webb Space Telescope (JWST). Supremely complicated endeavours, these space missions concentrate the efforts of thousands of engineers and researchers for decades, guaranteeing innovations through the challenge imposed by the harsh operating environment. Such innovations can however find their greatest impact far afield.

The advanced optics systems demanded by Hubble led to innovations in the focusing of light for computer chip etching creating faster, cheaper, and smaller microprocessors. The cameras (using charged coupled devices or CCDs) of Hubble, as well as the software techniques to process and enhance such revolutionary images, have found new uses (NASA, 2018) in medical fields such as mammograms. The translation of the technologies is aided by the generous and streamlined licencing arrangements (NASA, 2022) provided by NASA, actively encouraging the use of their patents far afield.

The growing scale of astronomy facilities, requiring ever more complex negotiations of funding and Intellectual Property, as well as the ongoing access to resources for operations, maintenance and upgrades (often at the expense of other priorities from the decadal plans) now rivals that of particle physics. The growth of the science has occurred with the maturing of the collaborations to advocate for political support and will in negotiating such arrangements, as well as for its funding. Astronomy, unlike particle physics, however, has a unique advantage. Amateur telescopes can still offer insights into space science, or permit discovery via citizen science initiatives, ensuring widespread awareness. Hence, increased public support for the costlier endeavours, that in turn can be leveraged for the government.

## 8.5.3  Distributed Opportunities

The widespread access to discovery, from archived data and open-source tools, to the relatively low technology barriers for finding new astronomical phenomena also ensures astronomy is an innovative science. Unlike a particle collider, a relatively cheap telescope or even a simple workstation for reprocessing openly available astronomy images means that no one entity is required to facilitate astronomy. Therefore, no one entity exists that can stifle or slow creativity/innovation through its bureaucracy, yet happily there are sufficiently large organisations to permit that institutional memory remains as explained earlier. Indeed, we see in the twenty-first century an evolution in this system with the integration and fostering of outside companies, from SpaceX in the USA to Fleet Space Technologies (FST) in Australia, that marry the organisational strengths of larger entities to the innovation and a cost drive that these smaller commercial entities exemplify.

Within the Australian context, the scale of astronomy is noticeable for its disproportionate size and enduring success, with sustained support for decades by all sides of governments, facilitating the development and innovation of new technologies. This support is in part a result of the community speaking with one voice to the government through the previously mentioned decadal plan.

Alongside this formal structure to foster innovation and the competition of ideas internally, without presenting competitive funding requests and confusion to external parties, the Australian community also greatly benefited from the CSIRO and its long-standing commitment to radio astronomy in particular. This has spanned the range of facilities from supporting the Apollo Moon landings with the Parkes Radio Telescope (now upgraded to be 10,000 times more powerful than when first built) to the latest Australian SKA Pathfinder and its breakthrough Phased Array Feeds (PAFs) that are now exported worldwide. This latter example well demonstrates the common path from a tentative, yet ambitious, scientific goal to the commercialisation of resulting technologies with the Quasar Satellite Technologies spinout of CSIRO as we will explore.

## 8.5.4  Radar to Radio

Australian astronomy has a competitive advantage with the nation affording pristine views of the Southern Hemisphere and the Milky Way galaxy stretching directly overhead for world-beating observational conditions. This opportunity, combined with a rich legacy of radio astronomy, will make Australia one of the true giants in the field of radio astrophysics. This interest in radio was first driven in Australia by radar systems and experience gained during the Second World War. Early pioneers in radio astrophysics, such as J.L. Pawsey, developed their skills as part of the defence effort in this new frequency domain with an auspicious first test site in 1940 at Dover Heights above Sydney (Forman, 1995). Since that time, primarily under the auspices of the CSIRO, radio astronomy has flourished in Australia with the likes of Parkes, which received the first televised signals of the Moon landing in 1969 that would then be broadcasted to the world (Sarkissian, 2001). In the decades after, this single dish would be subsequently upgraded to stay at the forefront of radio astronomy, making a series of breakthrough discoveries over the next half a century (Edwards, 2012). That technical focus provided by the CSIRO, and the intensity of scientific mission from the Australian Academy of Science, meant that leapfrog technologies could be realised through continued investment over decades, with such successes only encouraging Australia to consider a truly audacious undertaking.

In 1990, Peter Wilkinson introduced the concept of a *square* kilometre of telescope collecting area, to detect signals from the most distant hydrogen atoms in the observable universe (Wilkinson, 1991). The SKA, as mentioned earlier, was formally conceived by the International Union of Radio Science in 1993 (Ekers, 2012) in an effort to focus radio astronomy efforts worldwide towards an audacious new facility. The possibility of this scale of facility was discussed in the years prior; driven by the simple scientific goal to detect hydrogen within galaxies out to the very greatest cosmological distances in an ultimate galactic census. The counterpoint to this science was also the recognition at the time (Harwit, 1981) that scientific advances follow technical innovation, and to fail to plan and drive such innovation might doom the science of astronomy to stagnation. More than that, an active scientific community tends to be exponential in its advances (de Solla Price, 1984, 1986) ), which then requires the generation of new technologies to sustain such exponential growth rather than the refinement of existing technologies that plateau out (Ekers,

2012). The exponential growth in radio astronomy facilities' detection sensitivity has a doubling time of approximately three years (Ekers, 2012). On a log–log scale, such technology seems an inevitable extrapolation in time. However, it would be a decade before the international Memorandum of Understanding (MoU) was signed for the newly christened SKA and just over another decade for the SKA to be formalised as a legal entity in 2011 with an expectation that it might commence construction within the next decade. Yet this would transpire to have been optimistic; indeed, such delays to full operations were inevitable as Moore's Law[1] could not deliver the required computing power to correlate signals from so many telescopes until 2030 (Duffy, 2014).

A rich variety of technology paths were explored in precursor facilities, from MeerKAT in South Africa to China's Five-hundred-metre Aperture Spherical radio Telescope or FAST (Duffy et al., 2008), LOFAR in the Netherlands, and the Murchison Widefield Array (MWA) in Western Australia and the Australian SKA Pathfinder. Despite having wildly different designs, the majority of them shared the same design idea: connecting a lot of smaller telescopes to produce an unparalleled level of sensitivity and field of view of the sky.

For the Australian SKA Pathfinder, the critical technology lay not in the smaller telescopes, where a relatively mature design was utilised, but instead in the proposal of new receiver technologies for where the radio light was focused, at the feed. Rather than a single-pixel camera, or receiver, the feed would house a phased array of 188 receivers acting in concert. The abilities therein permitted 30 radio beams to be generated in a supercomputer, thereby monitoring a 30 times larger sky area. This was an enormous leap in radio astronomy capabilities, mapping more galaxies in greater detail than ever before (Duffy et al., 2012a, 2012b). The ability to monitor separate areas of the sky had enormous potential for satellite communications, and the technology was spun off into Quasar Satellite Technologies by CSIRO. Without such dedicated focus over decades to deliver a science instrument, the technology would not have been ready to coincide with the era of thousands of satellites that need to be tracked and communicated in low Earth orbit.

## 8.5.5   Infrared Comes of Age

Unlike radio astronomy, Australia's relatively low altitude observing sites meant that other wavelengths, such as optical and infrared (IR), did not have the competitive advantage of sitting above a large fraction of the atmosphere that sites in Hawaii or Chile offered. However, the field was able to remain internationally relevant by driving innovations, for example, wide-area sky surveys where the atmosphere is not as much of a limiting factor as with deeper surveys.

---

[1] Moore's Law, an empirical relationship postulated by Gordon Moore, states that the number of transistors that can be manufactured on an integrated circuit (which roughly equates to computing power) doubles every two years.

The process by which a site is selected for an astronomy facility is an involved process, an optimisation effort that considers the astronomical observing quality of the location (Aksaker et al., 2020) as well as the cost of installation or availability of existing infrastructure, and the potential for local engagement in the programme. The observing quality at optical and IR wavelengths explores the challenges from light pollution (termed radio frequency interference, or RFI, in radio astronomy), cloud cover, aerosol concentrations and altitude which allows a site to be above significant fractions of the air column (Daniyal and Hassan Kazmi, 2019). Politics can of course play a role in the selection of the site for a ground-based observatory, particularly where international collaborations are involved. Various features should be considered, such as sky transparency, protection of dark skies, altitude and climate, radio frequency interference that may be reconciled when deciding on the location of a site as was the case with the Square Kilometre Array (SKA), which is the largest and most sensitive radio telescope ever built. Both the South African and Australian sites were selected primarily due to their low radio frequency interference levels, large collecting areas to capture weak radio signals from space, and the ability to cover a wide range of radio frequencies among other site-specific advantages.

The broader global history of astronomy in the last century and a half resembles the Australian experience, with dedicated R&D efforts driven by defence needs or dramatic new frequencies opened up that have spurred innovation and supported collaboration. Indeed, for almost all of humankind's existence, our entire visual experience of the universe was through an electromagnetic window between about 400nm and 700nm—what is now known as the visible spectrum, to which the human eye is sensitive. It was only very recently, in the year 1800, that William Herschel noticed the presence of electromagnetic emissions beyond the reddest of visible wavelengths (Herschel, 1800), the first signs of a secret universe shining invisibly on, waiting to be measured and understood. But to do so, we would be entirely reliant on machines; Herschel's discovery of infrared radiation heralded an interplay between astronomy and instrumentation that continues to this day.

## 8.5.6 Infrared Universe

Infrared wavelengths occupy such a vast portion of the electromagnetic spectrum—between about 700nm and 1mm—that no single technology or device can cover the entire range. Regardless, what was evident from the beginning was the connection between infrared radiation and heat, with the first infrared detectors being purely thermal: Herschel's seminal work used a simple thermometer; by the 1820s, the discovery of the thermoelectric effect (Seebeck, 1822), followed by the invention of the thermocouple (Nobili, 1830), and later the more sensitive thermopile, had provided a way to convert incident heat into a measurable electric current. The next great leap came in 1878 when American astronomer Samuel Pierpont Langley invented the bolometer (Langley, 1881), its wire-thin sensor and high sensitivity allowing the study of dispersed infrared spectra for the first time. The need to invent new instruments to satisfy scientific curiosity—in Langley's case, fuelled by a lifelong fascination with the nature of the Sun—is something with which practically all observational astronomers will be familiar. As thermal infrared detectors continued to develop, so did our understanding of heat as a form of *light*. The scale of such experiments and the technologies being invented were however at the individual level, that could be supported by personal (or single benefactor) wealth. This would dramatically change towards the end of the 1800s and into the twentieth century.

The benefits of observing the infrared universe were clear to astronomers, yet detector technology throughout the early 1900s lacked the sensitivity required to measure all but the brightest sources. Following WWI and throughout WWII, military interests in heat detection technology snowballed, fuelling the development of new infrared sensors to enable night vision. Such programmes would ultimately lead to today's state-of-the-art scientific detector arrays for astronomical imaging and spectroscopy, a prime example of how science has directly benefited from defence spending (Lovell, 1977).

Perhaps the most significant breakthrough in infrared sensing came after WWII with the discovery of the variable-bandgap mercury cadmium telluride (HgCdTe, or 'MCT') alloy (Lawson et al., 1959). Unlike previous technologies, MCT responded to *photons* rather than heat. Even more important was that its spectral response was tuneable to narrow wavelength bands anywhere in a 1–30 micron range, allowing application-specific designs. This, along with their superior sensitivity and speed, is why MCT-based sensors still dominate the high-performance infrared sensor market decades later.

In a fortuitous technological coincidence, 1959 also saw the invention of the metal oxide semiconductor field-effect transistor (MOSFET), leading to the development of the complementary MOS (CMOS) fabrication process in 1963. While it's hard to overstate the importance of these inventions to the field of computing, they also catapulted optical detector technology to new heights. In 1969, Boyle and Smith invented the 'charge-coupled device' (CCD) to store and transport electrical charge across silicon using MOS technology (Boyle and Smith 1970), later winning them a share of the 2009 Nobel Prize in Physics.

In hindsight, it's easy to trace the path from the two-dimensional CCDs currently capturing images in various astronomical instruments around the world (many boasting over 16 million pixels), and to Boyle and Smith's one-dimensional row of capacitors with possible applications 'as a shift register, as an imaging device, as a display device, and in performing logic'. The fact that the inventors were researching memory technologies is a reminder that R&D has a habit of producing solutions in one person's lab to problems in someone else's; often, innovation is little more than connecting these worlds. Indeed, silicon's tendency to internally generate charge when exposed to ultraviolet- and visible-wavelength photons was seen by some as an inconvenience; because who wants to worry about making memory chips completely light-tight?

By the 1970s, the stage was set for infrared detector technology to make its next major leap, once again funded by the US military. With the advent of 'hybridisation' (Thom, 1977), it became possible to bond exotic photosensitive materials to silicon pixel arrays, such that silicon devices could still be used to capture and read-out the charge in each pixel, but a different material could be exploited for photo-generation. Suddenly, tuneable-bandgap MCT and MOS technology collided, and it's hard to imagine this happening so quickly had the benefits to surveillance and weaponry not been so substantial.

Fast-forward to today and many of our most productive facilities–such as the Sloan Digital Sky Survey and the Hubble Space Telescope, which together yielded over 15,000 papers with a six-figure sum of citations—would be starkly less capable without the defence-funded technology installed at their focal planes. Similarly, the newly launched JWST has added more than 66 million infrared-sensitive pixels to our space telescope arsenal, spread across 18 detector arrays from commercial product lines initially developed for military purposes.

We are currently experiencing another evolution of the science funding paradigm, as wealthy entrepreneurs take it upon themselves to answer 'big' questions on behalf of humankind. One example is the Breakthrough Foundation established by Russian-Israeli entrepreneur Yuri Milner (Breakthrough, 2021). A physicist by training, Milner nevertheless made his fortune as an information technology investor, with his portfolio including internet giants such as Facebook and Twitter. In fact, Facebook CEO Mark Zuckerberg sits on the board of Breakthrough 'Starshot'—a $100M initiative with the goal of sending a flyby mission to Alpha Centauri within a generation. Stephen Hawking was also a board member. Perhaps this hybrid model of private financial power and great scientific minds is what's required to leapfrog traditional government-funded R&D efforts. Indeed, the parallel 'Breakthrough Watch' initiative—aimed at identifying and characterising Earth-like planets around Alpha Centauri and other nearby stars—has already yielded a Nature-published success in the direct imaging of low-mass planets within the habitable zone of the star in question (Wagner et al., 2021). This was achieved with the 'VISIR' thermal infrared coronagraph on the 8 m VLT, its recent upgrade was funded by Breakthrough Watch.

While sophisticated observational techniques such as the pupil masking at the heart of the Breakthrough VISIR facility can tame the colossal dynamic range (relative brightness to faintness) of a star-and-planet image, other limitations are simply baked into the universe: the uncertainty in photon arrival times, for example, or the random thermal excitation of charge carriers in electronic circuits. These sources of noise plague the photon-starved astronomer perhaps more than any other, since they define just how faint a celestial object can be usefully observed in each time.

By far, the dominant noise source in modern-day infrared detector arrays is so-called *read noise* or *readout noise*. This is the penalty one pays for measuring the charge in any optoelectronic pixel and is the sum of the random fluctuations introduced by the chain of transistor and amplifier circuits required to probe the pixel itself. A good CMOS circuit introduces less noise, yet we haven't seen the best detector manufacturers consistently deliver devices with noise levels below around 18 electrons (equivalent to 18 photons if we assume perfect quantum efficiency). As such, there has been little relief beyond increasing exposure times or stacking multiple frames to leverage Gaussian statistics, both of which come at a literal cost in terms of telescope time.

But what if the trading signal-to-noise ratio with time is not acceptable? Modern challenges, including wave front sensing for adaptive optics, fringe tracking, and time-delay integration from scanning satellites, make no apologies for requiring sensitivity and speed. The Astro2020 decadal white paper on All-Sky Near Infrared Space Astrometry (Hobbs et al., 2019) acknowledges that only new technologies can make the proposed near-infrared version of the hugely successful Gaia mission feasible. One of the four candidate solutions for 'GaiaNIR'—and one of the most promising, given recent progress—is the linear-mode avalanche photodiode (LmAPD) array technology from commercial sensor manufacturer Leonardo MW (Baker et al., 2019). Notably, the development of this technology throughout the 2010s has been driven by (and for the most part, funded by) astronomical groups, with collaborators including ESO, the University of Hawaii, the European Space Agency (ESA), and the Australian National University.

While single-pixel avalanche photodiode detectors have existed for decades, LmAPD arrays are revolutionary in their two-dimensional format and linear response. They utilise bandgap-engineered MCT just like traditional infrared detectors, but with the addition of an 'avalanche gain' region between the photogeneration layer and the readout circuit. Applying an electric field (or 'bias') across this region results in a selectable multiplication of the signal before it is read out, therefore reducing the effective contribution of the readout noise floor. Demonstrated gains of several hundred have led to these devices being dubbed 'noise-free'.

The current focus is on increasing the size of these arrays from the 0.8-megapixel devices already on the market to the 4 megapixels required by next-generation instruments such as GaiaNIR mentioned before. Dark current—an accumulating and temperature-dependent background signal generated within the device—has also been dramatically improved, although this does not eliminate the general requirement to cool MCT-based detectors to around 80 K. It does, however, allow the sensitivity of LmAPDs to be exploited by traditional long-exposure regimes such as ground-based spectroscopy; indeed, this is one of many Australian goals for the deployment of the new generation of arrays, along with the demonstration of their capabilities in orbit (Gilbert et al., 2019).

Parallel developments in the infrared sensing landscape are myriad: Microwave Kinetic Inductance Detectors (MKIDs) promise per-pixel direct measurement of photon energy up to the mid-infrared region, capturing spectral information without the need for dispersive optics or filter wheels, albeit with an operating temperature of just 100 mK.[2] We see a resurgence of the humble bolometer in the development of

---

[2] 100 mK or 100 minikelvin is equivalent to −273.04999999999995 °C.

'microbolometer' focal plane arrays for thermal imaging, which can operate with no cooling at all. At a more fundamental level, work continues on new detector materials, from the attractive broadband response of germanium to metamaterials that appear to defy traditional physics.

While the new technologies at the cutting edge of infrared detector technology each have unique merits, history has taught us that their compatibility with current and future funding landscapes will be at least as decisive when it comes to their ultimate success.

## 8.6  Gravitational Waves and Big Science

The major breakthroughs achieved in the field of gravitational wave (GW) astrophysics beginning in late 2015 are no less remarkable than those of photon-based astronomy and make this scientific endeavour a prime candidate for answering or at least informing a number of the key questions posed in this chapter. In five years, the ability to conduct GW observations has gone from 'nearly impossible' to 'almost routine'; GWs are now detected weekly or even more frequently when GW detectors are running. That it took almost exactly 100 years from Einstein's postulation of the existence of GWs in 1916 (even Einstein didn't completely believe in their existence for a time; see Kennefick (2005) to when they were first detected is a testament to advances in key technologies (stabilised lasers and ultraprecise manufactured mirrors) and theoretical and computational advances in GR, as well as in the dedication and tenacity of clever scientists and engineers in harnessing those advances. Going from 'impossible' to 'routine' owes itself to many factors, chief among them innovations in GW detector technology and the size and scale of GW collaborations that have developed over the past 25 years.

Gravitational waves are unlike the more well-known types of electromagnetic radiation such as light, X-rays, or radio waves discussed earlier in this chapter. They come about as a direct consequence of GR, a geometric theory of gravity in which mass and space-time weakly interact with each other in such a way that attractive force experienced by objects possessing mass comes about because of the induced curvature (or warping) of space. The eminent theorist John Wheeler nicely summed up GR's essence in the following way—'spacetime tells matter how to move and matter tells spacetime how to curve' (Wheeler, 1998). GWs are emitted when any massive object accelerates and can (simplistically) be described as the stretching and squeezing of space-time ('ripples in space-time') resulting from the change in geometry as an object accelerates through space. More precisely, they physically manifest themselves as strains or changes in length per unit length which occur in directions perpendicular to the propagating wave. The LIGO gravitational-wave detector concept is illustrated in Figure 8.7.

LIGO uses a sophisticated variation of Michelson interferometry to detect differential strains produced by passing gravitational waves along orthogonal arms. LIGO exploits the physical properties of light and of space itself to detect and understand

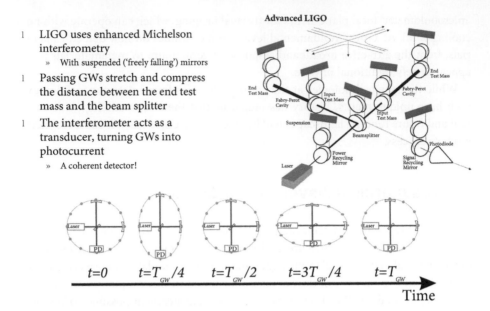

l   LIGO uses enhanced Michelson
     interferometry
     » With suspended ('freely falling') mirrors
l   Passing GWs stretch and compress
     the distance between the end test
     mass and the beam splitter
l   The interferometer acts as a
     transducer, turning GWs into
     photocurrent
     » A coherent detector!

$t=0$ $\quad$ $t=T_{GW}/4$ $\quad$ $t=T_{GW}/2$ $\quad$ $t=3T_{GW}/4$ $\quad$ $t=T_{GW}$

Time

**Figure 8.7** The LIGO gravitational-wave detector concept

*Source:* David Reitze, LIGO Laboratory, California Institute of Technology

the origins of gravitational waves using two enormous laser interferometers located 3000 kilometres apart.

The challenge of detecting GWs is a direct result of the incredibly weak nature of gravity relative to other fundamental forces. The coupling constant linking space-time (geometry) and stress-energy (matter) is $8\pi G/c^4$ where $G$ is Newton's gravitational constant and $c$ is the speed of light. This constant is *mind-bogglingly tiny*— about $10^{-43}$. It is simply inconceivable with current technology that a laboratory-scale GW generator could be built. To generate *detectable* GWs, very massive objects such as black holes and neutron stars moving in close orbit at significant fractions of the speed of light are required. GWs possess other important characteristics in that they travel at the speed of light but, unlike light, they propagate unimpeded through matter. This latter aspect implies they can carry information unavailable to other types of astrophysical messengers.

Fortunately, the immense challenge of detecting GWs also turns out to be a tremendous scientific opportunity for understanding the high energy universe. Because they emanate from the bulk relativistic motion of matter, GWs reveal unique information about the nature of the most energetic and violent events in the universe in ways that conventional astronomy cannot. They are a new window into the cosmos, providing completely new observational insights into the nature of black holes, neutron stars, and even the Big Bang.

## 8.6.1 History

The history of the development, construction, and operation of GW detectors provides some insights into the need to undertake ambitious large-scale projects to answer fundamental questions about the nature of the universe. That it took nearly 100 years from the time GWs were postulated to the first direct detection of gravitational waves is no surprise when considering the challenges posed by their feeble amplitudes as described above. The Laser Interferometer Gravitational-Wave Observatory (LIGO) in the US (Aasi et al., 2015) is used as an example in this section; however, the European-based Virgo gravitational wave detector (Acernese et al., 2014) took a very similar path towards its operation.

Figure 8.8 presents a detailed timeline for LIGO from its inception as a joint project led by Caltech and MIT in 1984 through the funding and construction phases of Initial and Advanced LIGO until the present. The first 'inception/funding' phase from 1984 through 1992 established the early collaboration between Caltech and MIT to jointly lead the research and development programme to design and prototype a large-scale interferometer in the United States. In 1984, less than 100 physicists were working in GW detector development in the United States, with a somewhat larger worldwide community. The transition from tabletop R&D to a large-scale construction project occurred between 1990 and 1994, with construction beginning in 1995. The LIGO Scientific Collaboration (LSC) was formed in 1997 with 24 institutions and 250 scientists and engineers from Europe, the US, and Australia. This was a key

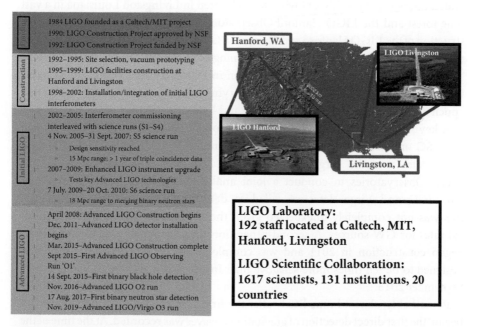

**Figure 8.8** A timeline of LIGO construction and operations

*Source:* David Reitze, LIGO Laboratory, California Institute of Technology

**Figure 8.9** LIGO Livingston aerial view
*Source:* Courtesy Caltech/MIT/LIGO Laboratory

step in beginning to grow the community. LIGO has two identical and widely separated interferometers. LIGO Livingston is located in Livingston Louisiana in a vast
pine forest and the LIGO Hanford Observatory is about 3000 km away. The construction of the LIGO Hanford and Livingston Observatories was completed in 2000,
followed by the installation and commissioning of the Initial LIGO interferometers.
The aerial view of LIGO Livingston is given in Figure 8.9. The Initial LIGO interferometers ran from 2002 through 2010, carrying out six observing runs during that
epoch at increasing sensitivities (Zweigzig and Lazzarini, 2007).

A key step in developing the global GW collaboration occurred in 2007 when
the LSC and Virgo Collaborations signed an MOU (LIGO-VIRGO, 2007) to
share data from the Virgo Observatory located near Pisa, Italy and the two
LIGO Observatories to conduct a joint analysis programme. Notably, the Initial LIGO and Virgo detectors failed to detect gravitational waves. However,
this was not completely unexpected given the uncertainty about the astrophysical rates for GW sources. A second-generation detector project, Advanced LIGO,
began construction in 2011 and was completed in 2015. Advanced LIGO was
designed to be ten times more sensitive than Initial LIGO and began operations in
September 2015.

Remarkably, almost immediately after the first Advanced LIGO observation run
began, the first direct detection of gravitational waves was recorded. At the time of the
first detection, the LSC Virgo had roughly 1500 scientists and engineers representing
more than 100 institutions worldwide.

## 8.6.2 Gravitational Waves: A Cascade of Scientific Breakthroughs

The first confirmed detection took place on 14 September 2015 when the two LIGO Observatories in Hanford, WA and Livingston, LA each recorded the signal from GWs emitted from a pair of black holes located approximately 1.3 billion light-years distant from Earth in the final stage of the inspiral and merger LIGO and Virgo collaborations (2016). Named GW150914, the detection was remarkable not only in that it proved that GWs could be detected, thus confirming in a direct way one of the remaining outstanding predictions of GR,[3] but also providing definitive proof that (i) astrophysical black holes are 'Kerr-like' and consistent with general relativity, (ii) that binary black hole systems exist in nature, (iii) that such systems can merge in Hubble time (the current age of the Universe), and (iv) that populations of black holes exist with masses greater than 20 solar masses (the observational limit set by x-ray astronomy).

Since 2015, nearly 90 binary black hole mergers have been detected by LIGO and Virgo. These detections have produced further significant insights into how black holes form and evolve. Very recent results from observations carried out in 2019 have reported the first confirmed observation of an intermediate-mass black hole (long thought to exist, but never previously observed) as well as confirmation that black holes can form in a theoretically forbidden region known as the pair-instability supernova mass gap (Abbott et al., 2020).

Taken together, these observations are already answering the question of how intermediate-mass black holes form and grow in size and are giving us tantalising hints that binary black hole mergers seed the formation of supermassive black holes found at the centres of most galaxies.

An equally spectacular detection occurred on 17 August 2017 from the inspiral and merger of a binary neutron star system located 140 million light years distant (Abbott et al., 2017). Figure 8.10 presents data obtained from the first (and up until now only) simultaneous 'multi-messenger' observations of GWs and a short gamma-ray burst (GRB) from the merger, named GW170817. The time-frequency spectrogram at the bottom of the plot shows the GW emissions during the inspiral phase right up to the instant the two neutron stars collide. The upper plots display the gamma-ray emissions in different frequency bands, which come approximately two seconds after the collision and are produced in the aftermath of the collision as jets emanating out of the orbital plane of the binary neutron star system.

The joint independent and simultaneous detection of GWs and GRBs from the same event led to further 'follow up' by nearly every X-ray, optical, infrared, and radio telescope in the world. The resulting optical counterpart left behind after the merger, a kilonova, was discovered within 12 hours of the LIGO-Virgo detection.

---

[3] A long-term series of radio observations of the evolution of the orbital period of the binary systems pulsar JSR1914+17 by Taylor, Hulse, and Weisberg (Weisberg et al., 2010) demonstrated that the rate of orbital decay was precisely that predicted by general relativity for a system radiating gravitational waves.

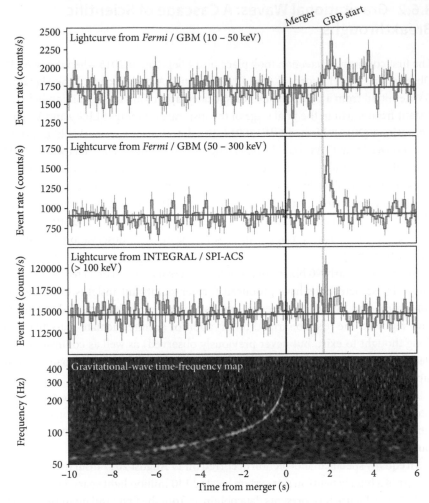

**Figure 8.10** Simultaneous detection of a gravitational wave and a gamma ray burst from binary neutron star merger GW170817

*Source:* Gravitational Waves and Gamma-Rays from a Binary Neutron Star Merger: GW170817 and GRB 170817A, B. P. Abbott et al 2017 ApJL 848 L13. © 2017. The American Astronomical Society.

The resulting astronomical observational campaign, one of the largest in the history of astronomy, lasted weeks (and years in the case of radio telescopes) and led to several breakthroughs in astronomy, nuclear physics, and cosmology. Key among them: the origins of heavy elements—those with atomic masses greater than 100 (such as silver, gold, platinum, and uranium)—were long speculated to take place in the neutron-rich environments occurring in the aftermath of binary neutron star mergers through rapid neutron capture (r-process nucleosynthesis).

Spectroscopic data from the kilonova in the optical and infrared electromagnetic spectral regions taken by optical telescopes provided stunning confirmation of the r-process nucleosynthesis hypothesis. The discovery of the kilonova also provides a precise measure of its distance, thereby allowing a new and completely independent determination of the Hubble constant which measures the expansion rate of the universe. This latter measurement is a hallmark of cosmology.

Unlike GW150914, the two LIGO Observatories were joined by the Virgo Observatory in the GW170817 detection. The participation of Virgo as the third observatory in this discovery was absolutely critical in enabling the much larger follow-up campaign, as three physically separated detectors are required to localise the position of the source in the sky. The discovery of GW170817, which inaugurated the era of multi-messenger astronomy with GWs, was only made possible by the joint collaborative efforts enabled by the LIGO-Virgo data sharing agreement signed in 2007. We will revisit the importance of the GW network below.

While the discoveries made thus far have provided deep insights into dynamical strong-field gravity, black holes, and how the high energy universe behaves, there remain a number of GW sources that have yet to be discovered. Isolated spinning neutron stars emit gravitational waves at twice their rotational frequencies if they have surface crustal deformations, internal hydrodynamic modes, or free precession resulting in elliptical deformations. These sources are expected to be exceedingly weak and only observable out to a few thousand light-years but may be detectable even though they do not produce radio emissions (which are termed pulsars), thus revealing a new population of neutron stars.

Primordial GWs were produced in the Big Bang and, unlike the Cosmic Microwave Background, do not interact with matter making them pristine probes of the earliest moments of the birth of the universe. Primordial GWs would manifest themselves as random (stochastic) background but would be exceedingly difficult to detect using interferometric detectors. More possibilities for GWs detection exist in the form of cosmic strings produced in the early universe as well as exotic quarks or bosonic stars. Most intriguingly, since ground-based observatories are the best probes of dynamic gravity in the strong-field regime, detected gravitational waveforms that deviate from predictions could provide hints to new physics beyond General Relativity.

## 8.6.3 The Complexity of Gravitational Wave Detection

Gravitational-wave detectors are 'Michelson' interferometers that have been scaled up and made supremely sensitive through numerous state-of-the-art technical enhancements. Advanced LIGO employs 4km long arms, along with Fabry-Perot cavities in the arms to increase the interaction time with a gravitational wave, as well as power recycling and signal cycling to further enhance the interferometers' sensitivity. To suppress ground motion, the primary 'test mass' mirrors are suspended in a four-stage pendulum configuration and further mounted to seismic isolation platforms. These actively sense and compensate for low-frequency seismic disturbances, resulting in a *12 order of magnitude suppression* of ground-noise disturbances in the 1–10Hz band.

Lasers producing 50W of frequency- and amplitude-stabilised light seed the interferometers; these lasers are the world's most stable and reliable lasers. Residual phase noise arising from the fundamental quantum nature of light due to the random arrival times of photons is further reduced by generating 'squeezed states' enabling sensing below the 'Heisenberg' limit. The use of squeezed light, a fundamentally quantum mechanical entity, in gravitational-wave detectors is yet another example of how innovations interact with and indeed enable astrophysical discoveries.

Each interferometer produces data at a rate of 50MB/sec over more than 200,000 independent data channels, producing 1.5PB of data per year. More than 350 servo loops are required to sense and control all aspects of the interferometer to maintain quite reliable interferometer operation. With the exception of the stabilised laser, the entire interferometer is housed in one of the world's largest ultra-high vacuum systems. Collectively, these innovations permit a strain sensitivity of approximately 4 x $10^{-24}/\text{Hz}^{1/2}$ in the interferometers' most sensitive frequency region. Displacements less than $10^{-18}$ m, roughly 1/1000 the diameter of a proton, are routinely recorded.

Like particle accelerators/colliders such as the LHC and detectors such as ATLAS and CMS, GW detectors are large-scale scientific infrastructures and fall squarely into the arena defined as 'Big Science'. The physical size of gravitational wave observatories is measured in kilometres; the scientific and engineering personnel needed to carry out the science programme numbers in the thousands.

## 8.6.4 The Criticality of Large-Scale Collaborations and Cooperation in GW Astrophysics

The overall complexity of the interferometric detectors and the data analysis methods requires a complex and diverse ecosystem of physicists, astronomers, mechanical/electrical/controls engineers, and computer scientists to design, build, commission, and operate the detectors, carry out the analyses, and interpret the data. Beyond this, there is one key aspect of GW astrophysics that simply cannot be overemphasised. The ability to detect GWs, locate their position and distance from Earth, and extract the underlying physical parameters from the detected events is greatly enhanced when multiple observatories carry out observing runs together in a cooperative manner. This was convincingly demonstrated with the binary neutron star merger GW170817 and is leading to a further expansion of the GW observational network.

In October 2019, the Kamioka Gravitational Wave Detector or KAGRA detector (Aso et al., 2013) located near Toyama, Japan formally joined LIGO and Virgo as the fourth gravitational-wave observatory. A fifth observatory, LIGO-India, is making progress towards construction with the expectation it will come online towards the end of this decade (Unnikrishnan, 2013).

Future observatories are envisioned to operate in the mid-2030s which will greatly increase the horizon to which we can detect GWs. The Einstein Telescope, to be built in Europe, is an underground gravitational-wave observatory with 10km long arms housing six interferometers (Punturo et al., 2010). Cosmic Explorer, planned for the US, will have 40km arm length (Reitze et al., 2020). Together, these will extend the range of detection out to cosmological distances (redshifts greater than 50) and be able to detect GWs throughout the entire universe from the time the first stars were formed. Bringing these observatories to fruition will undoubtedly allow us to answer longstanding questions about the high energy universe and, more importantly, lead us to ask new questions that arise from new and unanticipated discoveries.

## 8.7  Dark Matter and Big Science

Dark matter (DM) was already introduced in Chapter 5. Unlike normal matter, it is generally believed that DM does not interact through the electromagnetic force. DM does not absorb, reflect or emit light, making it extremely hard to detect. DM is thought to be everywhere in the universe, with its existence inferred by its gravitational effects on visible matter. DM is estimated to 'outweigh' visible matter roughly five times over, and according to the standard cosmological model ($\Lambda$CDM) makes up about a quarter of the mass-energy density of the universe (Particle Data Group, Zyla et al., 2020). Weakly interacting massive particles (WIMPs) are a popular DM and electroweak-scale DM candidate.

In addition to WIMPs, axions or more generally the so-called axion-like particles (ALPs) are a second class of DM candidate particles. The theory of axions was first introduced independently by Weinberg (Weinberg, 1978) and Wilczek (Wilczek, 1978), following the work of Peccei and Quinn, as a solution to the strong-CP problem in quantum chromodynamics (QCD), the theory of strong interaction which binds neutrons and protons inside atomic nuclei. Contrary to WIMPS, axions are very light particles and their mass can be in the range 1 MeV and $10^{-12}$ eV, which spans a very large range (Tanabashi et al., 2018). Despite their light mass, axions are also considered as 'cold' DM because they are believed to have been produced non-thermally in the early Universe.

If axions exist, they would be produced in large quantities in the solar interior. Once produced, axions escape from the star unimpeded and travel to the Earth, offering a tantalising opportunity for direct detection in the terrestrial so-called axion helioscopes (Sikivie, 1983). This utilises the conversion of solar axions back to photons in a strong laboratory magnet, with the resulting photons as X-rays that can be detected behind the magnet when it is pointing to the sun.

This has been the strategy followed by the CERN Axion Solar Telescope (CAST) using a decommissioned LHC test magnet that provides a 9T field inside the two 10m long, 5cm diameter, magnet bores. The CAST magnet is placed over a platform that allows it to be moved to track the sun for three hours per day. CAST performance has mostly relied on the availability of this first-class magnet. The Axion Dark Matter Experiment (ADMX) in Washington, USA takes a similar experimental approach (Du et al., 2018).

To substantially improve the CAST bounds and go deeper into unexplored axion parameter space requires a completely new infrastructure, like the one proposed by the International Axion Observatory (IAXO) at CERN and it will search for axions or axion-like particles emitted by the Sun with unprecedented sensitivity. IAXO is a next-generation axion helioscope aiming at a $10^4$ better signal-to-noise ratio compared to CAST, increasing the sensitivity to the axion-photon coupling more than one order of magnitude beyond the CAST bound, deep into unexplored ALP space and in particular probing QCD axion models down to the MeV scale (Irastorza, 2021).

### 8.7.1  Dark Matter Searches

The search for DM particles implies being able to measure some of their properties such as their mass, spin, and its interactions with Standard Model (SM) particles. Further, DM could be comprised of a single particle species or a combination of more than one (as is the case for the visible matter consisting of three generations of SM particles). Are these particles stable or simply long-lived particles and finally is the newly discovered Higgs boson connected to DM?

The possible masses for different classes of DM particles cover an enormous range. Furthermore, the strength of their interaction with SM particles extends to a range of 60 orders of magnitude. Thus, different experimental approaches for DM searches need to cover a large parameter space, calling for complementarity between different experimental techniques. There are three primary approaches:

### 8.7.2  Direct Searches for Dark Matter in Underground Detectors

Currently, there are many direct DM searches being carried out around the globe by different collaborations. The only trace of a DM WIMP-like particle passing through a detector would be the consequences of its hitting nuclei in the detector and changing their energy by a minuscule amount. DM detectors search for the tiny amounts of generated heat or recoil energy created when DM particles pass through (Froborg and Duffy, 2020). The detectors are designed to be very sensitive (and often operate at very low temperatures) in order to record the small heat or energy deposits from DM particles interacting inside the detector.

Despite over two decades of searches, there is so far no convincing evidence for WIMP dark matter from any direct detection experiment. The mass region below 6 GeV is studied with detectors with very low energy thresholds and/or lighter target nuclei; Cryogenic Rare Event Search with Superconducting Thermometers (CRESST) experiment in Italy, Expérience pour DEtecter Les WIMPs En Site Souterrain (EDELWEISS) in France, Super Cryogenic Dark Matter Search (SuperCDMS)

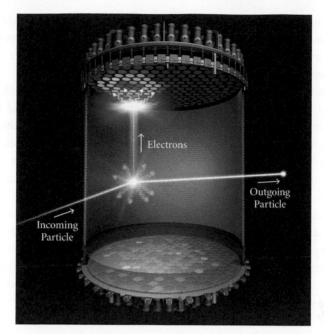

**Figure 8.11** LUX-Zeplin (LZ)—5 foot detector filled with 10 tonnes of liquid Xenon
*Source:* Illustration by Greg Stewart, SLAC National Accelerator Laboratory

in Canada or based on CCD technology to record the particle tracks (DArk Matter In CCDs in Modane DAMIC-M).

The other class of experiments involves detectors made of liquefied noble gases. A DM particle hitting an atom of e.g. xenon or argon can lead to a flash of characteristic scintillation light. These detectors cover the higher mass region above 6 GeV with very low background rates. Examples of detectors using liquid xenon include LUX-Zeplin in South Dakota, XENON1T/nT in the Grand Sasso Laboratory in Italy, and PandaX in Sichuan, China and they provide the tightest constraints on the interaction strength of WIMPs with nuclei. Proposed future noble liquid experiments include ZEPLIN and ArDM (see Figure 8.11).

## 8.7.3 Observatories for Indirect Dark Matter Searches

The second way to detect DM is through indirect detection in the sky or on Earth. In the indirect method, DM particles can be detected by observing the radiation produced when they annihilate in the galactic halo.

Although DM is non-electromagnetically interactive, DM particles nevertheless may collide and annihilate, releasing energy and other measurable particles in the process. Annihilation doesn't happen enough to significantly affect the overall density of DM in the universe, however, it might be enough to occasionally produce a measurable signal.

DM pair annihilation would lead to the production of detectable SM particles, which then decay into familiar products that could potentially be observed above the standard, astrophysical background. Examples are energetic cosmic rays, such as neutrinos, gamma rays, antiprotons, or positrons or pairs of photons.

The search for the products of DM annihilation is pursued with telescopes or detectors that look for particles and photons emitted by galaxies and the exotic objects that lie within them. Such observations might also illuminate the nature of DM. Since antimatter particles are relatively rare in the universe and since the distribution of photon energies could exhibit distinctive and identifiable properties, such detections could eventually be associated with DM. The spatial distribution of these particles might help distinguish DM annihilation products from more common astrophysical backgrounds. It should be noted that one of the main challenges in this approach is to correctly model the expected backgrounds from standard astrophysical processes.

The National Science Foundation (NSF) Cerro Tololo Inter-American Observatory (CTIO) is a complex of astronomical telescopes located near La Serena, Chile, is an example of an observatory carrying out such DM searches.

### 8.7.4 Collider Searches for Dark Matter

Finally, experiments at colliders such as the LHC look for DM particles that could be produced in the high energy proton collisions, where the kinetic energy of the colliding particles transforms into the mass of the newly created DM particles.

Since DM itself will escape the collider detector undetected, one needs to look for tell-tale signs of missing transverse momentum. There are two main types of DM searches at the LHC. One type is guided by new physics models, such as supersymmetry (SUSY) models where the lightest supersymmetric particle is a WIMP. Another type of search involving the missing-momentum signature is guided by simplified models that include a WIMP-like dark-matter particle and a mediator particle that would interact with the known ordinary particles. The mediator can be either a known particle, such as the Z boson or the Higgs boson or yet an unknown particle.

Despite the intense efforts, there is yet no evidence from the LHC for DM or any other type of new particle outside the SM. Nonetheless, more high-quality data is being collected, and the search is ongoing.

## 8.8 Conclusions

Astronomy and astrophysics have changed over the past 50 years from 'the lone astronomer on the mountain top pointing a telescope towards the heavens' to Big Science initiatives with hundreds or even thousands of collaborators working together on massive and highly complex instruments, be they ground-based telescopes, satellites, or GW (Gravitational Waves) and DM (Dark Matter) detectors. This knowledge evolution was fuelled not only by our desire to continually ask and find answers to the most fundamental questions about the origins, structure, evolution, behaviour, and nature of our universe, but also crucially by the ongoing development and uptake of a wide range of new technologies that can be used to address those fundamental questions. These inquiries do, in fact, bring human knowledge closer to understanding how our universe is structured and what our origins within it might be. These fields together contribute to the shared knowledge of many overlapping areas such as DM, GW, elementary particles, different forms of electromagnetic radiation, the study of neutrinos, the Sun, supernovae, and cosmic rays. With powerful instruments and techniques, human knowledge is expanding at an unprecedented scale.

Technological innovations play a foundational role in driving new astronomical discoveries, be they large-scale (with larger aperture primary mirrors in optical/infrared telescopes, massive radio telescope dishes, or multi-kilometre scale vacuum systems) or small-scale (MCT mid-infrared sensitive detectors, MOSFETs, lasers for adaptive optical guidestars). The complexity, cost, and audacity of these investigations, ranging from searching for Earth-like planets and the existence of life in nearby solar systems to measuring sub-nuclear scale displacements that capture the whips of passing GWs produced in cataclysmic collisions of black holes and neutron stars to identifying the elusive nature of DM, require large collaborations pioneered in high energy physics experiments. Large-scale collaborations are home to hundreds or perhaps thousands of physicists, engineers, and astronomers who produce an ever-increasing quantity of data. Indeed, the two first images of black holes by the Event Horizon Telescope (EHT) collaboration were the result of hitherto unimaginable Big Data collection and analysis from numerous observatories working together around the world.

The applications of mirror technologies and technologies associated with ground and space telescopes, the algorithms developed for data analysis, and the use of AI (artificial intelligence) in astronomy have many practical and documented outcomes. Radio astronomy has significantly contributed to our understanding of celestial objects and has led to many useful innovations such as GPS.

Big Science research is possible due to the concerted efforts of dedicated governments, institutions, and individuals, and it brings us one step closer to finding the answer to the question Einstein posed: 'Did God have any choice in the creation of the universe?' The search to understand our universe, at both the smallest and largest scales, as attempted by the particle physics and astrophysics communities, is a big collaborative leap into the unknown future for humanity.

# 9

# Big Science Medical Applications from Accelerator Physics

## Impact on Society

*Mitra Safavi-Naeini, Timothy P. Boyle, Suzie Sheehy, and Shantha Liyanage*

## 9.1 Introduction

Accelerators and detector technology have pushed the boundaries of science not only in advancing the frontiers of high energy physics, but also in helping us to advance medical science and improve the lives of billions of people. These accelerator technologies have significant 'human-centric' impacts. These technologies have revolutionised modern medicine both in terms of diagnosis and treatment, including through the development of particle beam therapy (proton beam, carbon ion beam, and alpha and direct electron beams). It can be argued that the use of radioisotopes and radiotherapy based on common electron LINACs[1] has had a higher impact than particle therapy (also known as particle beam therapy or hadron therapy) with high-energy accelerators.

Particle accelerators play an important role in many areas of medicine, from the production and application of radioisotopes for nuclear medicine for the diagnosis and treatment of disease (Starovoitova et al., 2014), to the clinical application of radiation for the treatment of cancer and other diseases that affect tissue growth (Kamada et al., 2015) such as plantar fibromatosis and Dupuytren's contracture (Bomford et al., 2002; Seegenschmiedt et al., 2012; Kadhum et al., 2017). Radiation therapy (i.e. where radiation is given by an external radiation source) and molecular radiotherapy (i.e. where a radiopharmaceutical is used to target a specific receptor or molecular site) are currently used to provide curative and palliative care to over 50% of all cancer patients (Lutz et al., 2014). Consequently, radiotherapy is the largest medical application of accelerators, with over 11,000 accelerators worldwide (IAEA, 2021b). Big Science initiatives such as those at CERN, Berkeley, and SLAC have paved the way for

---

[1] LINACs refers to a linear particle accelerator 'small, hospital based' electron accelerators and a type of *particle accelerator* that accelerates charged *subatomic particles* or *ions* to a high speed by subjecting them to a series of *oscillating electric potentials* along a *linear beamline*.

Mitra Safavi-Naeini et al., *Big Science Medical Applications from Accelerator Physics*. In: *Big Science, Innovation, and Societal Contributions*. Edited by: Shantha Liyanage, Markus Nordberg, and Marilena Streit-Bianchi,
Oxford University Press. © Mitra Safavi-Naeini et al., (2024). DOI: 10.1093/oso/9780198881193.003.0010

these developments. For example, CERN's GEMPix detector is a promising way to screen new radiopharmaceuticals in large cell and tissue libraries (Diamante, 2018).

More recently, the generation of more exotic forms of radiation for cancer therapy has been at the forefront of accelerator technology development and radiotherapy research. Charged particles (including protons, helium ions, and carbon ions), $\pi$ mesons and neutrons are applied with varying degrees of uptake and success. In the last 10 years there has been a significant growth in the number of carbon and proton accelerator facilities which offer several advantages over X-ray radiotherapy. These accelerators range from very compact, metre-long electron linear accelerators (LINACs) to large heavy ion synchrotrons with a circumference of over 100 metres. Adjacent spill-over technology from Big Science in detecting and measuring radiation has given rise to inventions such as the Chemo Camera in the 1950s that allowed radioisotopes produced by accelerator technology to be utilised in diagnostic imaging and treatment.

Many radioisotopes are produced for the diagnosis and treatment of diseases, using proton and ion cyclotrons. These are significant uses of accelerator-enabled technology in the medical field; however, in this instance, radiation application—not accelerator technology—is the human-centric factor. For this reason, in this chapter we aim to provide an overview of the current and emerging accelerator technologies used in medicine and explore beam-based accelerator technologies and the convergence of nuclear and other technologies that have failed to reach widespread adoption. This convergence is giving rise to novel collaborative innovation models for knowledge creation and dissemination.

## 9.2  Historical Context

The medical application of accelerators almost coincided with their invention. Within two weeks of discovering the emission of an *invisible light* from cathode ray tubes, Röntgen had taken the first photograph—rather an X-ray image—of his wife's hand. Röntgen chose not to patent the discovery and by January 1896 the world knew about X-rays. By the end of that year, X-ray images were being taken all over the world—a very fast adoption of new technology. In the same year, X-rays were used for the treatment of cancer by Amy Colbert and Victor de Pina (Alabama University, 2008). The discoveries of spontaneous radioactivity (Henri Becquerel in 1896, Marie and Pierre Curie in 1897), the electron (John Joseph Thompson in 1897), cosmic rays (Victor Francis Hess in 1912), proton, alpha, and beta radiation (Ernest Rutherford, up to 1919), and neutrons (James Chadwick in 1932) posed a challenge to create a ground-based laboratory source of charged particle beams for research investigations.

Early accelerators were based on 'static voltage' concepts, such as those proposed by Robert Van de Graff, and had a limited energy reach due to electrical breakdown (sparking). However, an alternative solution began in 1927 when Rolf

Widerøe, a Norwegian engineer, successfully invented the first linear accelerator: electrons were accelerated through radiofrequency (RF) fields on a straight path, i.e. a linear acceleration of electrons.[2] Widerøe (1928) publication inspired Ernest Lawrence (who won the Nobel Prize for Physics in 1939), the father of accelerator-based Big Science. Lawrence combined the RF idea with magnetic recirculation to invent the 'cyclotron' in 1929. The first of which was realised by Lawrence and Stanley Livingstone in 1931 enabling the acceleration of protons through 80,000 volts.

This provided a mechanism to produce radioactive isotopes in amounts surpassing their naturally occurring isotopic abundance. Lawrence and Livingstone published a seminal article on the production of usable quantities of radionuclides in 1932 (Lawrence and Livingstone, 1932). The development of subsequent cyclotrons at Berkeley and their construction, operation, and maintenance needed a team of specialised engineers, physicists, and chemists—a departure from a 'lone genius in a small laboratory' model.

In recent years, there has been an accelerated development of key technologies. For example, in the 1990s, a CERN-led study of synchrotron technology for particle therapy called the PIMMS project (Proton-Ion Medical Machine Study) provided the underlying principles and technical designs which resulted in the creation of Europe's two main hadron-therapy treatment centres. The CNAO (Centro Nazionale di Adroterapia Oncologica) in Italy and the MedAustron in Austria. These facilities use a particle accelerator to direct a beam of protons or heavy ions onto a tumour, depositing energy directly into the tumour while avoiding damage to surrounding tissues. This treatment is particularly effective for deeper, denser tumours, which would be out of reach using conventional methods (Hortala, 2021).

## 9.3 Teletherapy

The innovations of particle accelerators led a new generation of radiation pioneers into the application of radiation in medicine. While Sir William Henry Bragg published the rate of energy loss, which is also known as the Bragg Curve, in 1903 the increase in accelerator energy reach provided an opportunity to investigate and confirm the principles of his proposed model describing the energy-loss of ionising radiation along its trajectory in matter. The shape of this depth-dose profile of initially high energy ions shows that ionisation occurs at a greater depth or at a maximum energy deposition just before the projectile stops. This led Robert Wilson—physicist and founding director of the US-based Big Science facility Fermilab—to first propose the use of the Bragg peak for radiotherapy with protons (Wilson, 1946). Wilson's idea was taken up by the Lawrence brothers: John, an expert in nuclear medicine, together with Ernest. Cornelius Tobias and John Lawrence studied the biological effects of protons using this 4.7 metre synchrocyclotron in the late 1940s to early 1950s.

---

[2] This concept was proposed by Gustav Ising in 1924, but Widerøe is credited with the first working prototype.

The Lawrence brothers (or Lawrence boys, as they were affectionately known) treated the first patient with protons at the University of California, Berkeley in 1954, only eight years after Wilson's paper was published. The initial tumour targeted was a pituitary tumour in patients with metastatic breast cancer, easily located in 3-D using orthogonal plane X-ray films and rigid immobilisation of the cranium (Lawrence et al., 1958). The rationale was that a high proportion of human breast cancers are hormone dependent, and the elimination of pituitary hormones may include the regression of hormones in humans.

The limitations of LINAC design proposed in 1928—where the length of accelerators had to increase with the increased electron speed due to the fixed frequency— were resolved in the decades that followed. The first successful operation of an electron LINAC took place in 1947 at Stanford and at the Telecommunications Research Establishment in England (Blewett, 1979). William Hansen from Stanford developed an electron accelerator (Rhumbatron, named after a popular dance of that time) that utilised a resonant microwave cavity to increase the electrons' energy as they passed through each section. The subsequent development of high-power airborne and surface microwave radars used by the military during the Second World War paved the way for the development of a two-cavity oscillator magnetron by Russel and Sigurd Varian, who were working with Hansen at the Stanford Physics Department (Caryotakis, 1998). While the Stanford team focused their efforts on high energy physics research, across the Atlantic, an improved cavity Magnetron was developed by John Randall and Henry Boot, both part of Marcus Oliphant's group in Birmingham University, England. The British team proceeded rapidly to develop LINACs for cancer therapy by 1941; they succeeded in producing approximately 1 MeV electron beam at 10 cm, an order of three times the magnitude improvement with respect to their first prototype (Blanchard et al., 2013). Their work led to the design and commissioning of the first 8 MeV electron linear accelerator specific to radiotherapy at Hammersmith Hospital in London in 1953 which operated for more than three decades (Thwaites and Tuohy, 2006). Back in the US, a collaboration between Henry Kaplan of the Sandford Medical Department and Ed Ginzton of SLAC National Accelerator Laboratory reignited efforts in the application of electron LINACs for cancer therapy (Blewett, 1979). Varian Associates, a company cofounded in 1948 by Ginzton and the Varian brothers secured the rights to commercialise the klystron and developed the first small LINAC for external photon radiotherapy with the first patient treated with this machine in 1956 (Gauvin, 1995).

After more than 60 years and 50 million patients, LINACs have proliferated throughout hospital settings and are now the largest application of accelerators in medicine.

## 9.4 Molecular Radiotherapy and Diagnostic Imaging

The ability to prepare and isolate a large variety of enriched radioisotopes with known decay radiation accelerated their use in diagnostic imaging and molecular radiotherapy. The use of radioactive compounds for therapeutic purposes dates back to the early 1900s. Henri Alexandre Danlos and Eugene Bloch proposed the use of radium for the treatment of tuberculosis skin lesions in 1901 (Yeong et al., 2014). Frederich Proescher performed the first therapeutic intravenous injection of radium (Mackee, 1921). However, the birth of molecular radiotherapy can be traced back to George de Hevesy's discoveries (Nobel, 1943). Originally tasked by Rutherford to separate 'radium-D from all that lead', he proposed 'marking' lead with radium-D (lead-210). Together with Jorge Christiansen and Sven Lomholt, Hevesy performed the first radiotracer studies on animals in 1924 and thus the field of nuclear medicine was born (Obaldo and Hertz, 2021).

The first synthesis of a radioactive material, phosphorus-30, was performed in 1934, by Irene Joliot-Curie and Frederic Joliot serendipitously, by irradiating an aluminium target with alpha particles (Nobel, 1935). This was followed by the production of phosphorus-32, through the irradiation of sulphur targets with neutrons. In 1936, John Lawrence pioneered its clinical therapeutic application to treat leukaemia.

Meanwhile, Ernest Lawrence brought together a large team of young and enthusiastic scientists—physicists, chemists, engineers, medical doctors—and together they investigated the process of radioisotope production through high energy particle bombardment and devised plans for its potential applications. Lawrence himself took the initiative and enthusiastically supported the development of the first medical cyclotron. In 1937, it produced radioisotopes (iron-59) used for the studies of red blood cells (haemoglobin), with physics, medicine, chemistry, and biology applications represented in equal parts.

The first radionuclides for therapeutic use were produced in 1946 by the Oak Ridge National Laboratory (ORNL) and shipped to Barnard Free Skin and Cancer Hospital in St Louis by scientists in the Manhattan Project, to advocate for peaceful uses of atomic energy (atom for peace) (Creager, 2006). Abbott Laboratories began the distribution of radioisotopes in 1948. Nowadays, cyclotrons and nuclear reactors are the main production sources of radionuclides for medicine, with a more recent move towards the use of accelerator-based neutron sources to replace nuclear reactors.

## 9.5 Radiation and Human Health

Humans are constantly exposed to some level of radiation, which is inescapable: life itself has evolved in a sea of radiation. In 1987, Planel and his team conducted experiments in caves 200m below ground in the French Pyrénées on single-cell tissue cultures shielded against terrestrial background radiation, and observed a reduction in the proliferation rate (Planel et al., 1987). Today, experiments are still ongoing to quantify and explore the impact of low to high radiation levels and varying dose rates on living matter at microscopic and macroscopic levels. This is particularly relevant in the context of renewed interest in short and long distance space travel, which is fast becoming the driving force in understanding the impact of radiation on living matter (Lampe, 2017; Furukawa et al., 2020).

The impact of space radiation on crew safety was identified as one of three major 'red' risks by NASA's Human Research Program. Astronauts will be exposed to galactic cosmic rays (GCR) for up to three years if they take the shortest route to Mars, which is made up of a cocktail of photons, high energy protons, and light to heavy ions of all flavours, energies, and intensities. Recent epidemiological analysis of radiation on human health suggests that while cancer remains the dominant risk of GCR, cardiovascular, cognitive decrement and the central nervous system (CNS) should be included when calculating the radiation exposure induced death (REID).

Recent figures, published by NASA suggest that the radiation received by the crew will result in an increase of 10% to 20% in mortality and morbidity (Cucinotta et al., 2013; Patel et al., 2020). With five companies that are racing to make space tourism a reality and an annual market of at least US$20 billion, there is an increase in the number of humans primed for high-speed travel or short- to long-term residencies in low earth orbit (LEO). The radiation environment in LEO is very complex and is made up of GCRs, solar particle events and protons and electrons trapped in the Van Allen Belt. The interaction of these particles with the aircraft's outer shell produces a shower of low energy secondary particles. These particles typically travel a short distance in air or tissue and lose their energy primarily through the process of direct or indirect ionisation.

The current models of radiation focus on predicting the impact of radiation on the cell nucleus and mitochondrial DNA. Recently, other intra-cellular organelles and intercellular signalling pathways have been included. The biological effects of acute and chronic radiation exposures vary with the dose, dose-rate, and radiation flavour. An average background radiation dose of naturally occurring radiation that is received by an average person is of the order of 2 to 15 mSv[3] per year without causing any detectable harm. An exposure of 1 Sv/hour can result in radiation poisoning (resulting in nausea and vomiting). Several ongoing studies are underway on the effects of radiation on human health.

## 9.6  Radiation in Medicine

Particle accelerators are often considered in the context of Big Science. However, the analysis of research infrastructure and accelerator installations globally indicates that over half of the accelerators are involved in medical applications ranging from radioisotope production, radiotherapy, sterilisation, and medical research into new therapeutic techniques (Kokurewicz et al., 2020). Most accelerators currently being used for medical purposes are involved in the development of isotopes and tracers for diagnostic imaging. A clear majority of advancements and the use of accelerators

---

[3] The sievert (Sv) is a derived unit of ionising radiation dose in the International System of Units and is a measure of the health effect of low levels of ionising radiation on the human body.

in medical science are for the treatment of cancer patients. The use of radiation in medicine is now common around the world.

Diagnostic radiography, nuclear medicine, and radiation therapy are used routinely for the diagnosis and treatment of several epidemiologic conditions ranging from infectious diseases such as tuberculosis to non-communicable diseases such as cancer and cardiac conditions. According to distribution maps, an average of two radiation therapy units, three nuclear medicine specialists and 45 radiologists per million people with varying levels of access depending on the development status of each nation (see Figures 9.1, 9.2, and 9.3).

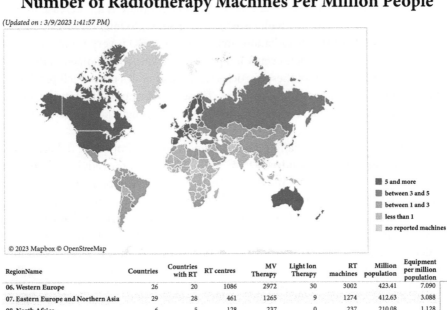

## Number of Radiotherapy Machines Per Million People

*(Updated on : 3/9/2023 1:41:57 PM)*

Legend:
- 5 and more
- between 3 and 5
- between 1 and 3
- less than 1
- no reported machines

© 2023 Mapbox © OpenStreeMap

| RegionName | Countries | Countries with RT | RT centres | MV Therapy | Light Ion Therapy | RT machines | Million population | Equipment per million population |
|---|---|---|---|---|---|---|---|---|
| 06. Western Europe | 26 | 20 | 1086 | 2972 | 30 | 3002 | 423.41 | 7.090 |
| 07. Eastern Europe and Northern Asia | 29 | 28 | 461 | 1265 | 9 | 1274 | 412.63 | 3.088 |
| 08. North Africa | 6 | 5 | 128 | 237 | 0 | 237 | 210.08 | 1.128 |
| 09. Middle Africa | 45 | 24 | 48 | 88 | 0 | 88 | 1,099.61 | 0.080 |
| 10. Southern Africa | 6 | 4 | 63 | 107 | 0 | 107 | 83.98 | 1.274 |
| 11. Middle East | 15 | 15 | 291 | 574 | 0 | 574 | 355.66 | 1.614 |
| 12. South Asia | 8 | 5 | 475 | 821 | 1 | 822 | 1,901.53 | 0.432 |
| 13. East Asia | 8 | 8 | 1948 | 3091 | 30 | 3121 | 1,663.70 | 1.876 |
| 14. Southeast Asia | 15 | 10 | 215 | 457 | 0 | 457 | 675.40 | 0.677 |
| 15. Southern and Western Pacific | 13 | 3 | 114 | 260 | 0 | 260 | 44.77 | 5.808 |

| IncomeGroup | Countries | Countries with RT | RT centres | MV Therapy | Light Ion Therapy | RT machines | Million population | Equipment per million population |
|---|---|---|---|---|---|---|---|---|
| Grand Total | 214 | 156 | 7745 | 15130 | 107 | 15237 | 7,895,79 | 1.930 |
| High income (H) | 75 | 62 | 4588 | 9449 | 96 | 9545 | 1,238.31 | 7.708 |
| Upper middle income (UM) | 53 | 41 | 2198 | 4023 | 10 | 4033 | 2,546.36 | 1.584 |
| Lower middle income (LM) | 53 | 37 | 934 | 1615 | 1 | 1616 | 3,391.62 | 0.476 |
| Low income (L) | 29 | 15 | 24 | 40 | 0 | 40 | 713.82 | 0.056 |
| Temporarily unclassified (NC) | 4 | 1 | 1 | 3 | 0 | 3 | 0.68 | 4.384 |

**Figure 9.1** Global distribution of radiotherapy centres

*Source:* ©OpenStreetMap, DIRAC, International Atomic Energy Agency (IAEA)

# NM Physicians (per 1 mil)

| Countries | Countries with NM Physicians | Regions | Population (mil) | Number of NM Physicians | NM Physicians (per 1 mil) |
|---|---|---|---|---|---|
| **212** | **140** | **6** | **7,674M** | **22,531** | **3** |

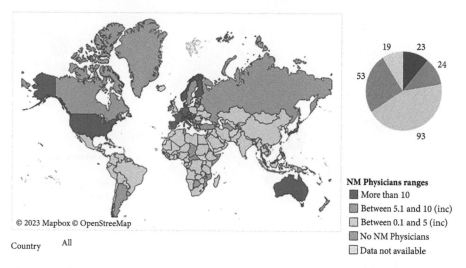

© 2023 Mapbox © OpenStreetMap

Country    All

**NM Physicians ranges**
- More than 10
- Between 5.1 and 10 (inc)
- Between 0.1 and 5 (inc)
- No NM Physicians
- Data not available

## Income Group

| Income Group | Countries | Countries with NM Physicians | Population (mil) | Number of NM Physicians | NM Physicians (per 1 mil) |
|---|---|---|---|---|---|
| High Income | 75 | 53 | 1,237M | 13,247 | 11 |
| Upper-Middle Income | 54 | 40 | 2,854M | 7,805 | 3 |
| Lower-Middle Income | 50 | 35 | 2,913M | 1,443 | 0 |
| Low Income | 30 | 12 | 669M | 36 | 0 |

## UN Regions

| UN Regions Name | Countries | Countries with NM Physicians | Population (mil) | Number of NM Physicians | NM Physicians (per 1 mil) |
|---|---|---|---|---|---|
| Australia/New Zealand | 2 | 2 | 30M | 425 | 14 |
| Central and Southern Asia | 14 | 9 | 1,993M | 971 | 0 |
| Eastern and South-Eastern Asia | 19 | 13 | 2,298M | 6,809 | 3 |
| Europe and Northrn America | 48 | 42 | 1,110M | 10,927 | 10 |
| Latin America and the Caribbean | 39 | 25 | 647M | 2,043 | 3 |
| Northern Africa and Western Asia | 25 | 23 | 517M | 1,231 | 2 |
| Oceania (excluding Australia an.. | 15 | 0 | 12M | 0 | 0 |
| Sub-Saharan Africa | 49 | 26 | 1,066M | 125 | 0 |

**Figure 9.2** Global distribution of nuclear medicine specialists per 1 million

*Source:* ©OpenStreetMap, DIRAC, International Atomic Energy Agency (IAEA)

# Radiologists (per 1 mil)

| Countries | Countries with Radiologists | Regions | Population (mil) | Number of Radiologists | Radiologists (per 1 mil) |
|---|---|---|---|---|---|
| 212 | 128 | 6 | 7,674M | 345,475 | 45 |

**Radiologists ranges**
- More than 100
- Between 50 and 100 (inc)
- Between 25 and 50 (inc)
- Between 10 and 25 (inc)
- Between 0 and 10 (inc)
- Data not available

Country   All

## Income Group

| Income Group | Countries | Countries with Radiolo.. | Population (mil) | Number of Radiologists | Radiologists (per 1 mil) |
|---|---|---|---|---|---|
| High Income | 75 | 46 | 1,237M | 114,701 | 93 |
| Upper-Middle Income | 54 | 38 | 2,854M | 185,172 | 65 |
| Lower-Middle Income | 50 | 32 | 2,913M | 44,808 | 15 |
| Low Income | 30 | 12 | 669M | 794 | 1 |

## UN Regions

| UN Regions Name | Countries | Countries with Radiologists | Population (mil) | Number of Radiologists | Radiologists (per 1 mil) |
|---|---|---|---|---|---|
| Australia/New Zealand | 2 | 2 | 30M | 2,491 | 82 |
| Central and Southern Asia | 14 | 9 | 1,993M | 27,518 | 14 |
| Eastern and South-Eastern Asia | 19 | 10 | 2,298M | 140,785 | 61 |
| Europe and Northrn America | 48 | 37 | 1,110M | 129,066 | 116 |
| Latin America and the Caribbean | 39 | 30 | 647M | 26,723 | 41 |
| Northern Africa and Western Asia | 25 | 15 | 517M | 16,761 | 32 |
| Oceania (excluding Australia an.. | 15 | 0 | 12M | | |
| Sub-Saharan Africa | 49 | 25 | 1,066M | 2,131 | 2 |

**Figure 9.3** Global distribution of radiologists per 1 million

*Source:* ©OpenStreetMap, DIRAC, International Atomic Energy Agency (IAEA)

Early and late side effects limit the radiation dose and might affect the patient's long-term health-related quality of life. The inherent properties of ionising radiation provide many benefits but can also cause potential harm. Its use within medical practice involves an informed judgement regarding the risk/benefit ratio. This judgement relies on a set of guidelines that are informed by a systemic approach informed by the physics of radiation, biophysics, and biochemical properties that are at play when radiation is interacting with human lives.

## 9.7 Accelerators and Their Direct and Indirect Impact on Health

There exists an intimate relationship between radiation type, radiation-dose fractionation, radiation-dose rate, spatial dose distribution, and the clinical outcome of radiation therapy. The advancement of accelerator technology is pushing the boundaries of each of these factors, exposing the flaws in our long-held assumptions.

Until the mid-1990s, the target-cell approach was the dominant biological model for discussing early and late side effects, in which side effects were attributed to direct cell killing, resulting in subsequent functional deficiencies. This model is still applied for predicting and managing early side effects. The multivariate nature of this phenomenon has been brought to light by more recent discoveries in radiobiology and molecular pathology, including the crucial roles that cytokine pathways play in the development of a latent misplaced systemic response and wound healing gone wrong, as well as the variability in patient response to different courses of radiotherapy.

There are several drugs that are used to minimise the incidence of latent radiation-induced conditions in patients (e.g. radioprotectors). Sensitisation of the malignant target to radiation using radiosensitisers is another growing field in cancer treatment. It is safe to assume that the increased interest in the development and use of therapeutics in radiation therapy will result in further improvements in the efficacy and utility of radiation therapy.

The physics of radiation is somewhat more reliable and offers a better future for radiation therapy (Bortfeld and Jeraj, 2011). Photon beams are the most commonly employed in radiotherapy. The therapeutic use of fast and thermal neutrons has ebbed and flowed, with an increased interest in the latter in recent times. Light ion beams including protons, alpha particles, and carbon ion beams are finding increasing utility with particle therapy systems in over 89 facilities worldwide (PTCOG, 2021). The usefulness of these radiation modalities with respect to their therapeutic applications can be assessed by comparing their microscopic energy deposition in matter.

**Table 9.1** Linear energy transfer (LET) and relative biological effectiveness (RBE) of therapeutically relevant radiation modalities

| Radiation type | LET (keV/μm) | RBE |
|---|---|---|
| Linac X-rays (6–15 MeV) | 0.3 | 0.8 |
| Beta particle (1 MeV) | 0.3 | 0.9 |
| Cobalt-60 γ-rays | 0.2 | 0.8–0.9 |
| 250 kVp X-rays (standard) | 2 | 1.0 |
| 150 MeV protons (therapy energies) | 0.5 | 1.1 |
| Neutrons | 0.5–100 | 1–2 |
| Alpha particles | 50–200 | 5–10 |
| Carbon ions (in spread out Bragg peak) | 40–90 | 2–5 |

The linear energy transfer (LET) is a physical parameter that quantifies the energy deposition density locally along the track of a particle and is measured in units of kiloelectron volts per micrometres (keV/μm). It provides the means of predicting the relative biological effectiveness, or RBE of the radiation modality. As LET increases, so does the RBE and peaking at 100 keV/μm, after which point the RBE decreases with any further increase in the LET value. LET allows us to account for the difference in the biological effectiveness of different kinds of radiation—to convert 'dose' to 'dose equivalent'. RBE is an important measure when quantifying the impact of radiation on mammalian DNA. RBE is relevant in therapeutic use of radiation or shielding from it. Table 9.1 shows the relationship between LET and RBE for different radiation types.

For imaging applications, a balance is struck between the image quality and the unwanted dose. With the former, there is a balance between absorption and scatter. These days, it is rare to find a computed tomography[4] (CT) scanner with megavoltage energy X-rays used for imaging. Current state-of-the-art CT machines operate in the kilovolt range of around 300 to 600 kV. When used in therapy, that packs a punch where the target is—if it originates within the body, it should stay there. Delivery of the maximal dose to the treatment region while sparing surrounding healthy tissue remains the main objective (see Figure 9.4).

External beam radiotherapy (EBRT) is the most common form of radiotherapy and the most common use of accelerators in medicine. In EBRT the patient is exposed to an external source of radiation via a beam of energy which is pointed towards a tumour site within the body. There are several techniques for external beam therapy, with X-ray and electron beams being the most commonly used.

The following sections will further discuss the main applications of accelerators used in the advancement of cancer treatments, starting with widely used radiotherapy, proton therapy, and heavy ion therapy. We will also touch upon neutron capture

---

[4] Computed tomography is a medical imaging technique which provides detailed images of the body noninvasively for diagnostic purposes.

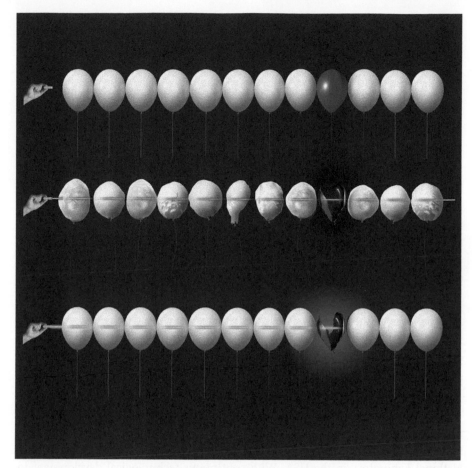

**Figure 9.4** Depth-dose profile for different types of radiation[5]
*Source:* ANSTO

therapy and advances in accelerator technology, which are enabling this technique to be more widely available.

## 9.8  X-ray and Electron Beam Radiotherapy

The most common techniques for external beam radiotherapy employ electron beam accelerators of varying energies to provide distinct treatments for separate indications. The technique involves directing a beam of radiation at a tumour mass from outside the body. Common radiotherapies include:

---

[5] How radiation interacts with matter – is an important factor to consider when planning its therapeutic application. Our aim is to pop the red balloon, or the tumour (top panel). Photons and electrons that deposit their maximal energy in the shallow part of their track (middle panel), light ions deposit the majority of their energy at the end of their range (bottom panel) – Images courtesy of Karl Mutimer, ANSTO.

- Orthovoltage X-rays, for the treatment of epidermal cancers such as skin cancer;
- Megavoltage X-rays for tumours located deeper within the body; specific indications include bladder, bowel, prostate, brain, and lung cancers;
- Megavoltage electron beams are used to treat epidermal cancers and interior tumour masses up to 5cm in depth.

Most linear accelerators used for radiotherapy can produce both X-rays and electron beams. For some indications such as skin cancer, electron beam radiation therapy has replaced orthovoltage (low energy) x-ray therapy which often has energy penetration through the epidermis to the underlying tissue. Electron beam therapy has the advantage that it delivers radiation primarily to the superficial layers of the skin avoiding unintended exposure to healthy cells. Electron beam radiation is very damaging to the tumour cells but is well tolerated by the surrounding normal skin cells.

## 9.9  Light Ion Therapy: Protons, Helium and Carbon Ion Therapy

The use of accelerated ions or protons for the therapeutic delivery of a radiation dose dates back to 1946 (Wilson, 1946). Ten years later, the first practical demonstration of highly accelerated particle therapy was performed at the Lawrence Berkley National Laboratory in the USA. With continuous technological development in accelerator technologies, the availability of proton and heavy ion facilities has progressively expanded. By 2021, there are 89 therapeutic proton or carbon ion facilities in the world with another 53 under construction and 25 in the planning stage of development. These are located in a variety of geographic regions from Argentina to Australia, Saudi Arabia, and Thailand (PTCOG, 2021). The use of accelerated ions has several advantages over photon and electron therapy:

- Protons and heavy ions deposit the majority of their kinetic energy at the end of their range, in the region known as the Bragg peak. By changing the energy of the incident particle, the position of the Bragg peak (range of the particle) can be controlled. This makes the use of protons and heavy ions well suited for the treatment of deep situated tumours.
- Collimated proton and heavy ion beams scatter less than beams of photons or electrons, resulting in a higher degree of dose conformity.
- Heavy ions have a higher LET than photons, which also sharply increases towards the end of the particle's range. Consequently, protons, and to a greater

extent heavy ions, will provide a higher RBE compared to photon and electron therapy. This makes the heavy ions well suited for the treatment of radio-resistant tumours.[6]

There are two common methods by which accelerated ions can be delivered for therapeutic purposes: active or raster-scanned and passive or passively scattered delivery (see Figure 9.5).

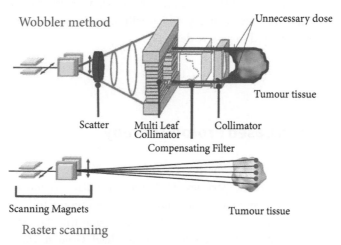

**Figure 9.5**  Passive (top) and active (bottom) beam delivery methods
*Source:* Image courtesy of Toshiba Japan

In active delivery particle therapy systems, the energy of a narrow monochromatic particle beam is changed by modulating the degree of acceleration via the accelerating source (typically a cyclotron for proton therapy or a synchrotron for heavy ion therapy) in order to change the depth of the Bragg peak. Additionally, the lateral position of the beam is changed by the use of two or more orthogonal magnetic fields at right angles to the beam such that the point of maximum dose can be adjusted in three dimensions (Matsubara et al., 2018). By contrast, in passive delivery systems, the energy spectrum of a broad, high energy, monoenergetic beam is spread via a ridge filter, producing a polyenergetic beam covering the desired depth range in the patient (a so-called spread out Bragg peak). Passively scattered beams are often laterally broadened with a combination of magnetic fields and dense (high Z) scattering materials. Further beam shaping devices such as a collimator and patient-specific bolus then laterally shape the beam and compensate for the exterior dimensions of the body to ensure conformity with the target site. The use of active beam delivery has several advantages over passive delivery including:

[6] Evidence suggests that Ions heavier than protons have both physical and radiobiological advantages over conventional X-rays- see for example: Durante, Debus, and Loeffler (2021).

1. More precise and conformal delivery of the dose, especially in the proximal region of the treatment region (Figure 9.5);
2. There is significantly less beam contamination in the form of fast neutrons compared to passively scattered delivery systems due to the absence of beam-shaping material. This will result in a lower dose outside of the treatment region, especially in the distal region;
3. It is possible to implement an adaptive treatment plan with active delivery, since there is no need to manufacture a patient-specific beam-shaping bolus;
4. Beam efficiency is very high since there is no beam-shaping material in the beam path during active delivery, which will result in reduced patient treatment times compared to passive delivery.

## 9.10  Cyclotron Based Proton Therapy

The use of proton treatments for the radiotherapy of cancer is not new, with the earliest uses described at Berkeley, USA and Uppsala, Sweden in 1954 and 1957 respectively. In fact, standard techniques in proton therapy today are derived from research and clinical transformation that occurred at the Harvard Cyclotron Laboratory in the USA by Herman Suit and Michael Goitein (Jongen, 2010). Additionally, there have been numerous innovations since Clatterbridge, UK, including pencil beam scanning, which has advanced knowledge regarding ocular treatments.

The protons are most often generated by an isochronous cyclotron. Compared with other accelerators, a cyclotron is simple to operate and maintain and relatively affordable, reliable, and compact (Smith, 2009; Degiovanni and Amaldi, 2015). In addition to the cyclotron, the proton treatment system also contains a detector for imaging capability, a treatment planning system for 3D mapping of the beam to the three-dimensional target volume, and treatment rooms where patients receive their dose.

There are two primary methods of delivering protons from the cyclotron.

- Passive scattering beam: The first commercially available technique where the proton beam is scattered by devices that interrupt the beam path. A focused beam is then reconstituted by using shaping devices such as collimators and compensators. This results in a homogenous radiation dose across the tumour/target mass (Radhe, 2017).
- Pencil beam: A relatively new technique that 'paints' the beam sequentially over a target volume. As the radiation is delivered layer by layer with high precision, the technique gives a much higher level of control and flexibility over the treatment (Radhe, 2017).

As mentioned before, proton radiotherapy is more often used than Helium or Carbon heavy ion radiotherapy in the treatment of cancers. While X-ray and photon radiotherapy make up the vast majority of treatments, proton therapy is often the preferred treatment for disease sites that respond favourably to high doses of radiation, around the location of the tumour mass requires a higher level of precision to minimise the unwanted radiation dose to healthy tissues adjacent to the tumour and reduce unwanted side effects (Levy and Blakely, 2009).

Specific indications for which proton therapy is suitable include: eye tumours, lower cranial tumours, head and neck cancer, breast cancer where the tumour is close to the heart, lungs or vagus nerve, lymphoma, prostate cancer, gastric melanomas, Hepatocellular carcinoma, and reduction of recurrent tumour mass.

While proton therapy offers several distinct advantages over X-ray radiotherapy, there are also several drawbacks and limitations. Particle therapy is generally considered advantageous for deep-seated tumours; however, to have full-body penetration depth the cyclotron must have an energy of around 250 MeV for protons, which is quite a large cyclotron compared to e.g. radioisotope production machines. However, this is not a 'limitation' in the therapy.

Cost is another major drawback. The cost of proton therapy is nearly three times the price of X-ray radiotherapy (Goitein and Jermann, 2003). Cost is becoming less of an issue as new accelerator technologies progress to market, and treatment protocols allow higher levels of radiation to be delivered in a single treatment session. These advances are expected to see the cost of treatment rapidly decrease in the future. The costs of proton therapy infrastructure are also considerably more expensive with the capital costs of a multiroom treatment facility starting at US$200 million (Raeburn, 2017).

## 9.11  Cyclotron and Synchrotron Technologies: Particle and Heavy Ion Therapy

Protons and heavy ions can also be accelerated by synchrotrons as an external beam radiotherapy for the treatment of cancer. As with cyclotron-based proton therapy, the particle beam exhibits a Bragg peak and loss of energy through the body allowing the optimum dose of radiation energy to be deposited within the target tumour mass and minimising the radiation dose to surrounding tissue. It should be noted that the cyclotron and synchrotron are not different treatments. but different technologies with different capabilities.

The particle therapy technique works by sending an external beam of high energy ionised particles to a tumour (Amaldi and Kraft, 2005). The charged particles cause irreparable damage to the DNA of the targeted cells within the tumour mass, which leads to apoptosis and cell death. Particle therapy, sometimes referred to as heavy ion therapy, uses accelerated ionised atoms such as neon, argon, silicon, and carbon. Particle therapy is a mature technology that is technically demanding and requires large energies to accelerate heavy ions to reach therapeutic efficiency.

The acceleration of heavy ions for treatment has the advantage of preferable dose distribution with less scatter than lighter particles due to the larger particle size. This results in very little loss of dose to healthy tissue. Heavy ions exhibit a much sharper dose fall off than protons in the longitudinal direction, resulting in greater beam concentration and targeting with a higher percentage of the particles stopping within the tumour mass and very few passing to the healthy tissue posterior to the tumour.

The most commonly employed heavy ion for particle therapy is carbon, and this technique is often termed carbon ion radiotherapy (CIRT) (Tsujii, 2017). CIRT is rapidly gaining attention for the treatment of a number of cancer types that present a solid tumour mass and are intractable hypoxic and/or radio-resistant cancers. Clinical trials have demonstrated CIRT to be effective in treating a number of poor prognosis cancers where x-ray based radiotherapy is no longer effective, including adenocarcinoma, adenoid cystic carcinoma, malignant melanoma, and various forms of sarcoma gastrointestinal and pancreatic tumours (Kamada et al., 2015). Since 2009, the Heidelberg Ion Beam Therapy Centre (HIT) has been treating with protons and carbon ions many cancer types such as salivary gland cancer (e.g. adenocystic carcinoma), ENT tumours, e.g. paranasal carcinoma, Chordoma /Chondrosarcoma of the base of the skull or the pelvis, prostate cancer, brain tumours, e.g. glioblastoma, glioma, meningioma, pylocytic astrocytoma, lung, pancreatic and liver cancer and recurrent rectal cancer. HIT has extensive experience in treating children and adolescents.

While particle therapy—especially carbon ion therapy—shows strong promise as a radiotherapy technique, there are several limitations and impediments to widespread uptake and adoption. As with proton therapy, the most serious impediment is the high initial capital cost. Currently the capital cost of a heavy ion treatment centre with a modest capacity to treat 1000 patients per year is roughly twice the capital cost of a proton therapy treatment centre of the same size. However, it should be noted that whilst the capital cost of both proton and heavy ion treatment facilities is large, the cost is still lower than the full cost of bringing a small molecule cancer therapeutic to market (Laine, 2016). There are also several new accelerator technologies currently under development which may reduce the size and cost of the most expensive synchrotron components, providing similar capability within a smaller footprint. These technologies which have the potential to democratise and decentralise particle therapies are discussed later in the chapter.

For both proton and particle therapies there is serious debate on the metrics of beam generation via cyclotron versus synchrotron. There are many considerations related to the use of accelerator technology for proton and particle therapy and these include capital outlay, size, reliability, maintainability, beam specification, as well as particle species requirements. Synchrotron technology is mature, efficient, and requires less radiation shielding due to its high beam extraction efficiency. Synchrotrons can also vary beam energy. Cyclotrons can be compact and the beams are acceptable and increasing in performance due to pencil beam techniques. Many cyclotrons operate at fixed energy, which some consider a downside (Fukumoto, 1995).

## 9.12  Neutron Capture Therapy

Neutron capture therapy is a radiotherapy modality in which the target is irradiated with thermal (or epithermal) neutrons. In this technique, externally produced neutrons are directed at the target volume with some being captured within the target region using tumour-specific molecules including certain isotopes with high thermal neutron capture cross sections. This has the effect of producing a high biological dose within the treatment region, since the neutron capture process results in the production of high-LET secondary particles. The specific capture products vary according to the neutron capture agent. The only neutron capture isotope presently in clinical use is $^{10}$B (Boron), however, $^{157}$Gd (Gadolinium) has a much higher thermal neutron cross-section and has a high potential for effective neutron capture therapy (Figure 9.6).

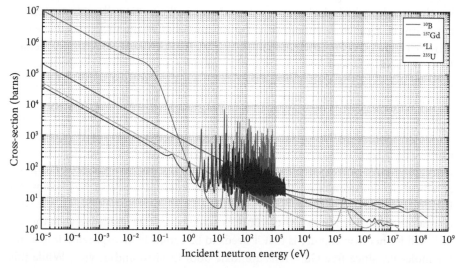

**Figure 9.6** Total neutron interaction cross-sections for isotopes used or proposed for therapeutic use

*Source:* Data sourced from TENDL 2019: https://tendl.web.psi.ch/tendl_2019/tendl2019.html.
Koning, A.J., Rochman, D., Sublet, J., Dzysiuk, N., Fleming, M., and van der Marck, S., 'TENDL: Complete Nuclear Data Library for Innovative Nuclear Science and Technology', Nuclear Data Sheets 155 (2019) 1

## 9.13  Neutron Capture Agents: Current Practice and Future Development

Two agents are currently in clinical use for the delivery of $^{10}$B to a treatment region: L-p-boronophenylalanine ($^{10}$B-BPA) and sodium mercaptoundecahydro-closo-dodecaborate ($^{10}$B-BSH) (Snyder et al., 1958; Mishima et al., 1989; Hatanaka, 1991). Both drugs have been used to treat patients suffering from glioblastoma multiforme and malignant melanoma. Significant differences in the normal and tumour concentrations are observed depending on the treatment site, the delivery method, and whether the BSH or BPA compound is attached to another transporter compound.

While BPA and BSH have been clinically used as neutron capture agents (NCA), new compounds continue to be developed with the aim of reducing the normal tissue concentrations and increasing the tumour retention duration (Luderer et al., 2015; Nedunchezhian, 2016; Barth et al., 2018). The research and development of new, 'generation 3', $^{10}$B drugs has been undertaken using new methods ranging from the use of boron loaded nucleosides (sugars), to targeting peptides for the transport of $^{10}$B. While many of the next-generation neutron capture agents are under active research and development, none have yet progressed to clinical use. The predominant focus of the studies has been on increasing the concentration of the agents in the tumour, increasing the tumour-to-normal tissue concentration ratio and increasing the time for which the boron is retained within the target cell, with the general consensus being that a tumour boron concentration of 30 ppm will be sufficient to provide a significant enough therapeutic dose during boron neutron capture therapy (BNCT). With renewed interest in the BNCT field it is anticipated that the development of generation 3 drugs will continue with a selection eventually making it to clinical use and superseding BPA and BSH as preferred neutron capture agents.

While most of the drug development research and development has been focused on boron-loaded agents $^{157}$Gd-based compounds have been proposed as neutron capture agents, either solely or in conjunction with boron-based agents. $^{157}$Gd has the highest cross-section of all stable isotopes and the process of neutron capture results in the emission of multiple Auger and/or internal conversion electrons. An important advantage of $^{157}$Gd over $^{10}$B is that a high tumour-to-normal tissue concentration ratio, often in excess of 100:1, is achievable. The challenge with the use of $^{157}$Gd for neutron capture is that it needs to be carefully incorporated into a stable molecule, since free $Gd^{3+}$ ions are toxic both in vitro and in vivo. While this has successfully been achieved with Magnetic Resonance Imaging (MRI) contrast agents, none of the currently available MRI contrast agents are suitable as neutron capture agents for neutron capture therapy as there is low tumour specificity and low uptake in the cell nucleus (and other radiosensitive organelles). Several new delivery methods or modifications to current contrast agents have been proposed with the aim of increasing tumour specificity, with tumour uptake of 101 ug Gd/g wet tumour achieved (Hofmann et al., 1999; Watanabe et al., 2002; Le et al., 2006; Ichikawa et al., 2007; Ho et al., 2018). Although these results show the potential of using current imaging agents for neutron capture, these techniques have not yet been used in a clinical setting. Further development of these agents is required, especially considering the higher tumour uptakes which have been demonstrated with alternative vectors.

A few groups have explored the use of gadolinium-loaded chitosan nanoparticles for neutron capture. While these studies demonstrate the potential of gadolinium nanoparticles for neutron capture therapy, one of the unique challenges of using gadolinium for this application is that the high LET Auger electrons have a very short range (of the order of nanometres). If the nanoparticles are too large, electrons from the centre of the nanoparticle will not escape to damage nearby cell structures. Therefore, only interactions at or close to the surface of the nanoparticle will yield a useful source of local high-LET radiation that can deposit energy within the target site. As a result, interactions between atoms on the surface in inner nanoparticles will compete with one another. Due to part of the thermal neutron fluence being attenuated but not depositing high LET radiation at the desired site, there will be a decrease in the total useful thermal neutron fluence.

An exciting new approach to atomic, small molecule drug development has been proposed (Rendina et al., 2020; Marfavi et al., 2022). This group has developed two promising next-generation $^{157}$Gd based neutron capture agents that are based on 1,4,7,10-Tetraazacyclododecane-1,4,7,10-tetraacetic acid (DOTA) and diethylenetriaminepentaacetic acid (DTPA) (Morrison et al., 2014). These agents have the distinct advantage that they can form strong bonds to the gadolinium atom, making the agent very biologically stable (i.e. the gadolinium will not become toxic free gadolinium). Additionally, it may be possible to tailor these agents to target specific biological pathways, to optimise the agent to maximise its specificity for certain tumours, including the organelles within the target cells, such as the mitochondrial membrane. This potentially will enable very high uptake into radiosensitive organelles, therefore achieving a correspondingly high radio sensitisation of the cell (potentially for photon activation therapy as well as neutron capture therapy). As such, tumour uptake with concentrations of up to many thousands of ppm has been reported. The next challenge for the DOTA NCA agents is to refine the production process before expanding it to human trials. This process will be similar to what was previously undertaken to progress the DOTA gadolinium imaging agents from theory to clinical use.

## 9.14  Neutron Sources

Like X-ray and electron beam therapies described previously, not all reactor derived neutrons are created equal and are categorised by their energy (En) as:

- thermal (En <0.5 eV);
- epithermal (0.5 eV <En <10 keV);
- fast (En >10 keV).

Thermal neutrons are commonly used in NCT because of their ability to initiate neutron capture reactions switching boron, such as the 10B(n,)7Li capture reaction. There are two common sources of neutrons for Neutron Capture Therapy (NCT): nuclear reactors and accelerator-based neutron generators. Nuclear reactor-based NCT relies on neutrons being generated in the core of a nuclear reactor (Wittig and Sauerwein, 2012). These neutrons are then moderated and filtered before being transported to the patient along a beam line. Reactor-based NCT has the advantage that very high thermal/epithermal neutron fluences are possible, which can significantly reduce patient treatment times.

Reactor-based NCT has significant limitations. The biggest limitation of reactor-based NCT is the extremely high start-up and running costs of a reactor-based NCT programme. Despite the promise that NCT shows and the benefits to patients of selective tissue treatment, despite decades of research, BNCT's use in the clinic is very limited. An alternative source of therapeutic neutrons is through accelerator-based reactions. In this process, protons or deuterium ions are accelerated to energies between 1 and 5~MeV which then collide with a target, typically either lithium or beryllium. The achievable neutron fluences are an order of magnitude less than those from reactor-based facilities, increasing patient treatment time, decreasing economic viability, and increasing the likelihood that a patient (or a region within the patient) will move during treatment. However, as new targets and accelerator designs are developed, the current trend in NCT is a shift towards accelerator-based neutron sources.

Recent efforts by several scientific research and industry groups have culminated in several BNCT clinics equipped with different types of charged particle accelerators and targets. Major developments and industry leads are listed below:

- Japan: Sumitomo Heavy Industries has installed NeuCure™ System, an accelerator-based neutron source in South Tohoku Clinic (Koriyama, Japan). A second clinic—Kansai BNCT Medical Centre—is being built in Osaka, Japan and will be equipped with SHI's NeuCure™ System;
- Japan: Mitsubishi Heavy Industry Co. together with the University of Tsukuba and the High Energy Accelerator Research Organisation, Japan Atomic Energy Agency, Hokkaido University and the Ibaraki Prefecture have produced an 8-MeV 5-mA linac with a beryllium target for the BNCT clinic in Tokai (Tsukuba, Ibaraki, Japan);
- Japan: Cancer Intelligence Care Systems, Inc. has produced another LINAC-based neutron source at the National Cancer Center in Tokyo;
- Finland/USA: Neutron Therapeutics Inc. (Danvers, MA, USA) has manufactured a direct-acting electrostatic accelerator Hyperion™ to be housed at the Helsinki University Hospital;
- China: Neuboron Medtech Ltd. (Nanjing, Jiangsu, China), TAE Life Sciences (Foothill Ranch, CA, USA), and the Budker Institute of Nuclear Physics (Novosibirsk, Russia) were commissioned to manufacture a tandem accelerator neutron source. The source will be housed at Xiamen Humanity Hospital (Xiamen, Fujian, China).
- Even with the rapid increase in the development of accelerator-based NCT technology there are several other challenges that still need to be addressed. These include:
  - dose and delivery optimisation of neutron capture agents (NCA);
  - development of specific targeting vector-based neutron capture agents;

  ∘ detectors that can measure and quantify real-time dosimetry at both the tumour mass and normal surrounding tissue.

The underlying physics of neutron interactions with matter is another factor that restricts the utility of neutron capture therapy. Externally produced neutrons have a limited penetration depth as illustrated in the depth-dose curves in Figure 9.4. Skin and blood, which both retain some of the neutron capture agents, are exposed to high levels of neutrons in order to provide enough neutrons to a target volume at a depth greater than 4 cm. Due to this restriction, skin cancers like malignant melanoma can only be treated using conventional neutron capture therapy techniques that rely on reactors or accelerators (Harling, 2009). NCT has shown promise in treating brain tumours and recurrent head and neck cancers, all of which rely on discovering mechanisms for creating a thermal neutron field within the body.

## 9.15  Emerging Technologies and Trends

The particle therapy market (comprising proton and heavy ion therapies) is experiencing consistent growth (compound annual growth rate of 9.3% over the forecast period of 2018–2023). There are a number of major factors driving this growth, including the global prevalence of cancer, the growing adoption of particle therapy in clinical trials, its advantages over other cancer treatments as well as the increasing number of particle therapy centres worldwide. The Asia Pacific region is expected to dominate the particle therapy market during the forecasted period (2018–2023) and the paediatric cancer segment is expected to account for the largest share of the particle therapy market by cancer type.

NCEPT is a convergence of neutron capture therapy and particle therapy to provide radiation dose amplification to metastatic and satellite tumour lesions. Neutron

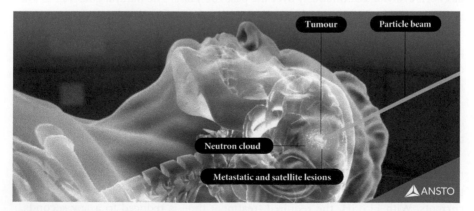

**Figure 9.7** Neutron capture enhanced particle therapy (NCEPT)
*Source:* ANSTO

capture enhanced particle therapy (NCEPT) is a transformative targeted radiotherapy technique proposed and developed at ANSTO in Australia for treating cancers of poor prognosis, including brain cancers such as DIPG, pancreatic cancer, and secondary melanoma. NCEPT delivers a double blow to cancer by exploiting internally generated thermal neutrons—a by- product of particle therapy—to deliver a targeted extra radiation dose directly to tumour cells (Figure 9.7). Significantly, NCEPT achieves this by combining the precision of particle therapy with the cancer-specific targeting capability of neutron capture therapy (NCT), using therapeutic agents already approved for use in humans. These agents were originally developed for neutron capture therapy (another type of external beam radiotherapy) that only treats very limited types of cancer. Interestingly, our very promising NCEPT results have stimulated a resurgence of research activity related to the development of new neutron capture agents targeting specific cancers.

NCEPT will improve patients' quality of life by reducing the side and late effects of radiation therapy, since the same therapeutic dose can be delivered to the tumour with less radiation damage to healthy tissue. As a bonus, NCEPT selectively targets cancer cells which may have spread to healthy tissue beyond the margins of the main tumour—even before this spread is large enough to be detected via diagnostic imaging. In late-presenting cancers (such as pancreatic and paediatric brain cancers), NCEPT may be the only viable treatment option.

Both proton and heavy ion therapies are increasingly becoming available throughout the world, with the first Australian facility (the Australian Bragg Centre for Proton Therapy) currently under construction in Adelaide. NCEPT can potentially be integrated into any of these proton or heavy ion therapy facilities.

## 9.16  The Future of Medical Accelerator Technology

We have seen in this chapter that the medical application of accelerator technology has been evolving continuously over the past 100 years. The natural question to ask is 'What does the future hold for medical applications of accelerators?'

In the last section, we explored an example of the convergence of two techniques: particle therapy and BNCT through NCEPT technology. This is one great example of a technology that increases the precision and accuracy of the treatment while minimising the impact on healthy tissue and ultimately reducing treatment related morbidity. The challenges are the core problems that innovators are seeking to solve in the development of new technologies and feature throughout the technology trends which will be reported in this section.

Along with NCEPT, one of the best examples is the FLASH Radiotherapy technique. FLASH is a technology that has the potential to reduce treatment times for photon and proton treatments by using very high dose rates that require large beam currents. With FLASH, the required treatment time is reduced to a few seconds, meaning that cancer cells are more selectively targeted preserving normal tissue and the technology also becomes accessible to a larger number of patients as a larger throughput is enabled by the shorter treatment times.

In addition to NCEPT and FLASH, other notable developments include intensity modulated radiotherapy (IMRT), stereotactic body radiation therapy (SBRT), and 4D imaging. The latter is interesting as advances in imaging technology allow a higher specificity in targeting tumour mass for external radiotherapy.

Supporting technologies like treatment planning systems are benefiting from the convergence of emergent technologies such as artificial intelligence, robotics and automation, computer vision, blockchain, and augmented reality. This convergence is allowing treatments to become increasingly personalised and tailored to individual patient needs.

A good example of this is the personalisation of radiotherapy doses based on the Positron Emission Tomography (PET) findings in prostate cancer and tumour volume mass determination from PET scans in head and neck cancer (Mak et al., 2011; Chang et al., 2012).

The PET technology widely used in hospitals dates back to the 1970s. Although PET scan was not invented at CERN, pioneering work carried out by CERN physicists Alan Jeavons and David Townsend made a major contribution to its development. In 1977 at CERN, Townsend working in the Data Handling Division together with Jeavons, who worked with the Charpak Group, decided to explore the possibility of merging and applying the novelties from information technology and detector development. Jeavons had developed a new detector, based on a high-density avalanche chamber, to take PET images. Townsend had developed the software to reconstruct the data from the detector and to turn it into an image.

They used a special software programme and bright new cameras based on a high-density avalanche chamber to produce the first image from a mouse injected with 5.7 µCi of $^{18}$F a positron emitter. From two chambers placed around the mouse they picked the pairs of emitted photons produced by the positron annihilation and were able to visualise the image (Figure 9.8).

This first image was a digital picture composed by numbers; later pictures were colour pixels and by comparing it with today's images one can appreciate the developments that are accompanying innovation technologies, concerning detectors, data acquisitions, and imaging reconstruction from prototypes to hospital applications. Progressing with all those steps in development and ideas of combining technologies, i.e. PET with CT, it is possible to see inside the body but also assess the metabolic functioning of tumours and the effects of treatments.

Today imaging is a very important and valuable adjacent technology as it converges with artificial intelligence for enhanced diagnosis and the selection of treatment protocols on a patient-by-patient basis. For example, these technologies will be able to determine a patient's suitability for radiotherapy versus other treatment options such as surgery, or immunotherapy.

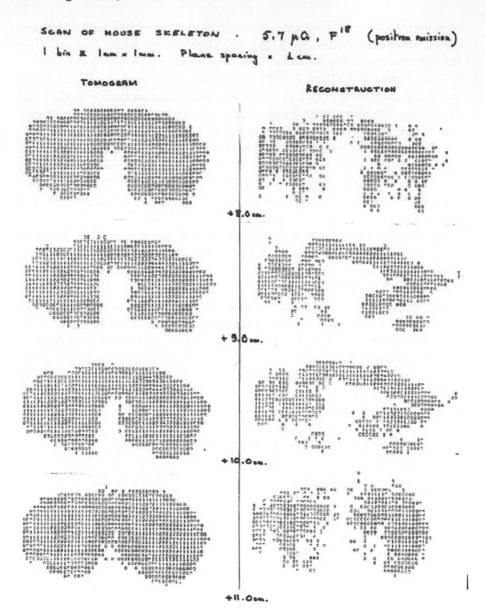

**Figure 9.8** The first PET image taken at CERN, in 1977, showing the skeleton of a mouse

*Source:* © CERN

In summary, three types of innovations appear to be driving the field forward:

1) Convergence of ideas or existing technologies;
2) New accelerator technologies enabling disruption (e.g. ADAM/LIGHT LINAC for proton therapy has this potential) or even totally new paradigms; and

3) New mechanisms including FLASH, change the requirements of the accelerator technology. e.g. VHEE with FLASH based on compact electron accelerators at around 100–250 MeV.

CHUV, CERN, and THERYQ are pioneering the development of a FLASH radiotherapy device based on CERN technology for the delivery of ultrafast, one-time high-dose irradiation with very high-energy electron (VHEE) radiation to treat cancers resistant to conventional treatments.

## 9.17 Conclusions

Chapter 9 demonstrates how Big Science investments can, sometimes over extended periods of time, create significant societal and environmental benefits in medicine by converting scientific discoveries and knowledge into practical artefacts, technologies, and methods.

Big Science technologies and expertise have been used in the field of medical research in diagnostics and treatment of cancer, neurological disorders, medical imaging, radiation oncology, and other medical applications. Therefore, Big Science contributions to human health and disease are extensive with notable contributions such as genome sequencing, imaging technology, accelerator technology applications, drug development, and the diagnosis of disease.

Accelerator and detector technologies were first developed nearly a century ago and proliferated into a wide range of applications utilising beams of electrons, positrons, protons, antiprotons, and heavy ions. As discussed in this chapter, the number of medical linear accelerators or LINACs around the world has grown exponentially for medical physics, radiation therapy, nuclear physics, industry applications, and research and development.

Radiotherapy is a form of cancer treatment that relies on photon beams generated by electrons in a straight line to irradiate cancer cells. The electron beam is used for direct irradiation but that can only reach superficial tumours. These methods differ from hadron therapy, which uses charged particles such as protons or heavier ion-like carbon to irradiate cancer cells directly. Neutron therapy is a form of radiation therapy that poses many challenges for physicists and medical practitioners. The availability, high cost, toxicity, radiation shielding, treatment plan and monitoring are among them. Research challenges include the development of new neutron sources including accelerator-based neutron sources and fast neutron therapy at hospitals.

Big science initiatives have resulted in many useful innovations that continue to grow and advance with industry involvement. Detectors are also needed in medical applications to monitor treatments and imaging techniques. Medipix is given as an example of a successful family of semiconductor pixel detectors developed by CERN in collaboration with industrial partners for imaging and spectroscopy which has tremendous use for medical diagnosis. Another important technology derived from Big Science is the Positron Emission Tomography (PET) scanner, which is used to create detailed functional images using radioactive tracers to identify and follow the development of diseases such as cancer and neurological disorders. The use of PET-CT for diagnostic and follow-up treatment, of gene receptor scintigraphy, of theranostic both for diagnostic and treatment, as well as molecular targeted cancer therapy, allows enormous progress in the treatment and control of the disease.

Innovation processes involve multiple stakeholders and face significant unpredictability in the early stages of the development of equipment. There is a long learning curve to developing standard medical practices that can be implemented and are acceptable to the medical community. Research into medical fields and introducing novel innovations is a complex and long-term process that involves expensive clinical studies and approval processes.

Innovative approaches demand working outside comfort zones and ongoing training, better communication, and collaboration, and, more importantly, human interaction between particle physicists and the medical community. Delays or lags in technology diffusion can be minimised by establishing strategic collaboration among the medical industry, medical scientists, and the Big Science community. Technology transfer initiatives and groups, such as CERN and EBI-EMBL, have used open innovation frameworks for the rapid dissemination of knowledge and information useful for medical applications.

Furthermore, the translation of the development of detector technologies for medical imaging is a prime example of how scientists have and still need to collaborate in specific fields like physics, medicine, and information technology for the good of all human beings and contribute to the field of medical science. Continuous improvements in medical accelerator technologies will enhance the quality of life and wellbeing of patients.

# PART 3

# ORGANISATIONAL AND SOCIETAL IMPLICATIONS

Big Science, like all other investments requires intelligent organisations and collaborations to become the binding force that makes creative and learning organisations and motivating individuals. Big Science facilities and experiments require the use of highly advanced technologies and well-thought-out concepts. Part 3 of this book examines the organisational complexity and wide range of factors that link science with social agendas.

# 10

# Big Science as a Complex Human Enterprise

*Beatrice Bressan, Anita Kocsis, Pablo Garcia Tello, and Shantha Liyanage*

## 10.1 Introduction

Open science is the fabric of making scientific research and research data available to every user irrespective of their role and contribution to research endeavours. The design or the rubrics of the scientific ethos and process constitute two starting reference realms for undertaking the complex task of analysing the contribution of Big Science to society. Open science facilitates ready access to scientific research and the wider dissemination of data and information. Such open science objectives are to enhance and accelerate learning and innovation and benefit society. This, in turn, increases trust in science and builds the credibility and reliability of scientific enterprises, invigorating a better understanding of the role of science in society.

This chapter offers yet another set of interlinked facets aiming to illustrate that scientific activities are complex and multifaceted human enterprises that escape closed and rigid definitional frameworks.

Thus, the first section starts by offering a view of science, society, and values in a historical context. Given the inescapable interdependence between science and society, this historical reflection also includes a social dimension. In this sense, the central issue being thought through is how science should advance in society and which social values should be cherished. Although answers to this complex question are far from straightforward, it is clear that science should engage holistically with the many aspects of engaging the norms and values of society embedded, for example, in politics, the arts, the economy, wellbeing, etc.

Using these factors as a foundation, the second section examines Big Science from the viewpoint of aiding in the emancipation of humanity, or the less fortunate and disadvantaged people of the world. Big Science produces both direct and indirect social benefits, particularly when it comes to major challenges and issues like climate change.

In light of the upcoming challenges to our society's future, the third section examines how scientific and artistic endeavours can energise one another by fostering

Beatrice Bressan et al., *Big Science as a Complex Human Enterprise*. In: *Big Science, Innovation, and Societal Contributions*. Edited by: Shantha Liyanage, Markus Nordberg, and Marilena Streit-Bianchi, Oxford University Press. © Beatrice Bressan et al., (2024). DOI: 10.1093/oso/9780198881193.003.0011 interdisciplinary forums for discussion and experimentation. Humanity needs to search for opportunities to launch ambitious and ground-breaking projects in order to respond in a way that disrupts systems and technology. The final section considers whether strictly neoclassical economic frameworks are adequate or even suitable for valuing Big Science given how diverse it is and especially how historically entwined it is with society.

## 10.2 The Social Value of Big Science

Given the various paths from fundamental science to experimental development, contributions to society can be viewed as a common denominator linked to the characteristics of research organisations. Indeed, the increasing high-technology requirements of research provide a fertile ground for technology and knowledge transfer, promoting the injection of science into all levels of daily life in a variety of ways.

Consider, for example, quantum entanglement, a physics phenomenon based on quantum theory. Who could have imagined the practical applications of cryptography and computing would result in the formation of companies to protect information sharing?

When organisations dealing with fundamental sciences permit 'freedom' to think, do, and discuss freely, it is possible to find fundamental research as a common denominator for technological development and applications. If so, is it not possible to provide non-science organisations with some basic guidance that will help them achieve this 'freedom' and teach them how to use it for themselves? The answer to these questions was well illustrated by Sir Ben Lockspeiser, the first President of the CERN Council in 1954, who stated: 'Scientific research lives and flourishes in an atmosphere of freedom—freedom to doubt, freedom to inquire and freedom to discover. These are the conditions under which this new laboratory has been established' (Lockspeiser, 1954).

In the case of CERN, high energy physics stimulates the continuous production of innovative technological development. In the quest to find out what matter is made of and how its different components interact, these organisations need to develop highly sophisticated instruments, in which technology as well as required performance often exceed the available industrial know-how. This is why, since its creation in 1954, CERN has pursued the tradition of collaborative partnerships with industry and making CERN's technologies available to third parties. In the LHC experiments, almost half of the participants are from non-member states of CERN. As a result, the technological learning from high energy physics has spilled over worldwide.

This worldwide spill over concerns not only technological learning but also the development of frontier technologies required in Big Science and their utilisation in fields other than those they were originally developed for. Scientific organisations are constantly improving their capabilities for making real-time observations, interactive data analysis, and automated processes.

The LHC experiments, astrophysics experiments, and gravitational waves laboratories have knowledge reservoirs that bring significant benefits to society. The recent

Covid-19 was an example of how knowledge reservoirs in physics and medicine at EMBL-EBI (PDBe-KB Covid-19 Data Portal) and the Repository of Corona Disease Research Community made available through Zenodo, which is a multi-disciplinary open repository maintained by CERN, were linked.

Each year hundreds of young people join organisations like CERN, as students, fellows, associates or staff members taking up their first job. This continuous flow of people, who come to these research centres, are trained by working with experts, and then return to their home countries, exemplifies knowledge and technology transfer via people. However, when it comes to industry, the potential of organisations such as CERN may be underutilised. It would be possible to enhance the spectrum of their technological impact by paying attention to their technological learning management.

Several studies provide evidence that the socialisation of participants in meetings leads to the acquisition of skills in various areas (Bressan, 2004; Bressan et al., 2008; Bressan, 2010). The development of interests through interaction with colleagues is a critical element in the learning process. The learning processes extend from tacit knowledge, which is essentially personal and hard to share, to explicit knowledge, which can be easily shared.

Individual and organisational learning is a core asset of research organisations, the latter being the social process by which a group of people collectively improve their capacities to produce an outcome. The creation of organisational knowledge amplifies the knowledge that is created by individuals who spread it at the group level through dialogue, discussion, experience, or observation (Nonana and Takeuchi, 1995).

Big Science research organisations must provide a context in which individuals can hold both formal and informal discussions to steer new ideas and foster collective learning if they are to be effective in the process of knowledge translation. This type of knowledge generation is regarded by economists and sociologists as significant, because such processes support organisational and technological innovation relevant to industry and society.

Scholarly studies (Kogut et al., 1992; Grant, 1996; Spender, 1996; Autio et al., 2003, 2003b, and 2011) have confirmed that knowledge acquisition in a multicultural environment is linked to interactions between social capital components (social interaction, quality of relationships, and network connectivity) and competitive advantage (development of inventions and uniqueness of technology).

Large experiments, such as those at the LHC, serve as the hub of an institutional and organisational network at CERN. Interactions between individuals and experiments, enabled by the collaboration's organisational structure and the frequent use of modern communication tools such as emails and websites are important routes for knowledge translation.

The fertile environment such as the LHC experiments fosters a dynamic, interactive, and simultaneous exchange of knowledge both inside and outside the collaborations, allowing individuals to create and expand knowledge through their social networks while also involving industry at various stages of project development.

Thus, when working in scientific environments, if the development of personal skills is well managed, used, and catalysed to target individual development, it is possible to improve labour-market opportunities. Individuals with a sense of entrepreneurship may want to consider working for a company that promotes learning and innovation in science enterprises, which can also be used for social business purposes (Bressan et al., 2008).

Nobel laureate Muhammad Yunus, a Bangladeshi social entrepreneur, banker, economist and civil society leader founded the Grameen Bank and he was awarded the Nobel Peace Prize in 2006, 'for their efforts through microcredit to create economic and social development from below'. The social business income is reinvested in the business itself with the aim of increasing social impacts. Yunus' philosophy was to profess the social benefit of the social enterprise and he demonstrated that it is possible to develop a social enterprise built on the selective transfer of knowledge and technology.

Such transfers can foster innovative solutions to promote good governance and develop strategies to address emerging global security challenges and the risks of over globalisation leading to inequity and social unrest. Besides Big Science initiatives, there are a myriad of other actions needed to foster innovative solutions to promote good governance and strategy to address emerging global security and economic challenges. The risks of globalisation, wars, pandemics, human rights violations, and poverty always have drawbacks.[1] In Yunus' words, 'a charity dollar has only one life; a social business dollar can be invested over and over again' (Yunus, 2009; 2011).

In doing that, Big Science organisations such as CERN in partnership with other governmental organisations such as the UN can promote emancipation processes leading to enhanced cooperation and operability for developing an intertwined framework among global members and stakeholders to make knowledge actionable from local to global. Under the 2030 UN agenda, the 17 Sustainable Development Goals (SDGs) could be a place to look for some inspiration.[2] It is possible to find actionable applications of such fertile collaborations in disadvantaged countries as well.

In his recent book on The Kyoto Post-Covid Manifesto for Global Economics, Hill and his colleagues (Hill et al., 2022) observe the underlying dynamic of the majority of contemporary global economics, self-interest, but then demonstrate the power of drawing our relationships instead from the wellspring of what makes our *humanity* work—*sharing*. The Kyoto Post-Covid Manifesto moves on to show how this alternate sharing dynamic can be built into existing and new institutional and exchange relations through 'Humanity-centred transformation'. The result is that the organisations and their exchange relationships become more productive because the participants now live and benefit from 'an increasingly broad culture of trust and cooperation rather than divisive self-interest and greed' (Hill et al., 2022: 352).

---

[1] W.L. Christman, *Global Resilience: The Manifesto* (forthcoming).
[2] CERN has observer status in the UN General Assembly.
https://home.cern/news/press-release/cern/cern-granted-status-observer-united-nations-general-assembly.

In this context, one example has been the realisation of a local IT social enterprise, a spin-off of UNRWA[3] based in Gaza, Palestine, where almost 70% of the approximately 1.000 students, who yearly graduate in Information and Communication Technology (ICT) programmes from local universities have difficulty finding a job[4] despite the rapid contextual industrial growth in this sector.[5]

Funded by the Korean government to provide short-term employment and a learning environment for young ICT graduates, this non-profit initiative aims to establish a local ICT service development business park as a hub for overseas ICT outsourcing solutions to a wide range of clients. In agreement with UNRWA, CERN invited the management of the UNRWA start-up to specific open-source software sessions in its IT Department, representing Gaza's possible solution from academia to the private sector, in order to fulfil the young social enterprise's mission and establish its high-potential socio-economic impact. The project allowed its staff to acquire the necessary knowledge and know-how to enrich their services and activities to better satisfy customers' needs and increase market segment opportunities. This is another example of how Big Science organisations can address social problems at the grass roots level.

Such initiatives between research laboratories and intergovernmental organisations like CERN and the UN will not only increase the social value of basic science but will also serve as a guidepost for future young leaders who will be able to build social businesses in order to positively impact their communities and society and to foster a new resilience culture.

As the former CERN Director General Rolf-Dieter Heuer said, 'Science has a responsibility to bring itself to the mainstream of popular culture, to engage in and shape public debate about major issues that are science based. It has the responsibility to make itself accountable, particularly if it is publicly funded. And it has a duty to work to the highest possible ethical standards. Science underpins almost every aspect of modern life, be it economic, social, cultural, or humanitarian, and it is blind to race, gender, language, and religion. In short, science represents the best in humanity' (Heuer, 2018).

## 10.3  Art, Design, and Science Colliders—Creating New, Young Leonardo's

The cultural convergence of art, science, and technology today is well represented by a community of international institutions that is a platform for collaboration. Exemplars such as the establishment of the Leonardo institution, the journal and book series founded (1967–1968) by Frank Malina, an aeronautical engineer and a kinetic sculptor, were important catalysts for ongoing social, political and environmental

---

[3] UNRWA: United Nations Relief and Works Agency for Palestine Refugees in the Near East.
[4] Mercy Corps Labour Market Needs Analysis for the Digital Economy, 2013.
[5] World Bank Ad Hoc Liaison Committee Report, covering the first nine months of 2012, estimated that the sector contributed 0.02% to Gaza's GDP growth; UNESCO Socio-Economic Report: Overview of the Palestinian Economy, 2013.

debates through the nexus of art–science interaction. One of the two key goals, as outlined by Leonardo, the International Society for the Arts, Sciences and Technology, MIT Press, and the affiliated French organisation Association Leonardo, is 'to create a forum and meeting places where artists, scientists, and engineers can meet, exchange ideas, and, where appropriate, collaborate', to tackle the 'hard problems' and bring about new agendas in science and opportunities for technological innovation.[6]

A generation later, a plethora of international organisations including science research centres like CERN, cultural institutions, universities, government, and private funding bodies are a locus for such forums.[7] This forum is a catalyst for new disciplinary collaborations, emerging theories, sharing of practices, and increased dissemination (true and false) of data across a spectrum of science. A common occurrence is the emergence of inter and transdisciplinary practices (Van Noorden, 2015; Bliemel and van der Bijl-Brouwer, 2018) through experimentation and invention.

Forums such as the European Digital Art and Science Network[8] initiated by Ars Electronica invite artists and scientists together to connect arts and cultural institutions with ESA (European Space Agency), CERN, and the ESO (European Southern Observatory). The forums are catalysts for co-innovation; multi and inter-disciplinary, cross-sectoral serendipitous engagement, and research. Spaces such as studios, workshops, galleries, cafes, and laboratories, encourage structured serendipity facilitating unlikely interaction.

The Design Factory Global Network, mentioned in Chapter 6, is one example of a change agent that uses networks and nodes made up of cross-sectoral, international universities and research organisations to structure serendipity and innovation across disciplinary silos (Björklund et al., 2019). Other forums include the European Commission's 'Science with and for Society' (SWAFS, 2020) full cycle innovation programmes structured for diverse societal actors; researchers, citizens, policymakers, businesses, and non-governmental organisations.[9]

ATTRACT[10] mission shifts gears to accelerate the conversion of opportunities gleaned from Big Science's lengthy timescales and advances structured serendipity to systematise serendipity for ground-breaking applications (Wareham et al., 2022).

With the help of Big Science initiatives, such as the CERN experiments, ATTRACT aims to create the next generation of scientific tools that will enable the emergence of new businesses (see Figure 10.1).

Despite their differences, both arts and science share a common goal in the above-mentioned programmes to combine the contributions that artists, designers, engineers, and scientists can make to the challenges of our time:

---

[6] https://mitpress.mit.edu/books/series/leonardo.
[7] http://userwww.sfsu.edu/infoarts/links/wilson.artlinks.org.html.
[8] https://en.unesco.org/creativity/policy-monitoring-platform/european-digital-art-science.
[9] https://horizon-swafs2020.b2match.io/.
[10] https://attract-eu.com/about-attract-phase-2/.

Both value the careful observation of their environments to gather information through the senses.

Both value creativity.

Both propose to introduce change, innovation, or improvement over what already exists.

Both use abstract models to understand the world.

Both aspire to creative works that have universal relevance.

(Wilson, 2002: 18)

Forums like residencies in scientific organisations, museum outreach programmes, academic institutions, and cultural organisations that have been boosted by digital social networks have allowed experts and non-experts alike to interact across the fields of art, design, science, and technology in order to use the research done by others and look into alternative contributions to their own research. The forums, such as the Australian Network for Art and Technology[11] for art sciences research have potentially influenced the idiosyncrasies of science production experiences. Increasing interest 'among scientists in interdisciplinary projects at the interface between art and culture'[12] may see the inclusion of aesthetic approaches to resolving a problem or at the very least offer multiple avenues for communication. Such is the case in the fight against plasticised fish and the unacceptable status of the 'plastisphere'. The exhibition and book, *Mare Plasticum* (Streit-Bianchi et al., 2020) reaffirm scientific purpose as an art intervention drawing on marine biologists, ecotoxicologists, oceanographers, mathematicians/modellers, chemists, and physicists

**Figure 10.1** ATTRACT programmes connect industry and research organisations
*Source:* ATTRACT

---

[11] https://www.anat.org.au/about/.
[12] https://en.unesco.org/creativity/policy-monitoring-platform/european-digital-art-science.

motivated by the urgency to act. Aesthetic and experiential approaches to science that are more than or extend beyond the purpose or utility of the science as transferable to other or unintended outcomes are less likely to emerge from a single discipline.

Reflecting on the design science research lens, what can we learn from science's integrated approach to the analytical, logical, theoretical, and synthetic aspects—as well as the practical value and conversion in the real world and the aesthetic—from a design science research perspective? What can we learn about the subjective experience of science? Which includes pleasure and flow (Baskerville et al. 2018)?

A snapshot of exemplars reaffirms the arduous journey from fundamental science towards unexpected permutations as a result of historic forums including:

- www.jodi.org, 1995 (Joan Heemskirk and Dirk Paesmans), responding to the aesthetics of the then World Wide Web (WWW), revealed abstract, non-functional designs commenting on slippages in truth and misinformation represented by a non-user design. Jodi.org, like many other information arts and design contributions built a digital language, disrupting print and providing new paradigms for omnipresent visualisation in today's common work tasks such as maps, 3D digital worlds, the use of gamification and database aesthetics;
- *Software engineer network collectives*: a hive model approach to open software use, also generated new interface visualisation semantics and interaction born from social networking, code sharing, and publishing. This community of practice democratises open-source development for rapid ideation, project development, and distinct repertoires of practice and language. Functions also have new graphic semiotics and hybrid languages, such as *fork, pull, request, and re-use*, which describe in very general terms a process where you can copy, extract, re-use, redevelop, add on to another string of code and improve, live test, and generate something new;
- *A tradition of citizen science*: through digital networks improved accessibility as volunteer computing clouds, volunteer thinking; projects like Seti@home (extra-terrestrial life search), Einstein@home (gravitational wave detection), Folding@home (protein folding), and ClimatePrediction.net (large-scale modelling of Earth's climate), (Bonney et al., 2014).

The above examples underscore the sentiment of working for a better world, which is increasingly found at the intersections of disciplines (Malina, 2008). He names a future generation of creative individuals, thinkers, and builders, 'new Leonardo's' at the intersections and disciplinary boundaries. Malina's report on four decades of contributions at the intersection of art and science saw the emergence of new Leonardo's, individuals and teams responsible for a new cultural fabric who both develop meaningful art, drive new agendas in science and disrupt technological innovation to address hard problems (Malina, 1997; Malina, 2008).

These creative individuals include the new Leonardo's who are pursuing basic science for both fundamental research and the rising demand for influencing major, intricate, and urgent social innovation challenges. Today, there may be a new generation of Leonardo's who will seek inspiration by crossing disciplinary boundaries, investigating alternative methodologies, engaging with and developing novel aesthetics, and perhaps even influencing their discipline. New Leonardo's may at the boundaries or outside specific disciplines create new or hybrid disciplinary fields, inventions, and experiments to respond to their world and future. New levels of access on professional, social, and disciplinary fronts afford new Leonardo's access to technical, scientific, digital, aesthetic, and cultural differences. The convergence of disciplines, 'Alt.Art-Sci' by a 'networked' community of practice conversant with big data, distributed networks, and digital culture, is motivated out of purpose as hackers, makers, and people's science movements are motivated by communities of interest (Malina, 2011). The acceleration of social development is influenced by the digital, technical, and network world, where social media redefines new words in the dictionary, disrupting, and influencing applications of creativity, the experiential, and narratives for art and science (Rosa, 2010).

New Leonardo's look through new lenses and open up to new tools and new research methods. The new Leonardo's, the contributors to basic science, may similarly draw on other sources of inspiration or disciplinary expertise. CERN's research data analysis platform was created as a result of leveraging code and data recycling, as led by the REANA project (Beck et al., 2020: 20).

Futuristics suggest that in the transition from traditional Big Science to modern Big Science, it is essential to focus on innovation for global challenges, economic growth, and sustainability in addition to science (Hallonsten, 2016).

Part of this mandate is to instigate breakthrough innovation in society, which necessitates the investigation into socio-cultural models, resulting in art, design, and science synergies. Art- science synergies have shifted beyond the socialisation of science for breakthrough innovation through design-driven innovation research (Verganti, 2008). Design methodologies are incorporated invariably into science experiments consciously or unconsciously as part of Big Science construction process. This concept trickles down to the values of technology applications in society through the focus on 'how signs, languages, and symbolic elements are shaped and diffuse society' (Verganti, 2008: 442).

As Big Science innovation projects explore the meaning and value of technological applications for society, design methodologies integrated into science experiments with 'how signs, languages and symbolic elements are shaped and diffused in society' (Verganti, 2008: 442) in cultural production.

There are many questions that need answers or solutions. What peculiarities might have emerged as a result of science synergies involving Big Science, deep technology[13] and enmeshed sciences as well as information, communication, and/or computational communication technologies?

As Big Science includes theory, laboratory experiments, Gedanken or thought experiments, and computer experiments (Mitchell, 2009; Li Vigni, 2021), how have

---

[13] https://www.ansto.gov.au/work-with-us/nandin.

interactive, experiential, and visual modes of science evolved in both fundamental and applied forms? What possible influences, for example, of computer modelling and simulation have influenced information design and vice versa? For example, in order to codify and widely share and distribute data, New Leonardos, who are digital natives, have integrated interaction, visualisation, simulation, and gamification as a kind of communication aesthetic (Reas and Maeda, 2004).

What can we learn about the techniques and art of Big Science, which influences disciplines at the boundaries through aesthetic approaches and methods?

New Leonardo's are well understood for contributing to artistic responses to science, however, they may equally be found in scientific responses to art and design. What can we learn from the New Leonardo's, the researchers, scientists, and advocates known as moonshots whose teams are for the mission itself purely to seek and solve the mission? Moonshot missions, synonymous with the challenges of the moon landing, are likened to ambitious, radical, exploratory and ground-breaking projects or missions. What aspects of the craft, specifically the theory, the lab, the mind, and computer experiments, have an impact on the internal and external collaborations of moonshot missions?

Today's generation of new Leonardo's contributes to cultural diversity and draw on multiple digital network touchpoints for information. It is at the intersections of different disciplines where co-innovation is more likely such as the Manhattan Project, the Human Genome Project (HGP) and Large Hadron Collider (LHC) experiments at CERN (Beck et al., 2020). Arts at CERN programme, invites multiple disciplinary contributions in the world's largest laboratory of particle physics to explore 'notions of creativity, human ingenuity and curiosity'.[14]

Exploration of basic science through a diverse disciplinary lens of art, design, science, and technology, has demystified the discourse of basic science (Latour, 1987), socialising institutional co-operation and increasing mutual respect for interaction across disciplines. However, the new Leonardo's, working in basic science, are drivers and producers of a highly specialised discipline, grappling with the building blocks of the universe, engaged in a language of head splitting theories that are challenging to learn let alone comprehend for the layperson. They may yet benefit from art and science forums offering artefacts, methods, and narratives, reinforcing the value and practice of experimental or novel methods and resulting aesthetics. The age of accelerations, digital convergence, artificial intelligence, and social networks has seen new image schemes, (Johnson, 2008) and aesthetics derived from the visualisation of complex large amounts of data; thus, building systems of symbols and narratives. Artists, software engineers, information designers, and scientists build software and form networks to find 'unusual relationships between events and images' (Wilson, 2002: 19). The artefacts born out of computational science include information design, computational aesthetics, and design thinking approaches to human–computer interaction providing an arsenal for new

[14] https://arts.cern/.

Leonardo's to experience methods that include ambiguity, serendipity, and accident (Maeda, 2019).

Could the interdisciplinary dialogue and experimentation forums envision by Leonardo and now exemplified by places like IdeaSquare and CERN (see Chapters 6 and 14) foster young Leonardos, mitigating Negroponte's risk of a big idea famine (Negroponte, 2018)? Big Science questions require 'long time horizons and high risk' (Negroponte, 2018), and ongoing benefits from art, design, and science forums in the pursuit of basic science are part of a cultural discourse.

The new Leonardo's are more likely to be enriched, 'by standing on each other's shoulders' (Negroponte, 2018) and shoulders from diverse disciplines to create a culture that explores questions about the universe we live in and how to apply this knowledge.

Normative actions in science can likely gain new insights if they stand on the shoulders of idiosyncratic arts. Answering the call in the first volume of Leonardo is to 'ask not what the sciences can do for the arts, ask what the arts can do for the sciences' (Malina, 2008). Today, as the importance of the problem context rather than the discipline grows, debates about disciplinary heterogeneity continue. Clues towards transdisciplinary thinking for emerging and disciplinary subspecialties 'may emerge in response to paradigmatic shifts, scientific spill-overs, increases in available data or the development of new types of data' (OECD, 2020: 29).

Finally, the difficulty in communicating meaning through the experience of Big Science is diffusion, which will facilitate through rich dialogue explored in art, design, and science forums, regardless of disciplinary differences. When discussing the difficulties of building the new collider, Tara Shears, a physicist at the University of Liverpool, writes that 'we do know that the only way to find answers is by experiment and the only place to find them is where we haven't been able to look yet'. This statement highlights the tensions that exist in any discipline between facts, context, and the connections that make information meaningful (Castelvecchi and Gibney, 2020).

Scientific and cultural endeavour calls for new Leonardo's from all walks of life to invest in the pursuit of understanding science and ourselves. Many fields and interactions between art, science, and design have shown the value of viewing things via an aesthetic, subjective, abstract, and emotive lens.

## 10.4  Valuing Science and the Need for a New Paradigm

One outstanding characteristic of fundamental research is the substantial time lag between discoveries and their materialisation into tangible and pragmatic benefits for society (Stefik and Stefik, 2004; Lehman and Stanley, 2015). Let us briefly illustrate the case with Schrödinger's equation below (Fleisch, 2020). It is not the aim to enter into physics or mathematical considerations of Schrödinger's equation here, but we cannot resist the temptation of writing it down in its modern form.

Thus, if you are not a technically trained reader, please consider the strange signs below in the same way as an abstract painting:

$$H\psi = E\psi$$

(H=double derivative. E= energy of e- $\Psi$= wave function representing e-)

It's interesting that all of the supposedly quantum particles' locations are controlled by this 'intellectual' thing, which is mysterious but elegantly symmetric. One of them is the electron, which is the ultimate 'actor' behind the operation of all the electronics devices we use today, such as your laptop and smartphone. It was proposed in 1925 by the Austrian-Irish physicist Erwin Schrödinger. The first working device fully based on his equation was the point-contact transistor invented in 1947 by the American physicists John Bardeen and Walter Brattain while working under the American physicist William Shockley at Bell Labs. Nevertheless, the realisation of the first integrated circuit did not come until 1958, thanks to many different contributions from scientists of diverse disciplines, especially Jack Kilby while working at Texas Instruments and Robert Noyce who co-founded Fairchild Semiconductor and Intel Corporation. Nonetheless, the personal computer did not become a mass-market consumer electronic device until 1977, when the first microcomputers were released (Orton, 2004).

As a result, there has been a 52-year gap between Schrödinger's audacious leap of creativity and imagination and the first market-ready electronic devices. Today, the electronics industry segments itself into various sectors. Just as an example, in 2018, into semiconductors generated annual sales of over \$480 billion in 2018.[15] The largest sector, e-commerce, spawned over \$30 trillion in 2017.[16]

Certainly not bad for a fundamental physics equation!

The incredible journey from Schrödinger's equation to the first microcomputers and beyond, teaches several lessons. Here, we highlight two of them. The first is that the pathway from Fundamental Science discoveries through abstract thinking towards tangible and concrete market innovations is highly non-linear and serendipitous. The second is that it is necessary, in view of this non-linearity, to have a long-term economic perspective mind frame. Unfortunately, the traditional economic mind-set, tools, and paradigms are not always suitable in this respect. Therefore, as this chapter title suggests, there is a need for new paradigms in the realm of science, and more in particular, Big Science valuation. One of the, arguably, most illustrative examples is the so-called discount rates.[17]

We use here a context for discount rates as the one considered in Discounted Cash Flow analysis (DCF). DCF models are powerful, however even the most sophisticated DCF calculations have shortcomings as they are purely mechanical and the

---

[15] Semiconductor Industry Association. 5 February 2018. https://www.semiconductors.org/.
[16] United Nations Conference on Trade and Development. 29 March 2019. https://unctad.org/en/Pages/Home.aspx.
[17] Economics of Big Science was extensively discussed, please refer to https://cds.cern.ch/record/2744400?ln=en; for discount rates in various engineering economics books and http://sciencebusiness.net/sites/default/files/archive/eventsarchive/OpenScience/BigScience.pdf.

quantitative valuation tool is subject to the axiom 'garbage in, garbage out'. Big Science projects are extremely complex, large, and have various components that work as a system, which has many variables and uncertainties and often runs for decades. Therefore, it is difficult to accurately predict future cash flows and costs associated with such projects. However, DCF is a widely used valuation method, based on the concept of 'time value of money'[18] for some R&D projects. DCF is an economic analysis tool for estimating the value of an investment based on its expected future cash flows. It helps assessing the viability of a project or an investment by calculating the present value of expected future cash flows using a discount rate.

It would be beyond the purpose of this section to enter into a full analysis of DCF. Therefore, the intention is to illustrate, with a somehow detailed example in the DFC context and especially for the not-so-familiar reader, the unsuitability of classical economic tools for long-term thinking. The concept of discount rates serves the purpose since, as it is worth noticing, acknowledged economic experts have expressed concerns since early on. For example, Hardin states that:

> The economic theory of discounting is a completely rational theory. For short periods of time, it gives answers that seem intuitively right. For longer periods, we are not so sure.

(Hardin, 1981)

For getting our heads around discount rates, it is useful considering two economic concepts that we will illustrate below with examples as explained in *Revealed Time Preferences* and *Opportunity Costs* (Drupp et al., 2018) as explained in future value. The future value of a sum of money today is calculated by multiplying the amount of cash by a function of the expected rate of return over the expected time period. Future value works in the opposite way, discounting future cash flows to their present value.

Another important concept is the opportunity cost. In the economics language, how much an investment pays in the future relative to other potential uses of the same invested money is known as its opportunity cost.

The joined consideration of the *Revealed Time Preference* and the *Opportunity Costs* establishes in economic terms the discount of the value of future benefits. An intuitive way of thinking about by imagining yourself as an investor. Any returns you will receive in the future recede each year by some percentage with respect to their value today. The longer they materialise in the future, the more they decline with respect to their value today.

The discount rate is the percentage that a return (or benefit) declines in value each year into the future. Let us illustrate the inadequacy of discount rate reasoning when long-term thinking is required. In practical terms, discounting means that a gain or loss in 50 years, for example, would be valued, using a relatively low discount rate at 4%, at only 14% of its value now. Transposed to problems faced by the world,

---

[18] For a detailed discussion on these topics relevant to Big Science projects see Florio (2019).

imagine an environmental related damage cost of $1 billion in one hundred years from now. The use of discounting, again at 4%, means that such a loss would appear as $140 million in any economic appraisal including a traditional cost-benefit analysis related to the mitigation measures and related investments today enabling future environmental preservation (Pearce et al., 2003).

As we mentioned before, one of the main characteristics of fundamental research is its non-linearity and serendipity with respect to predicting what its future benefits will be worth in economic terms, how much those will be and when in the future they will materialise in the form of new innovations. Applying the strict logic of traditional economic concepts such as discount rates will lead us to the conclusion that fundamental research is worth very little as an investment today. Nevertheless, as we have exemplified with the case of Schrödinger's equations, the long-term benefits can be staggering. Therefore, a dilemma appears right in front of us.

A recent study conducted by Florio (2019), suggests cost-benefit analysis is more suitable and offers a systematic analysis of the benefits in terms of the social agents involved. The benefits to scientists, students, and postdoctoral researchers as well as the effect on firms of knowledge spill overs and the benefits to users of information technology and science-based innovation can be considered in such an analysis to show the benefits of funding fundamental knowledge creation. Perhaps the solution to this dilemma of valuing Basic Research starts by reconsidering the vision of what Big Science (and especially fundamental science) carries towards all of us, as humankind. Many academics offer intriguing starting points that might serve as the foundation for such an integrative vision.

First is considering scientific research, scientific knowledge, and technology as global public goods, to be cultivated for the benefit of humanity and accessible to all (Bishop, 2015). In this sense, science is placed on with equal footing as other socioeconomic rights such as education and healthcare. It considers science and technology beyond their purely utilitarian value, emphasising their intrinsic value as a means of expressing our human nature, and personality, and facilitating international understanding. The essence of this approach is that scientific and technological knowledge should be accessible to all, as a human right.

Second is considering knowledge in general and (fundamental) science in particular, as a shared resource within a 'commons' context (Hess and Ostrom, 2007). Such a context is defined as a complex knowledge-sharing ecosystem formed by diverse groups subject into social dilemmas. Therefore, knowledge, in its intangible form of ideas, thoughts, or wisdom, would fall in the category of a public inclusive good. For example, one person's use of knowledge (such as Einstein's theory of relativity) does not subtract from another person's capacity for using it as well. New governance modes and models are necessary for organising efficiently and openly the production, access, use, and preservation of knowledge. This is key, especially with the current and upcoming digital technologies and paradigms such as the internet and the World Wide Web (WWW).

Third is considering science as an Option Generator (Boisot et al., 2011). In this sense, the value of science is viewed as consisting of two parts, broadening the scope of purely conventional economic paradigms. The first part does consider science's net present value in relation to an identifiable and definable stream of future benefits. The second one is the option part, which fundamentally broadens the traditional economic paradigms by reflecting potential future opportunities that scientific knowledge might create at a later date that are not yet known or clearly defined. The 'vagueness' overcomes, at least partially, the above-described rigid paradigm of discount rates. For example, omitting the option value carried by fundamental science will seriously distort its appraisal process.[19] This distortion will detrimentally favour what is investable in the short-term (e.g. favourable with respect to a certain discount rate standard) over what is potentially achievable and capable of transforming established paradigms and markets in the short term.

## 10.5 Conclusions

This chapter examines the multifaceted and entangled relationship between Big Science and society. Varied perspectives on the complexity, richness, and multifaceted nature of Big Science relationships require analysis using multiple lenses and flexible valuation frameworks. Therefore, the challenge is open to the collaboration of novel and flexible paradigm models that can capture both ontological and epistemological aspects of scientific activity. The ideologies of individual actors and communities need to be factored in. A potential starting point could be to ask oneself why and how fundamental scientific activities pursued in Big Science embed human traits like curiosity and imagination? How is Big Science able to build collaborative communities? Who is willing to contribute to both collective and individual values and ambitions?

In times where there is an excessive focus on the practicality, productivity, and efficiency of science and technology, human beings leave little room for the type of curiosity and serendipity that lead to the discovery of transformational ideas and paradigms. Moreover, as the Covid-19 pandemic teaches us, the knowledge generated by Big Science is not a luxury but essential for tackling not only this global emergency but many others including climatic and planetary changes.

Perhaps, we may intermittently revisit Abraham Flexner's great essay *The Usefulness of Useless Knowledge* (Flexner, 2017), to remind ourselves about the dangerous tendency to forgo pure curiosity in favour of excessive pragmatism:

> We make ourselves no promises, but we cherish the hope that the unobstructed pursuit of useless knowledge will prove to have consequences in the future as in the past.

---

[19] This option value is intrinsically necessary precisely to consider the non-linear and serendipitous nature of the process taking a Fundamental Science discovery to generate market value.

# 11

# Big Science and Social Responsibility of the Digital World

*Ruediger Wink, Alberto Di Meglio, Marilena Streit-Bianchi, and Shantha Liyanage*

## 11.1  Big Science, Big Data, and Computing

Big Science is often synonymous with 'big data'. 'Big data' generally describes large amounts of data, high data rates, or particularly complicated or unstructured data (Heiss, 2019). The ways data are produced, processed, distributed, made accessible, and analysed are important parts of data management processes in Big Science. Those who are familiar with the Large Hadron Collider (LHC) at CERN know that particles collide in the LHC detectors approximately 1 billion times per second generating about one petabyte (1 million gigabytes) of collision data per second (Gaillard, 2017).

Not all data is useful, and it takes a painstaking effort to determine what data is useful and what is not so useful. Usefulness of data and information is determined by the types of questions asked—not any question, but the right type of question. However, even with advanced filtering techniques, enormous amount of about 330 million petabytes of scientific data from past and present high energy physics (HEP) experiments had been created at CERN by the beginning of 2019 (CERN, 2021b). With the massive amount of data generated by numerous collisions, scientists must be able to go through a process to determine how a rare process differs from a common one. Reliable statistical analysis and probability studies are necessary. An open-source tool set called ROOT[1] developed at CERN and Fermilab computes vast amounts of data very efficiently (See Figure 11.1).

Similarly, the collaboration of eight telescope observatories around the globe to generate the first image of a black hole 55 million light years away from Earth depended on capacities to manage huge amounts of data, as every night of observation generated 1 petabyte of data, which could not be sent via the internet but had to be transported as hard drives from place to place (Castelvecchi, 2019). In

[1] The ROOT system provides a data analysis framework and consists of a set of OO frameworks with all the functionality needed to handle and analyse large amounts of data in a very efficient way (see https://home.cern/news/news/computing/big-data-takes-root).

Ruediger Wink et al., *Big Science and Social Responsibility of the Digital World*. In: *Big Science, Innovation, and Societal Contributions*. Edited by: Shantha Liyanage, Markus Nordberg, and Marilena Streit-Bianchi, Oxford University Press.
© Ruediger Wink et al., (2024). DOI: 10.1093/oso/9780198881193.003.0012

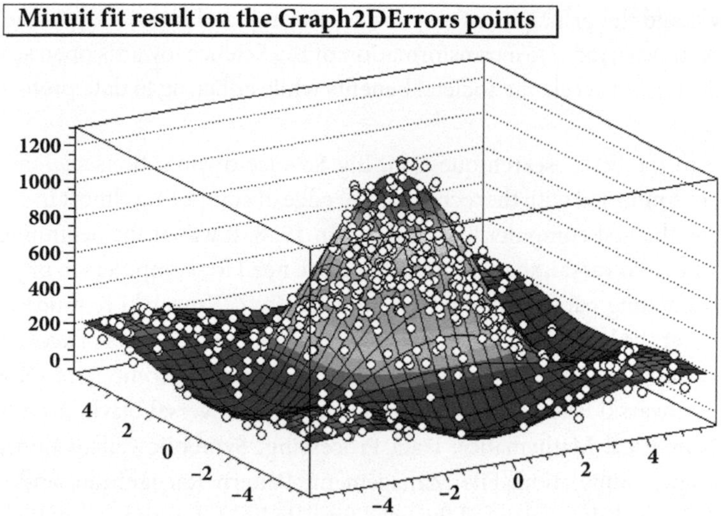

**Figure 11.1** An example of a plot created using the ROOT tool
*Source:* © CERN

fact, Big Science experiments produce extremely high volumes of data. The planned high luminosity LHC (HL-LHC) is expected to produce an annual data volume of approximately one exabyte[2] and the antennas of the Square Kilometre Array (SKA) will produce more than 100 terabytes per second on site and virtually online (Heiss, 2019).

This chapter focuses on three important connections between Big Science, big data, and overall technological developments that impact societal needs in the digital world: the long-lasting experiences in collaborations between the high energy physics community and Information Communication Technology (ICT) companies; the management of big data among scientific communities; and the transfer of knowledge and skills in big data infrastructures in an Open Science and open innovation context to initiate faster and more innovative solutions such as development of vaccines for Covid-19 pandemic. Consequentially, three research questions covered here are:

- Which organisational principles and elements were installed by Big Science organisations such as CERN, ESO, EMBL to maximise the mutual benefits from technological development in computing for the scientific community as well as for commercial partners from the ICT industry?
- Which principles help to organise and manage big data in Big Science projects, and how can these principles meet the expectations of policy and society?

[2] An exabyte is the equivalent of one quintillion bytes, one billion gigabytes, or one million terabytes (TB). In context with other units of digital data and storage: 8 bits equals one byte. 1,024 bytes equal one kilobyte (KB).

- How could the principles of responsive research and innovation in the digital age be transferred into a transformation of Big Science towards open science to maximise and accelerate societal benefits while adhering to data protocols?

Starting with the first research question, Big Science organisations such as CERN were closely connected with the technological edge of computing almost from its earliest days, as the first computer was installed in 1958. It was at the beginning of the 1970s that Lew Kowarski, shared his views on the need for computers in physics and wrote a stimulating paper entitled 'Computers: Why?'. Kowarski reminded us that 'we are only at the beginning to discover and explore the new ways of acquiring scientific knowledge which have been opened by the advent of computers' (Kowarski, 1972: 59). Kowarski listed eight applications for 'the universal black box at CERN', namely: Numerical Mathematics, Data Processing, Symbolic Calculations, Computer Graphics, Simulation, File Management, Pattern Recognition, and Process Control. This work demonstrated the need for data links, forerunners of high-speed networks, to transmit data between small online computers and larger ones in the central computer room.

Already starting in the 1960s, exchanging data between computers and external networking between scientific institutes were core challenges for high energy physics (Hemmer, 2018). The invention of the World Wide Web (WWW) by CERN was the solution to the urgent need of high energy physics community to share data via the Transmission Control Protocol/Internet Protocol (TCP/IP)[3] in a structured way. It is one of the most prominent examples of Big Science and Big computing (Hemmer, 2018). It led to experiences with the transition from big mainframe computers to decentralised networks of small computers at Fermilab (Melchor, 2021). CERN established the Worldwide LHC Computing Grid (WLCG), which allowed scientists to access and analyse data from anywhere in the world.

From an economic point of view, Big Science leads to big gains in knowledge productivity as more researchers with their ideas get access to useful knowledge. Generating, filtering, sharing, preserving, processing, and contextualising data is an important part of these options for scalability. This is one reason, why the close and early collaboration with software and computer firms was so important for the high energy physics community as well as for broader scientific communities in medicine and biology.

The high energy physics (HEP) community became familiar, relatively early on, with the use of common data formats, communication standards, and sharing practices. Due to historical experience in collaboration among the HEP community and ICT companies, formal organisational structures have been established (Zanella, 1990; Williams, 2004). For other scientific disciplines such as oceanography, human genome project, climate change, the transition towards common data management required considerable adjustment to social and organisational practices, making it

---

[3] CERN named Ben Segal as its first 'TCP/IP Coordinator' and the TCP/IP protocols (as Internet protocols were then called) were introduced at CERN in early 1990s, inside a Berkeley Unix system.

even more difficult to exploit the benefits of common research (Bos et al., 2007; Meyer, 2009). CERN and ESO are at the forefront of using big data infrastructure such as grid computing which facilitates faster dissemination, analysis, and retrieval of the enormous amount of data being produced by the LHC experiments and ESO detectors (see image of CERN computer centre in Figure 11.2). Such sophisticated infrastructure is not readily available for small laboratories and is a major drawback in analysing research data.

CERN's experiences with computing were transferred into biological research communities in 1994, when Paolo Zanella became the Director of the European Bioinformatics Institute (EBI) after having been the leader of the Computing and Data Handling Division at CERN during the period when major outcomes occurred with the start of computing online, computer automatic event recognition, PET development, and the World Wide Web (1976–1989). Zanella reported: 'The discovery of a new gene or the determination of the genomic variations related to a particular illness may have an impact on bio-medical research and on the pharmaceutical industry. The delay between a discovery and its effects on healthcare is shorter. This has to do with the strength of research performed by Industry and with the size of molecules, a billion times larger than that of quarks and leptons, thus resulting in less expensive research and development' (Zanella, 2014: 155),

In February 2001, both Nature and Science published the initial sequencing and analysis of the human genome. Nature published data from the international human genomic mapping consortium, which included several thousand Human

**Figure 11.2** CERN's Computing Centre
*Source:* © CERN

Genome Project (HGP) researchers led by Francis Collins (The International Human Genomic Mapping Consortium, 2001). Science published data from Craig Venter's joint private academic venture (Venter et al., 2001). The use of big data in biology, followed by a significant media launch and the sharing of bioinformatic information, was an important and unavoidable lever in many areas of research.

Similarly, the use of big data has been recognised as a critical prerequisite in neuroscience for better understanding of mouse and human brain functionalities (Koch and Jones, 2016).

An important step towards further collaboration between Big Science and leading global ICT firms was achieved by the foundation of CERN openlab in 2001. CERN openlab, which will be described in detail below, is a public–private partnership with different levels of intensity in the way that companies are integrated as members, ranging from partners (with at least a three-year commitment to a common programme of work), contributors (with a formal collaboration in joint tactical projects for one to three years) to associates with a formal collaboration on a specific joint targeted project (Di Meglio et al., 2017).

Current partners include the companies Google, Oracle, Micron, Siemens, and Intel. When companies partner with CERN openlab, they have access to challenging requests for their technological advancements. CERN Openlab's expertise stimulates the development of new ICT infrastructures and technologies, and it provides ideal testing conditions for new technologies in a demanding, pre-competitive environment (Grey, 2003; Hemmer, 2018).

Accordingly, CERN combines the roles of a lead user who defines avantgarde and specified needs and directions for technological solutions, a new technology testbed with challenging tasks and applications, and co-developers who collaborate with industrial labs to produce new and leading-edge technological solutions.

In CERN's White Paper to describe the priorities for the sixth three-year phase ending at the end of 2020, it has identified four major R&D priorities for CERN openlab (Di Meglio et al., 2017; Albrecht et al., 2017).

- Data centre technologies and infrastructures dealing with storage and processing needs for even extremely large scales of data generated in new scientific experiments;
- Computing performance and software to modernise coding techniques and optimise the exploitation of features offered by modern hardware architectures;
- Machine learning and data analytics to maximise the value to be generated from data while optimising resource usage;
- Applications in other scientific fields with large quantities of data and ICT challenges are comparable to high energy physics, like life sciences, medicine, astrophysics, and urban/environmental planning.

In 2020, CERN management announced the quantum technology initiative (QTI) as a new three-year activity to focus on further investments and research in quantum technologies, in particular quantum computing, with huge potential to be used in high energy physics as well as comparable other scientific fields (Di Meglio et al., 2020; Melchor, 2021). QTI has defined a medium- and long-term roadmap and research programme in collaboration with the HEP and quantum-technology research communities.

Again, the existing structures in CERN openlab with industry partners and other research institutes provide an excellent precondition for developing and exploiting the full potential of these new dimensions of computing (Di Meglio, 2021), and potential fields of applications in high energy physics (Tavernelli and Barkoutsos, 2021, on the industry perspective from IBM). CERN openlab is an important case example of Big Science and big data initiatives.

## 11.2  CERN openlab: Partnership in Scientific and Technological Innovation

### 11.2.1  What Is CERN openlab?

CERN openlab is a unique public–private partnership that works to accelerate the development of cutting-edge ICT solutions for the worldwide LHC community and wider scientific research. Through CERN openlab, CERN collaborates with leading ICT companies and research institutes. Within this framework, CERN provides access to its complex ICT infrastructure and its scientific and engineering experiences—in some cases this collaboration even extends to institutes worldwide. Testing in CERN's demanding environment provides the ICT industry collaborators with valuable feedback on their products and outcomes, while enabling CERN to assess the merits of new technologies in their early stages of development for possible future uses. In a similar way, research laboratories and academic institutes worldwide can join forces with CERN scientists and technologists to advance knowledge in computer and data sciences for large-scale scientific applications. This framework also offers a common ground for carrying out advanced research and development activities with more than one company or institute, thus accelerating innovation through collaboration and cooperation.

### 11.2.2  Brief History

CERN openlab was established in 2001 at the start of the construction of the Large Hadron Collider (LHC) at CERN to provide a framework through which CERN could collaborate with leading ICT companies to accelerate the development of cutting-edge ICT solutions needed by the HEP community. The complexity of the scientific instruments and infrastructures needed by CERN and the LHC experiments, presented extreme challenges and provided the ideal environment to carry out joint R&D projects and evaluation of new technologies in large-scale operations.

The joint collaborations were at the beginning aligned along a sequence of three-year phases, a balanced duration long enough to go beyond simple short-term investigation, but still short enough to ensure impact and deployment of results in production within reasonable life-cycle expectations of the technologies being assessed. The transition between phases would also provide a natural boundary to update the research priorities and collect requirements from the research community. Today the phase mechanism is still in place, CERN openlab has entered its seventh phase in January 2021. However, the projects are not necessarily aligned anymore with the phases in order to follow more closely the rhythms of the LHC experiments, and computing infrastructure upgrades, and the evolution of technologies and products.

The first four phases of CERN openlab between 2001 and 2013 were primarily focused on industrial collaborations in ICT and infrastructure technologies, from networks for distributed computing to computing platforms, from storage and databases to security for control systems.

The typical model for collaborations was based on a small number of large companies able to shape the broad technology landscape, such as Intel, Oracle, or IBM. As worldwide distributed infrastructures based on cloud computers became more established and reliable and new computing paradigms such as artificial intelligence and quantum computing started showing promising potential for scientific research, CERN openlab collaboration gradually extended into more academic research in computer and data science. In 2022, CERN openlab runs more than 30 different collaborative projects with both international companies and academic institutes worldwide.

## 11.2.3 Collaboration Principles: Win–Win Scenario

The CERN openlab collaborations are based on a few principles, or 'rules of engagement', designed to maximise the chances of achieving results and keep the parties engaged and committed for the duration of the projects. Every conversation is based on the rule that collaborations are purely focused on joint R&D and innovation. Explicit commercial interests must be left out of the discussion and taken up with the appropriate procurement services.

The second important rule is that all parties engaged in a project must have something to gain from the collaboration. For CERN and the physics experiment, the interest is to get access as early as possible to innovative technologies, be part of the development process and have direct channels to suggest requirements and improvements, and possibly get access to sponsorships and funding opportunities for students and researchers. For the companies, the interest must be related to the opportunity of assessing their technologies in CERN's challenging environment or applying them to computationally intensive use cases, so that they can be stress-tested and improved even before becoming products or services on the market. In addition, the association of a company brand with CERN brings tangible marketing value that companies have successfully exploited in many collaborations through case studies or joint participation at events and conferences.

Finally, the rule that 'there is no such thing as a free meal' must be respected. In order to ensure the continued commitment of all parties to the achievement of the agreed objectives, they must invest resources in the collaboration. Such resources include monetary contributions from the companies to the public activities organised by CERN openlab, such as conferences or training programmes or communications; contributions in the form of grants for dedicated researchers in the joint projects; and 'in-kind' contributions such as time of people, access to infrastructure or services, and hardware samples; and dedicated events at CERN to showcase the collaboration with a company and attract interest in the technologies under development.

The lifecycle of any new project goes through several steps, passing through a transition phase from the first introduction to the successful development of partnership. The typical sequence of the CERN openlab steps of the collaboration process is represented in Figure 11.3. The process starts always from with a technical requirement or, better yet, a challenging scientific or technological problem, that both the scientific community at CERN and the R&D teams in a company can relate to. The definition of the problem is followed by technical due diligence, which usually involves technical experts reviewing and brainstorming the technical importance of the problem and solution. If the discussion leads to positive outcomes, the project is deemed implementable on agreed timelines, objectives, and available resources. Finally, the project proceeded to form a collaboration agreement and contract to formally launch it with the approval of legal and financial services.

The parties can decide that the potential results are not worth the investment and explore other directions or part ways. When a project is finally technically and legally validated, it becomes binding on both sides. However, the process ensures that by then the parties share a clear understanding of the potential impact and 'return on investment'.

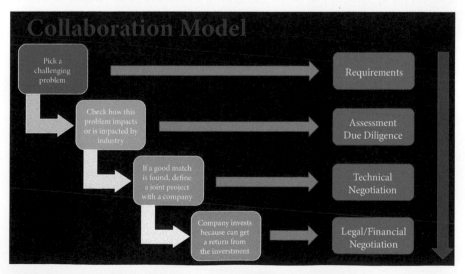

**Figure 11.3** The CERN openlab steps of the collaboration process
*Source:* Created by Alberto Di Meglio

During the first 20 years of the history of CERN openlab, on only two occasions was a collaboration terminated before the expected end of a project; once because of a lack of agreement on the intellectual property rights (IPR) of the project results, and once because the chosen technology did not live up to expectations very early in the project lifetime and no suitable alternatives could be found. These could have been avoided by a thorough assessment during the preparation phases.

## 11.2.4  Innovation, Knowledge Sharing, and Impacts

Although the primary goals of CERN openlab is to support the scientific research done at CERN in high energy physics applications, innovation in ICT infrastructure, computers, and data science is of general applicability and potential benefit in many other domains. Innovative multidisciplinary research projects have been developed during the past few years in a variety of disciplines, including earth observation, social and economic sciences, life science and medical research, and earth science.

The main principles remain the same, the starting point of every conversation is a common interest or challenge among the different parties. However, in this type of collaboration there is a strong element of knowledge-sharing across disciplines. The change of perspective that can be provided in multidisciplinary collaborations has proven to be very often the spark that ignites innovation. As an example, a project to 'modernise' the code used in radiation transport problems for high energy physics detectors, that is to optimise it for newer generations of high-performance multi-core architectures, sparked an interest in applying similar optimisation techniques in the simulation of biological cell development. A few initial projects were defined to assess the applicability of cancer cell growth simulation or to validate models of the development of eye retina structures. The activities attracted the interest of new experts, and a formal open-source community called BioDynaMo (Biology Dynamic Modeller) was formed in 2018. Applications to in-silico simulation of immunotherapy applications for the treatment of Alzheimer's disease with industry followed in 2019. In 2020, the agent-based simulation engine used to model cell dynamics was adapted to simulate epidemiologic models of Covid-19 propagation. Today Big Science systems are being adapted to model socio-economic conditions of city areas and populations and predict the possible impact of investments in transport, education, or healthcare systems.

The challenge for Big Science data is to converge different approaches on common problems and facilitate multidisciplinary teams to respond and collaborate on long term and complex questions. Such data convergent designs and models are useful because scholars from different disciplines are accustomed to using different research methodologies, whether qualitative, quantitative, simulation based, or involving interventions and action research (Smart et al., 2012). As discussed in Chapter 8, astrophysics organisations in particular have adopted policies and approaches to sharing data and facilities in an Open Science flat form.

The world has recently witnessed the rise of novel concepts and technologies that have the potential to revolutionise not only scientific inquiry but also industry and society. Artificial intelligence and new computing systems based on quantum technologies have moved from the laboratory to concrete applications, with a strong projected impact on physics research, chemistry, biology, finance, and the social sciences. However, questions of governance, fairness, and ethics cannot be ignored. CERN and its culture of collaboration between research, industry, and society can be a bridge between different disciplines and communities and a catalyst for open, fair innovation.

## 11.3  Challenges in Managing Big Data in Big Science

With the increased scale of data created in Big Science projects the management of processes to identify useful information within data, make information available along collaborative communities, communicate hypotheses and empirical results and disseminate insights to society became an incredibly complex task (Bicarregui et al., 2013, on practical challenges to implement data management processes in Big Science projects). With the term 'fourth paradigm of science' (Hey et al., 2009), data driven scientific research was even described as a new and different way of scientific exploration added to methods from empirical evidence, scientific theory, and computational science.

Big Science organisations and experiments have become increasingly data driven. These data programs now play a critical role in enabling and accelerating Open Science which open up a more collaborative culture that enables many possibilities for the open sharing and use of data, information and knowledge within and beyond the scientific community that generates such data.

Machine learning as well as artificial intelligence tasks as part of deep learning have been recognised as important tools to accelerate and extend the possibilities of identifying patterns and using big data sets (Hey et al., 2020). Experts in artificial intelligence and machine learning, however, still emphasise the limitations of these tools, when it comes to tasks like reasoning, planning, and acting in complex environments (Kersting and Meyer, 2018).

In general, the collaborative ethos with long traditions of sharing data as well as the strong awareness of the importance of excellent data management processes in Big Science projects are recognised as important preconditions for data management (Bicarregui et al., 2013). In the following, we look at the example of LIGO (Laser Interferometer Gravitational-Wave Observatory) in the US and its data management plan to describe the key ingredients of a systematic organisation of these processes. The Data Management Plan (DMP) of LIGO (2017) is a deliverable as stipulated

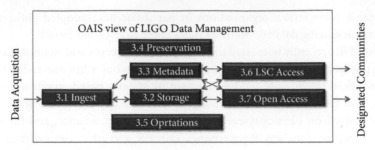

**Figure 11.4** LIGO Data Management Plan
*Source:* Courtesy Caltech/MIT/LIGO Laboratory

in the Cooperative Agreement with the National Science Foundation for the operation of LIGO (see Figure 11.4). The data plan describes how data from the LIGO instruments will be collated, stored, and made available to a range of communities of users of the data and should be preserved for future uses.

LIGO is working closely with the Gravitational Wave International Committee (GWIC) to engage with international partners to coordinate broader access to all gravitational wave data. The LIGO Scientific Collaboration (LSC) provides an avenue for the global community to be involved in the use and analysis of LIGO data.

LSC is a group of scientists focused on the direction of gravitational waves. Gravitational waves are used to explore the fundamental physics of gravity and the emerging field of gravitational wave science as a tool for astronomical discovery and the study of phenomena such as dark matter and dark energy. Such scientific collaborations are significant investments at the scale of Big Science operations. The data collection and analysis consist of multidimensional processes where information from all sources is shared and enriched within the organisation's protocols.

In the case of LIGO, open data and preservation are done in two phases: Phase 1 involves detection or discovery phases and Phase 2 involves observational phases. During Phase 1, data are released after careful vetting and in batches that detect, for example, the initial discoveries of gravitational waves. These data are continuously corrected and upgraded with better calibration and understanding of instrumental artefacts. As the data management plan explained, the data release to the broader research community makes the most sense for validated gravitational wave data surrounding confirmed discoveries. Detected events in this phase are released with sufficient data so that the broader community is able to comprehend the reported signals. In the observational phase, LSC work closely with the broader research community to establish data requirements, then build and field-test data access methods and tools. These methods and tools are provided to the broader community in Phase 2. The understanding of the data is continuously improved as the field progresses from initial discoveries to the more mature exploration of the astrophysical content of gravitational wave signals.

All Big Science organisations have sophisticated systems for releasing data for scientific research. For example, the LIGO Open Science Centre was created to build data products suitable for public release. LIGO's Open Archival Information System (OAIS) represents the storage, access, and preservation of data management for the scientific community. In this way, Big Science collaborations develop a system of production, dissemination, and storing of data in purposeful and systematic ways. Big Science organisations like CERN, LIGO, and Australian Nuclear Science and Technology Organisation (ANSTO) present real challenges for the creation and dissemination of useful knowledge and information. At the core of high energy physics experiments at LHC, CERN, and astrophysics detectors are the needs to generate data that provide useful information for the scientists. An extensive analysis of publicly available data provides more meaningful insights and knowledge that can be interpreted into new discoveries.

There is always a lag time before data is released to the public. Based on the experience of LIGO, the process of annotating the data that includes identifying artefacts, tagging hardware injections, providing accurate calibrations and performing analyses can take up to 18 months. At this point, an 18-month delay before public release is required for vetted gravitational wave data, at least for the first data from Advanced LIGO (LIGO, 2017: 7). Generally, there is a 12-month delay in data releases, during which time mission scientists have access to the data for analysis, vetting, and clean up before it is made available to the broader community. Similarly, it took two years for the virtual Event Horizon Telescope (EHT) project to start the first Earth-spanning observation campaign until the release of the first image of a black hole (Castelvecchi, 2019).

These data management plans are a reaction to challenges created by the complexity of big data as well as legal and political requirements. Government initiatives on open data, the privacy of personal data, and responsible research have a significant impact on data management issues because most Big Science projects require coordination between multiple governments and are dependent on public funding. A typical example of this is CERN and its wide range of international collaborators (Bicarregui et al., 2013). A closer examination of selected government initiatives is described in the next section.

## 11.4  Social Responsibility in the Digital Age

During the last three decades, technological progress in data generation, storage, processing, and assessment has stimulated a huge variety of communities in different scientific fields outside high energy physics towards the development of international collaboration projects paving the way towards Big Science (Ulnicane, 2020; Cramer et al., 2020b). Typical fields include molecular biology and neurosciences as well as climate research or spatial environmental planning. These directions towards investments in common basic research infrastructures were accompanied by intensified debates on the expected results from these investments. In the USA, controversies over government spending on basic research culminated in 2010, when the House Majority Leader issued a website called 'you cut' allowing citizens to vote for cutting Federal programmes, including basic research (Fahrenthold, 2013). Political decisions on the allocation of public R&D have been increasingly based on partisan political rationales (see von Schomberg, 2013).

In the European Union, 'responsible research and innovation' became the most influential catchphrase to define political expectations for basic research programmes in the context of Big Science (on the various dimensions of this term Burget et al., 2017; Ulnicane, 2020b). Within its EU R&D Framework Program 'Horizon 2020' the EU defined responsible research and innovation as the main focus. However, the underlying principles and their relevance in the context of digitalisation remained vague at the beginning of the programme. Four specific fields of challenges occurring in the digital age were identified by different authors and public authorities (Stahl, 2013; Elliot and Resnik, 2014 and 2019; Resnik and Elliot, 2016; Filipovic, 2018; Fothergill et al., 2019; Inverardi, 2019; NQIT, 2019; European Commission, 2020):

- Issues of data generation, (re-)use, storage, protection, and privacy;
- Issues of intellectual property;
- Issues of transparency to maintain trust in the verifiability of scientific results based on artificial intelligence and machine learning; and
- Issues of inclusive education and access to scientific data and information.

In the next section, recent developments and challenges in public health surveillance are introduced as a typical example for scientific fields, where 'responsible research and innovation' serves as important ingredients of strategies to cope with new opportunities and (technological as well as regulatory) challenges in the digital world.

## 11.5 Public Health Surveillance in the Digital World

Public Health Surveillance is 'the continuous, systematic collection, analysis and interpretation of health-related data' (WHO, 2021). Public health surveillance needs led to the definition and establishment of new research directions at the end of the 1990s. Thanks to the development of data management, storage, availability, and the inception of the cloud, an important step from conception to implementation in healthcare has been made possible. The development of online health surveillance platforms was facilitated by the ability to combine and coordinate distributed and heterogeneous computational resources, by gaining access to more advanced information systems and distributed services, by enjoying increased computing and storage capacity, and by being able to use more powerful computers and storage devices.

Since these first emerging steps, big data lakes have been available and used around the world. Numerous epidemiological and scientific queries on knowledge and treatments of specific diseases are now possible due to the implementation of technological findings from high energy physics experiments. One of the outcomes worth mentioning is the scientific grid infrastructure, GÉANT, the broadband and highspeed network for research and education.[4]

Generally speaking, deep-techs such as big data, blockchain, and artificial intelligence are now combined and at the core of research and innovation in many strategic activities of national and private business-oriented institutions and organisations: healthcare services, personalised prevention, the management of healthcare, legal security of digital transactions, or archiving all benefit from disrupting technological approaches. As an example, the recent collaboration at CERN openlab with Beys Research, resulted in a tripartite research team developing a platform based on the blockchain to support European research centres and hospitals in constituting 'Living Labs' where sensitive research data can securely be shared (Frisoni et al. 2011).

Beyond the Living Lab project, disruptive quantum computing has also become a field of collaboration in the public health context. This is the case for Quantumacy,[5] a project launched in January 2021 by Beys Research and CERN openlab.

Quantumacy stands for QUANTUM key distribution for large-scale informational self- determination, privacy awareness, and the distributed processing of personal sensitive medical data.Quantumacy aspires to support the transition to confidential computing and informational self-determination by integrating relevant technologies into core Quantum Key Distribution (QKD)-enabled services requiring key issuance and distribution, resulting in a new approach to making data available and processing under the so-called polymorphic privacy principle.

To that end, Quantumacy is developing a proof-of-concept platform in which existing encryption techniques are combined and extended to use the open QKD infrastructure across the platform's different layers, allowing for testing on concrete healthcare use-cases.

Combining privacy-preserving and enabling technologies with blockchain to secure data registration and exchanges across the entire data lifecycle, along with

---

[4] GÉANT develops and operates a range of connectivity, cloud, and identity services that ensure a safe and secure environment for researchers, educators, and students. See https://geant.org/.

[5] Quantumacy is a privacy-preserving data analytics platform combining the security of Quantum Key Distribution (QKD) protocols QKD protocols and links with state-of-the art homomorphic capabilities to execute machine learning and deep-learning workloads across a distributed federated-learning infrastructure (CERN, 2022, https://quantumacy.cern.ch/home).

full audit trails, while relying on a QKD-enabled network infrastructure, will result in unprecedentedly secure, reliable, and unfalsifiable information systems in the future.

Cancer screening, distant analysis of medical dossiers, and personalised medicine are areas where recent improvements in data analysis and sharing have acted as catalyst for innovative solutions in the health sector. The dawn of all those developments including the assessment of the genome of individuals, the functional genomics developments, and following data collection, storage, and sharing for personalised medicine brings with it many extremely important ethical questions in terms of security and privacy. Diseases with high societal and economic impact, such as cancer, neurodegenerative conditions like Alzheimer diseases, Parkinson diseases, or Soluble Liver Antigen (SLA) pathologies require innovative research methodologies using digitalisation for treatment, follow-up, data analysis, and epidemiological assessment (Fogel and Kvedar, 2018).

In Europe, software developments that handle personal data must adhere to the General Data Protection Regulation (GDPR). These regulations provide data protection principles and new individual rights in business processes. Different approaches to privacy and confidentiality preservation are being explored today. An emerging concept in the field is the Private or Confidential Computing, whereby data is protected (e.g. by means of encryption or anonymisation) across all stages of the data life cycle.

Several commercial cloud providers such as Google or Microsoft have recently proposed pilots using encrypted virtual machines. The future of large-scale data analysis systems, especially in areas where sensitive data is processed, requires new secure ways of protecting data. Typically, a Confidential Computing platform protects four areas of activities:

(1) *Data at rest*: protect data stored in repositories—typically using suitable encryption systems;

(2) *Data in transit*: protecting data as it is transferred across the network, typically by encrypting the contents, the channel, or both;

(3) *Data processing*: protecting data and models as they are analysed/trained/inferred, using techniques such as Secure Multi Party Computation protocols (SMPC) or ML/DL-compatible encryption schemes (homomorphic encryption, functional encryption); and

(4) *Operations*: recording and auditing transactions, managing AuthN/AuthZ,[6] provenance, etc., typically using technologies like blockchain ledgers, x509 security frameworks,[7] etc.

Against this backdrop of growing needs and opportunities to collect and connect big data for public health surveillance while ensuring the protection of individual personal data in the digital world, CERN, and other Big Science organisations' efforts

---

[6] AuthN is short for authentication and AuthZ is short for authorisation.

[7] *X.509* 'certificates are used in many Internet protocols, including TLS/SSL, which is the basis for HTTPS, the *secure* protocol for browsing the web'.

to build Open Science platforms which serve as critical building blocks for innovative and rapid public health strategies, are explained in the following section.

## 11.6  Open Science and Covid-19 (CERN and EMBL)

Ideally, all high energy physicists wish to share information openly about their research as enshrined in the values of CERN's original charter. CERN values recognise the universal importance of the fundamental scientific research knowledge produced at CERN to make this knowledge available to everybody for the benefit of the society.

The core foundation of Open Science[8] is making publicly funded research data results available for the public, education and development, and the swift transfer of scientific knowledge. Initiatives have been growing including Open Science access, Open Data, Open Sources and Open Science projects. All of these initiatives aim to disseminate knowledge for the public good. The dissemination of knowledge has been facilitated by the availability of tools and means that facilitate the dissemination, sharing and deposition of big data and information. CERN *Open Science Policy*, adopted in October 2022, provides provisions for all research publications and experimental data, and research software and hardware are to be made publicly available under the new policy, which also incorporates the previous rules for open access, open data, and open source software and hardware. It is important to note that the term 'Open Science' is not just about opening science in a way that makes knowledge accessible without cost for everyone (Naim et al., 2020).

CERN promoted volunteer initiatives such as Rosetta@home and Folding@home, as well as LHC experiment technologies such as the File Transfer System (FTS) and Rucio, a modular, scalable solution for searching high energy physics data files in distributed data centres for monitoring and analysis. Furthermore Zenodo, the open-data repository built and operated by CERN and OpenAIRE to ensure that everyone can join in Open Science, has been extended to research communities collecting research output and information for Covid-19 and SARS-CoV-2, as we will describe at the end of this section.

CERN and EMBL, as members of the European Intergovernmental Research Organisation forum (EIROForum, https://www.eiroforum.org/), have extensive experience in operating data-services for the international community over several decades and are key actors in data intensive science, curating some of Europe's most important and popular datasets. All members of the EIROforum are committed to ensuring their openly available scientific datasets are curated over the long term and maintained through certified digital repositories. Thus, CERN and EMBL, together with the other EIROforum members have stimulated the Open Science movement across Europe and have shown a strong involvement in the European Open Science

---

[8] Open Science is a term often coined to Steve Mann in 1998 (Wright, 2020), but also already mentioned by Chubin (Chubin, 1985).

Cloud. Open access to research data proved to be not only a selected benefit for the relevant scientific communities dispersed around the globe but also an important stimulus for interdisciplinary science and transversal dynamic research and education. The outcome of such Open Science initiatives has had and will continue to have implications and outcomes for society worldwide. We will now look closer at the variety of initiatives at CERN and EMBL.

## 11.6.1 CERN

CERN's convention states: 'The organisation shall have no concern with work for military requirements and the results of its experimental and theoretical work shall be published or otherwise made generally available.'

Open Access has a long tradition at CERN. Since its inception in 1954, particle physicists at CERN have shared their results with fellow scientists across the world by distributing pre-prints. With the introduction of *arXiv* in the 1990s electronic distribution of pre-prints became the norm and since 2014, the Sponsoring Consortium for Open Access Publishing in Particle Physics *(SCOAP)*, hosted at CERN, has made around 90% of all high energy physics peer reviewed journal articles available on open access.

CERN pioneered the largest open access initiative of research data from high energy physics in the world. It promoted the dissemination of research results by hosting SCOAP[3] with over 3000 libraries and research institutes from 46 funding agencies and research institutions from 46 countries and intergovernmental organisations like the European Commission and UNESCO. The aims of these activities go far beyond particle physics. From December 2020, the establishment of a new open data policy to further support Open Science on scientific level 3 data (the type required for scientific studies) from LHC experiments will be released alongside the software and documentation needed to use the data. The data is made accessible via the CERN Open Data Portal.

## 11.6.2 EMBL and EMBL-EBI

The European Molecular Biology Laboratory (EMBL) was founded in 1973 with the key mission of carrying out basic research in the life sciences. EMBL endorses Open Science as an integral part of its research activities, as underlined already in its Establishing Agreement of 1973, Article 2§1:

> The Laboratory shall promote co-operation among the European States in fundamental research, in the development of advanced instrumentation and in advanced teaching in molecular biology as well as in other areas of research essentially related thereto, and to this end shall concentrate its activities on work not

normally or easily carried out in national institutions. The results of the experimental and theoretical work of the Laboratory shall be published or otherwise made generally available.

EMBL-EBI is the European Bioinformatics Institute, based at the Welcome Genome Campus, Hinxton, Cambridgeshire, UK. It is one of the sites of the European Molecular Biology Laboratory (EMBL) and a world-renowned computational biology institute. Since its inception the newly created EMBL has needed leading-edge computing resources to deal with data collection, analysis, and sharing from different projects carried out by the many Member States within their projects. EMBL's biological data resources on nucleotide (EMBLBank) and protein databanks were initially hosted in Heidelberg until the migration of data and services to the site in the UK, thus establishing the EMBL-EBI. Biocomputing activities and computers are integral part of the research activities throughout EMBL. EMBL-EBI, leading computational biological research since 1993, is 'The home for big data in Biology'. Their data portal allows the user to browse data, perform analysis, and share research among the molecular biology community. Another important mission for EMBL-EBI is the delivery of training in data-driven life sciences.

Open Targets is a public–private partnership that uses genomic data for drug target identification with the aim of helping scientists develop new, safe, and effective drugs. The Open Targets Platform and the Open Targets Genetic Portal are the two sectors containing repositories. In the Platform, information is available for some types of illnesses spanning from RNA expression, genetic, drugs, text mining to animal studies, whereas the Genetic Portal enables the user to explore variant, gene, and trait associations. This fundamental knowledge will allow cohort-based, personalised medicine, and personalised treatment to become part of our daily lives, for which, all research groups need and use computational approaches to drive new experiments. EMBL-EBI's Industry Program has been a key mission since its founding, allowing regular contacts between research groups and pharmaceutical and biotech companies giving outcomes of paramount importance. The results obtained from the joint projects carried out in such contexts are also publicly available.

In addition, EMBL has commitments to Open Science in its Open Science Policy and as a signatory to DORA: the San Francisco Declaration on Research Assessment (https://sfdora.org) upholding the principles and processes defined in the declaration in all its sites.

## 11.6.3 EOSC

Alongside its commitment to Open Science, EMBL together with CERN and other members of EIROforum is involved in the continuous development of the *European Open Science Cloud* (EOSC). EOSC is a joint initiative between the European Commission, the EU member states and associated countries, and other stakeholders to provide open access to publicly funded research data across scientific domains and without geographical boundaries, including services addressing the whole research data life cycle. Developing iteratively, it aims at finding, accessing, combining, analysing, processing, and storing data in line with Open Science and FAIR principles (findable, accessible, interoperable, reusable). By bringing all available information and data together, it will maximise the impact of publicly funded European research and boost applications.

EOSC can be considered an important catalyst for Open Science and a major player in implementing FAIR data. Integrating EMBL's and CERN's rich open data resources with EOSC aims not only to further underline the organisations' commitment to Open Science but can also act as a catalyst to widen the use of their data resources and make them more accessible beyond the respective physics and life sciences domains. EOSC acts as the missing link between EIROforum's own data resources and those data sets provided by other disciplines, that are not readily accessible, thus fostering transdisciplinary research. Beyond the federation of data, EOSC enables more effective coordination of existing and future research infrastructures and e-infrastructures at the European level with the opportunity to connect with major centres and national initiatives in the European Member States and beyond. Consequently, EOSC aims to enable the federation of the organisations' existing and future computing services to support the distributed analysis of large-scale data. Additionally, EMBL, CERN, and others will be able to provide and also benefit from coordinated training, information and dissemination opportunities, particularly related to data science and FAIR data management.

Even before the EOSC initiative had started, CERN and EMBL were two founding members involved in the Helix Nebula Science Cloud initiative. A forerunner for EOSC, the initiative established a successful pilot cloud platform linking together commercial cloud service providers with the IT resources of ten leading European research centres in the areas of astronomy, high energy physics, life sciences, and photon/neutron sciences.

Throughout its different phases, CERN and EMBL have significantly contributed to the development of EOSC, e.g. through coordination of and participation in key EOSC projects of the European Framework Programmes Horizon 2020 and Horizon Europe (EOSC-Pilot, EOSCHub, EOSCEnhance, EOSC-Life, Covid-19 Data Portal, and others). Additionally, both organisations have been heavily involved in shaping the implementation of EOSC, playing major roles at the EOSC governance level. EMBL was represented on the EOSC Executive Board in 2019–2020 and co-led the Board's Working Group on EOSC Sustainability, which significantly contributed to implementing the EOSC Association. CERN is a member of the EOSC Association directorate.

The EOSC Association AISBL was established to sign a co-programmed European Partnership with the European Commission under the Horizon Europe Framework Programme. The EOSC Association is expected to play a crucial role in EOSC's development by enabling funding by the Commission as well as contributions by

EU Member States and Associated Countries throughout the duration of Horizon Europe through a Strategic Research and Innovation Agenda (SRIA). Another important aspect is that the EOSC Association brings together a rapidly growing number of stakeholders as Association members and both EMBL and CERN became full members of the EOSC Association at the first General Assembly in 2020, joined by other EIROforum organisations. Both organisations continue to be involved in different strategic efforts to shape EOSC, e.g. through the SRIA and by participation in the Association's Advisory Groups.

This engagement has not only brought additional visibility in areas where e.g. EMBL was less prominently known before, but has significantly shaped the direction in which EOSC is moving forward. Setting up EOSC is an activity involving a large variety of communities with different maturity levels in their data and services. EMBL, CERN, and other Big Science centres are forerunners with regards to cloud-based, large-scale, and globally distributed research activities. Aligning EMBL and CERN with EOSC is therefore both a challenge and an opportunity for both organisations to act as role models for others.

## 11.6.4 Covid-19

The Covid-19 outbreak and continuing pandemic called for rapid and synchronised interactions by sharing data and research results across borders. Open Science and the tools and infrastructures already developed to share scientific information and data rapidly responded to this critical and important need facilitating the sharing, analysis, monitoring, and assessment of breakthrough advancements to explore and find possible solutions. In April 2020, EMBL-EBI launched the Covid-19 Data Portal to include a wide range of data types including genomics, protein and microscope data, as well as a repository of scientific literature. One can be impressed by the rapidity acquired in analysing the virus variants, but this is the result of the strong involvement of EMBL in European Open Science in the development and sharing of many years of previous genomic research and ad hoc development of technologies, as well as the availability and freely shared international scientific information in a dedicated and rapidly designed secure portal. CERN also got involved in ventilator and computer technology to deal with the Covid-19 pandemic. Following the request for actions from the European Commission, Zenodo and OpenAIRE joined their competencies and collaborated to make scientific information available via the OpenAIRE Covid-19 Research Gateway. The OpenAIRE Covid-19 Research Gateway provides a single point of access where publications, data, software, and other research outcomes are made available due to the collaboration with pan European research infrastructures and national and international alliances. Trusted sources and scientific material with bibliographical references are made freely available. All these initiatives to accelerate and facilitate the exchange of leading-edge scientific knowledge formed the basis for the relatively fast responses to the challenges caused by the global pandemic.

## 11.7  Big Science's Contributions to Societal Challenges

At the beginning of this chapter, we introduced three research questions to guide us through our observations:

- which organisational principles and elements were installed by CERN to maximise the mutual benefits from technological development in computing for the scientific community as well as the commercial partners from the IT industry?;
- which principles help to organise and manage big data in Big Science projects, and how can these principles meet the expectations of policy and society?; and
- how could the principles of responsive research and innovation in the digital age be transferred into a transformation of Big Science towards Open Science to maximise and accelerate societal benefits while protecting personal data?

CERN openlab provides key messages to answer the first question. Transparent processes and joint agreements with commercial partners on exclusively pre-commercial activities and mutual benefits served as foundations for joint and collaborative research on leading-edge technologies for the digital age. The example of the data management plan at LIGO took us to the answer to the second question.

Despite massive and continuously growing volumes of new data, it is called for new data management protocols to facilitate scientific progress. These data protocols also stimulate data management and linkage systems in other scientific fields such as medicine and other social research environments by improving tools to filter and combine meaningful data and forming open data infrastructures.

Our examples from public health surveillance, particularly their applications to research during the current Covid-19 pandemic, helped to answer the third research question. The emergence of open data infrastructures to be used in transdisciplinary environments and the guarantee of protection for personal data by the FAIR principles formed the preconditions for fast reactions to unprecedented research questions during the pandemic.

Especially in times of sovereign budget cuts and austerity policies, funding Big Science projects comes under high scrutiny. The digital age cannot be imagined without the important technological and social developments initiated by collaborations between scientific communities, industrial partners from the IT sector as well as governments and political organisations. From the first steps of computing to collaborations using the Internet and the future of quantum computing, Big Science organisations such as CERN and ESO provide stimuli for new technological developments, create challenging innovation environments to test leading-edge appliances and stimulate creative ideas for new forms of use and applications. The case of CERN openlab reveals the potential of clearly focused organisational frameworks for Big Science–big industry collaborations in the context of pre-market developments and testing. Programmes, processes, and learning environments were developed to maximise the positive impact on technological infrastructures for scientific research as well as on innovative output in the IT sector.

Simultaneously, the transition towards big data led to new challenges for responsive scientific research. These examples from public health emphasise the importance of Open Science infrastructures using experiences from scientific communities in high energy physics as well as bioinformatics. These infrastructures do not only help to follow the pathways of well-ordered science but also to initiate fast common scientific efforts in urgent crises like the Covid-19 pandemic. They also serve as a model for how the FAIR principles of dealing with big data (findable, accessible, interoperable, reusable) can be implemented in a transparent way. These experiences form the basis for broad applications of novel regulatory approaches in commercial uses of big data. Again, the experiences of Big Science communities can help to understand and imagine how the protection of privacy in everyday digital applications based on big data can be achieved in a meaningful way. A challenge, however, remains for any societal use of open data infrastructures, as the collaborative ethos and mindsets of Big Science projects are missing in commercial and competitive contexts.

## 11.8 Conclusions

We live in a digital society that requires fast and rapid changes due to the way we access, share, and use data. The LHC, LIGO, and other large telescopes produce enormous amounts of data. The present and next-generation Big Science instruments are even going to be laden with massive data generators. The quest for collecting, storing, and analysing data is a challenging task.

Volumes of data, data analysis, and data management in Big Science from HEP to astronomy and molecular biology became ever more important and highly sophisticated. As detection, computing, and digital technology advances, large volumes of data can be captured and stored and need to be disseminated to collaborations located worldwide. Therefore, in Big Science organisations and experiments, well designed research, data handling methodologies and coordinated approaches to carry out and execute the experiments as well as transfer the collected information to the stakeholders are vital and necessary to establish from the inception.

It is important to outline the joint partnerships that Big Science has established with leading IT companies and other research institutes, with openlab at CERN and the industry programmes at EMBL. It is impossible to predict how newly developed algorithms, rare signal discrimination, and methodologies to reconstruct images used in high energy physics and astrophysics will affect our daily lives in the near future. Furthermore, the more intriguing question is how the massive volume of data produced in cosmology or by the LHC experiments can be used and contribute to the broader physics and astrophysics communities and/or educational purposes by companies for practical use and public benefit. The LIGO and ESO for example, have data protocols for more public access to all their data within accepted data management protocols.

Another challenge for IT is how to design memories that are fast, large, and responsive. The ATLAS at CERN has already generated 140 petabytes of data, distributed between 100 different computing centres, with most of it concentrated in 10 large computing centres like CERN and Brookhaven. New large-scale astrophysics projects are producing data and information at unprecedented scales and facing computing challenges at the 'exascale' level and beyond. Astrophysics now uses large data surveys like the Dark Energy Survey, Sloan Digital Sky Survey, and will generate petabytes of data from LIGO, telescopes such as the Square Kilometre Array and the Large Synoptic Survey Telescope. The high energy physics community and astrophysicists must define data collection and storage and use strategies to deal with the massive amounts of data that these infrastructures will generate over the next few years.

Furthermore, this calls for the harmonisation of data usage and ongoing efforts are being made by national governments and the European community. In EMBL's Open Science initiatives, for example in the development of the Covid-19 Data Portal or in the development of the European Open Science Cloud (EOSC), the EU member states and associated countries have collaborated on open data usage.

Big data and Big Science are closely coupled and have the potential to revolutionise the way we organise and conduct research. Open access to publicly funded research data across scientific domains and without geographical boundaries, including services addressing the whole research data life cycle, is increasingly becoming valuable for science-making inroads to social, educational, and economic development. Rapid progress in artificial intelligence with neural networks will also assist societies in making use of data from Big Science projects.

# 12

# Well-ordered Big Science, Innovation, and Social Entrepreneurship

*Faiz Shah, Beatrice Bressan, Pablo Garcia Tello, Marilena Streit-Bianchi, and Shantha Liyanage*

## 12.1 Introduction

Big Science, usually means 'Big Dollars, Big Machines, Big Collaboration' and contributes to the advancement of knowledge in significant ways. Big Science is commonly associated with particle accelerators and large telescopes, but it also includes Big social, environmental, and information technology concerns such as climate change, human genome research, and artificial intelligence.

Several countries and research communities have collaborated towards achieving scientific advances that no single country or research group in one country could have produced on their own. Such collaborations have unique challenges, but left alone, the scientific community has to deal with them. Knowledge exchanges follow common academic and research practices regardless of ties to politics or geography. In the real world, however, the dynamics of knowledge exchanges are often tempered with political alignments, business interests, and legal considerations.

Volatile socially disruptive occurrences in a number of countries have recently increased, creating unrest and suspicion among and within the international scientific community. A new polarisation in global politics is resulting in even lower levels of collaborative knowledge-sharing, which will have a negative impact on how scientists conduct Big Science and transfer its beneficial outcomes to society.

The Covid-19 pandemic has exacerbated these disruptive tendencies even further. It might take even longer to close the knowledge transfer gap as physical isolation, ongoing mobility restrictions, and socioeconomic survival become more urgent issues that need to be addressed by governments and institutions (including those involved in research and outreach) during the recovery phases.

Ironically, the recovery from Covid-19 may make it the perfect opportunity to invest in Big Science discoveries that benefit humanity greatly and improve quality of life.

This chapter highlights examples of Big Science that trigger innovations that impact social wellbeing. Social, technological, and organisational innovation—together with entrepreneurship to close the gap between fundamental research

Faiz Shah et al., *Well-ordered Big Science, Innovation, and Social Entrepreneurship*. In: *Big Science, Innovation, and Societal Contributions*. Edited by: Shantha Liyanage, Markus Nordberg, and Marilena Streit-Bianchi, Oxford University Press.
© Faiz Shah et al., (2024). DOI: 10.1093/oso/9780198881193.003.0013

and its application to greater benefits for society—are explained. The discussion shifts from establishing the contextual relationship between scientific research and quality of life to examples of how successes in breakthrough Big Science have led to grassroots solutions. These examples show not only earth-shattering scientific and technological solutions but also explain how social innovations can engineer grassroots responses.

## 12.2  Role of Entrepreneurship in Shaping People and Nations

The progress and prosperity of human civilisation have always been linked to wealth and industry. The evolution of society, from tribes to communities to nation-states, is a chronicle of individual initiatives and self-interest in the pursuit of wealth.

Such self-interest propels the nation's wealth, ensuring peace, stability, and social welfare. In essence, society's history has always been the tale of entrepreneurship, with particular emphasis on the creation and preservation of wealth.

The economic prosperity of countries is inseparable from their social development. The inclusion and involvement that have come to define wealthy societies are those in which all individuals share a relatively more equal distribution of the benefits of development. High levels of healthcare and education for citizens, as well as state guarantees of security and equal rights to citizenship, are some of the indicators of social well-being. Economic prosperity, on the other hand, is closely related to science and technology development. With such development, we witness a well-developed workforce, gainful employment opportunities and disposable incomes, a higher standard of living, efficient public infrastructure and public services, good health care, welfare, and education systems that can be sustainable in a viable economy.

The concept of balanced growth, also known as 'the big push', was highlighted in post-war studies by Nurkse (1953), Scitovsky (1954), and Fleming (1955). According to this concept, economies that are in between the pre-industrial and industrialised paradigms can advance into a higher gear by merely absorbing technological advantages that have been developed at a high cost by others.

India's leap into the digital economy, Bangladesh's move into light manufacturing, and Thailand's evolution as the 'Detroit of Asia' are examples where entire populations have experienced quality of life changes because of this technological leap (Murphy et al., 1989: 1004). Science, technology, and innovation play a crucial role in this transformation.

Investing in Big Science has a significant influence on this kind of technology transfer with investments in fundamental knowledge and human capacity building. However, Big Science development necessitates large investments in technology, infrastructure, and expertise, which in turn bestows influence based on control and exclusivity.

Naturally, countries with relatively high Gross Domestic Product (GDP) per capita are able to participate freely in such Big Science investment projects. Those nations with large research funding and technical expertise are able to contribute to such projects. Despite these challenges, Asia, South America, and some African countries are able to participate in some of the Big Science activities on a limited scale with international collaboration programmes (Praderie, 1996), for example, international projects such as ESO's Square Kilometre Array (SKA) telescope located in South Africa with additional stations located in eight other African countries,[1] known as the African VLBI Network (AVN). These radio telescope stations are part of the African Very Long Baseline Interferometry (VLBI) to produce high-resolution images of celestial objects. Another example is Chile, in South America, where ESO's flagship ground-based Paranal observatory, the Very Large Telescope (VLT) and several other telescopes are located in the Atacama Desert in Chile (Figure 12.1). This location is regarded as the best astronomical observation site by astrophysicists.

Pakistan and India became the first non-European associate members of CERN in 2015 and 2017, respectively, after decades of scientific collaboration (CERN, 2017b). By allowing scientists to take part in prestigious scientific investigations like the LHC, such scientific collaborations helped increase capacity building in developing nations.

**Figure 12.1** ESO's Very Large Telescope (VLT) at Paranal, Chile
*Source:* G.Hüdepohl (atacamaphoto.com)/ESO

---

[1] SKA project involves eight African countries Botswana, Ghana, Kenya, Madagascar, Mauritius, Mozambique, Namibia, and Zambia.

Building scientific collaborations is a key mandate of CERN and ESO and has led to increasing global scientific capabilities and cultivating science diplomacy among nations. Global research infrastructures such as the Synchrotron-light for Experimental Science and Applications in the Middle East (SESAME), the International Space Station (ISS), the Abdus Salam International Centre for Theoretical Physics (ICTP), and the European Council for Nuclear Research (CERN) are shining examples of how to bring increased collaboration among various nations irrespective of their wealth, status, and scientific capability.

Equally, men and women with high intelligence were able to comprehend the origins of the cosmos and make contributions to human understanding of the living environment. Among them are exceptional individuals like Ernest Rutherford, Albert Einstein, and more recently Stephen Hawkins, a brilliant physicist and cosmologist who contributed to work on black holes and the origins of the Universe, including Hawking radiation (Figure 12.2).

Historically, it has been shown that public investment and policies are necessary to promote Big Science infrastructure for socioeconomic development. Such investment drives technological innovations that benefit nations and improve the quality of life. A notable example, already covered in several chapters, is the development of the World Wide Web (WWW), a component of the ubiquitous internet that was imagined and created at CERN (Berners-Lee and Cailliau, 1990). The WWW or Web provided open access to information-sharing among researchers around the world.

**Figure 12.2** Professor Stephen Hawking's visit to the Large Hadron Collider (LHC) tunnel in 2006

*Source:* © CERN

Such information sharing has facilitated the continuous development of magnetic resonance imaging (MRI) and computerised tomography (CT) scans benefiting millions of people (Rinck, 2008; Cirilli, 2021).

Research knowledge, when effectively commercialised, generates enormous private wealth. It may be possible such wealth generated, as Yunus claimed, can be confined to a privileged few (Yunus and Weber, 2017). Apffel-Marglin and Marglin (2015) illustrate the unintended consequences of Big Science investments that span developmental failures, environmental degradation, and social fragmentation.

The dramatic transformation of China's economy and society over the past three decades serves as a compelling example of how significant investments in Big Science and knowledge infrastructure, when made within a highly centralised policy framework and state-controlled economy, can create human capital that is globally competitive in tandem with accelerated economic growth.

China is closing the gap with the USA and Europe on research investments. China holds significant stakes in important international initiatives like the Square Kilometre Array (SKA)[2] and China's 185-million-dollar single-dish Five-hundred-metre Aperture Spherical Telescope (FAST) telescope technology transfer from invention to innovation.

When taken into perspective, the example of centralised national growth may not be unique to China, but the size and scope of its achievement are interesting. Emerging evidence suggests that a society that is opposed to widely held views on popular representation can build citizen participation without mass electoral processes and share the fruits of economic development without a free market.

In a way, China's socioeconomic trajectory stands in contrast to another well-studied national development approach exemplified by Bangladesh. Here, small-scale private enterprise-driven growth in the national income is matched by a simultaneous rise in self-help strategies including microcredit, social services, delivery enterprises, and the developing idea of Social Business[3] championed by Bangladeshi Nobel winner Muhammad Yunus. These two cases from China and Bangladesh illustrate what the Yunus global network refers to as 'Enterprise-led Development', a strategy that shapes public participation and resource allocation to promote socioeconomic development while being supported by socially responsible entrepreneurship within the confines of legal safeguards and a clear focus on eradicating poverty.

As a result, human development indicators for Bangladesh show promise compared to its neighbours, and the Grameen Bank has become an international model for microcredit for a poverty alleviation strategy which may be as complex as Big Science problems (Bernasek, 2003). In contrast to Myrdal's (1968) Malthusian assessment of poverty in Asia, Asian economies have had 50 years of continuous GDP

---

[2] The SKA project gathers top experts and policymakers to build the world's largest radio telescope whose image quality will exceed that of the Hubble Space Telescope. https://www.skatelescope.org/the-ska-project/.

[3] Muhammad Yunus and Weber (2007; 2017) describes Social Business as a market-competitive non-dividend enterprise created solely to address human problems. A specific sub-set of social enterprise it is designed to counter wealth concentration.

development, driven by technological discoveries imported from the Western world while utilising low production costs. In comparison, Latin America and Africa, where knowledge transfer has been slower, show a low rate of growth.

Bangladesh and China demonstrate two similarities in their respective development journeys. Both have controlled population growth while increasing mass education levels proportionally. Alongside investments in human capital, Bangladesh and China, each in their own way, have invested in knowledge transfer within their means and context. Despite the structural constraints in two countries, they have increased the pace of development, and made resources available for improving quality of life (Nayyar, 2019). According to Myrdal's (1968) view, development must be internally based, deliberately persuaded, and nurtured. Since then, the spontaneous growth-inducing stimulus of relatively free and expanding international trade has faded in the current socioeconomic context.

Enterprise as a driver of growth has been well-established for decades. This enterprise concept is still evolving. Enterprise-led development straddles a continuum between business and philanthropy. This notion has been fundamental to Yunus' model of Social Business. Here, resources, otherwise meant for philanthropic purposes, are used as seed capital to start social impact businesses that eventually become successful and continue to channel profits into further expanding organisational scope and impact.

Similarly, Big Science investments are funds largely generated by the public and are not necessarily motivated by profit. When such investments are profitable, they open the door to commercialisation for the general public. According to Yunus' school of thought, social business investments in Big Science make sense since they share the same dynamics of solving complicated issues and combining dispersed knowledge into technologies that improve the quality of life of the most vulnerable populations.

## 12.3 The Role of Big Science in Social Construction

Science in general has always played an important role in the transformation of societies. A Pakistani theoretical physicist, Professor Abdus Salam, made an important contribution to Big Science. Salam and his colleagues (Sheldon Glashow and Steven Weinberg) received the Nobel Prize in Physics in 1979 for their work on the electroweak unification theory, a theory confirmed by the discovery of the 'neutral currents' in 1973 at CERN. The discovery at the Gargamelle bubble chamber was also the first experimental indication of the existence of the Z boson observed at CERN in 1983. Professor Salam contributed to the advancement of science and its applications to society, and he played a key role in the creation of the World Academy of Sciences (TWAS) and the Abdus Salam International Centre for Theoretical Physics (ICTP). The former served as a bridge for the transfer of scientific knowledge between developed and developing nations, and the latter overcame the 'Iron Curtain' that divided Europe (Del Rosso, 2014b).

Upon reflection, the swift socioeconomic development of Asia confirms the close connection between science, enterprise, and development, even if only indirectly.

All previous examples cited above, including China, Pakistan, India, Bangladesh, and Vietnam, negate Myrdal's expressed apprehension about how a number of countries would be constantly at risk due to economic stagnation, corruption, and poor governance. Theories of breaking economic stagnation in countries like India and China address a combination of progressive measures of export-oriented growth, investment in human capital, market-oriented reforms, and an open economy that combines science and technology development.

Big Science continues to play a key role in building economic independence, even as most nations cannot afford to invest in expensive knowledge infrastructure on their own. Collaborative platforms such as CERN, ICTP, TWAS, and SKA, provide an opportunity to collaborate and derive direct and indirect dividends from frontier research. The opportunities are enormous to enrich information and data sciences, medicine, education, poverty alleviation, and agricultural development (Reynaud, 2005).

If wealth is an essential component of social well-being and human development, its equitable distribution across all social strata is the goal, and successful enterprise in its many evolving forms is the channel, then science can be argued to be the catalyst. Without business, which has self-interest as its fundamental principle and is driven by the human tendency for 'fair and deliberate exchange',[4] the phenomena of wealth creation and, in fact, its concentration would not exist. Entrepreneurs would have fewer options to disrupt markets and alter value propositions in order to generate money if an open science regime were inaccessible.

Muhammad Yunus (Yunus and Weber, 2007) is quoted on how he built upon his formal expertise in economics, 'standing it on its head' to create a brand-new micro-banking ecosystem that initially served the marginalised, but now is a mainstream product across the banking industry. The Grameen network may be comparable to the network of Big Science collaborators, and it has achieved success in healthcare, infrastructure, software and communications, and professional education, building on the established theory and practice of science and its application, to design 'pro-poor' enterprise solutions to the world's most pressing problems.

## 12.3.1 Enterprise as Social Equity and Social Transformation

Enterprise has inherent potential as a social equity mechanism, provided it conducts itself with social responsibility and meets the core condition of 'fair exchange'. All else being equal, a fair business transaction is itself a levelling mechanism from the social anthropology perspective (Eller, 2010), because even the richest person cannot buy what the poorest may not wish to sell unless a fair exchange is agreed upon.

---

[4] Excerpted from Smith's *Wealth of Nations* (2000, Modern Library): Nobody ever saw a dog make a fair and deliberate exchange of one bone for another with another dog. Nobody ever saw one animal by its gestures and natural cries signify to another, this is mine, that yours; I am willing to give this for that.

Big Science organisations such as CERN and ESO have different approaches to knowledge transfer. These organisations follow the idea that the knowledge and technology developed should be made readily available to society in order to benefit the public. Profit considerations lie at the root of inequity (Yunus, 2017). As we have seen since the rise of the corporation, the propensity to place profits above social values and the abdication of personal accountability in favour of a corporate structure create a dynamic of power and privilege where profits may impose an enormous cost on society and the natural environment. Much of the criticism levelled at corporations stems from an unbridled pursuit of profits, which is frequently aided by regulatory safeguards that, until recently, made few demands on corporate citizenship and responsible business.

CERN and ESO technology transfer does not always target industry. Licencing is also used as a tool to share knowledge and techniques with other research institutes at no cost. Training and education are often used to foster entrepreneurship and prepare future generations of physicists who are capable of utilising intellectual property and engaging in start-up ventures.

The modern entrepreneur understands that bargains are ingrained in human nature and that profits are a natural outcome of value exchange. Not much is discussed about the importance of maintaining social and environmental responsibility, which is a tangible part of the value exchange. Utting (2007) examined the equity and equality aspects of doing business from the perspective of corporate social responsibility (CSR), which is being articulated by an increasing number of companies as an effort to redress the imbalance in wealth and power resulting from the conduct of business. Indicators such as the working environment, workers' rights, community engagement, and stakeholder interest show that companies invest increasingly in social and environmental remediation. However, these efforts need reformation to improve empowerment, redistribution of resources, quality of life, and equity.

The recent Covid-19 pandemic has disrupted the progress of Big Science like in many other areas of the economy. Covid had disproportionate impacts on different nations, with poor nations having to rely on rich nation for health, economic, and social support. However, Covid also provided an opportunity to demonstrate equitable access to vaccine and help needy nations. The development of the Covid vaccines at record time in several countries by selected pharmaceutical companies. Keeping in line with Open Science initiatives, more than 25 Nobel Prize winners, including Mohammed Yunus call for the Covid-19 vaccine to be declared a public good. Initiatives such as Covax, the Global Alliance for Vaccines (GAVI), the Coalition for Epidemic Preparedness Innovations (CEPI), and the Covid-19 Tools Access Accelerator (ACT) have responded to efforts to increase vaccine availability, distribution, sharing doses, and redress intellectual property rights to ensure that all nations have access to the vaccine to combat the global pandemic.

Enterprise is thus intricately woven into notions of community development and social advancement. Shared prosperity, social security, equal opportunity, and democratic freedoms are all central to the role of enterprise as a driver of equity. Social entrepreneurship and impact investments are new terms that describe the role society demands from businesses. As governments create incentives for socially responsible businesses and consumer groups rally for more social and environmentally accountable behaviour by entrepreneurs, a visible shift may be underway towards the original role of business as an equity-based mechanism for fair exchange and redistributing wealth. When Big Science generates breakthrough research that is commercialised, enterprises can generate years of profits from it. The current debate seeks to reclaim the social role expected from business and re-introduce the concept of equity that has been lost as a result of free-market profit-seeking incentives.

Big Science discovery and commercial success are not straightforward for a number of reasons. First, fundamental research is directed towards answering scientific questions and is not usually driven by the purpose of seeking commercial success. Second, the capital-intensive nature of Big Science is often distributed across numerous collaborators, making it difficult for lenders to translate into cost recovery over the short term. Third, given the amount of public resources often deployed in the service of Big Science, it is often difficult to justify them to policymakers inclined towards tangible rewards. Fourth, Big Science research can often only be commercialised with significant public subsidies, at least in the start-up phases.

Governments and corporate funders are already encouraging researchers to find ways to reduce the seeming gap between Big Science and entrepreneurial solutions. It remains a challenge, but there is evidence that commercialisation of research outcomes is a priority for leading enablers such as the European Innovation Council Accelerator and the European Institute of Innovation and Technology (Romasanta et al., 2021). CERN, for example, provides training for young researchers who later join many companies, demonstrating immediate contributions to the economy.

## 12.3.2 Reliable Knowledge, Trust, and Entrepreneurship

Fundamental scientific research, as the bastion of knowledge creation and a fountainhead of reliable ideas, continues to be the best source of business ideas fuelling entrepreneurial inspiration, even though it may not be as obvious to the ordinary citizen as might be expected.

The key to entrepreneurship at CERN is building connectivity. Connectivity between world-renowned scientists, engineers, and practical staff who are able to identify scientific and technological problems not only to advance concepts but also provide solutions that benefit society. This requires building a culture of entrepreneurship across Big Science organisations.

The general public leans towards those they feel they can trust. Building trust in evidence-based research in general and cohesion and respect among social actors and institutions is integral to the perception of well-ordered science and a hallmark of the success of liberal democratic societies. The goal of well-ordered Big Science is to ensure research is conducted for the public good and carried out in a transparent, efficient, and responsible manner to realise the goals of society.

At the individual level, physicists seem to have limited interest in generating public goods and research commercialisation other than the diffusion of useful knowledge. The garnering of public trust and support is necessary to sustain Big Science enterprises. As in the case of high-energy physics and astrophysics, the average citizen may rarely connect to scientific values and even not see their relevance in daily life, until innovations such as biomedical technologies, cancer cures, and imaging technology provide the link to Big Science investments.

Big Science community responds to 'grand challenges', such as environmental threats like climate change, demographic, health, and well-being concerns, and to the difficulties of generating sustainable and inclusive growth. These challenges are associated with 'wicked problems' requiring the Big Science community's strategic action as they are complex, systemic, knowledge-intensive, interconnected, and requiring the insights of many scholars and epistemic groups. This research challenges the imagination of ordinary citizens. CERN has launched citizen science projects to inform and engage the general public. Citizen science projects directly involve the public in the scientific process, and they provide meaningful engagement between science and society.

A more recent success story is Estonia, which was until 1991 a part of the Soviet Union. Estonia has established a reputation for rapid innovation piggybacked on technology infrastructure, earning the nickname *E-stonia*. In 1996, the government-owned technology investment body, the Tiger Leap Foundation, led reform towards a digital economy with large investments in ICT infrastructure. A decade later, ten corporate entities have joined the Look@World Foundation, a public–private partnership aimed at bringing digital technologies to all citizens on the right side of the digital divide. As of 2016, 91 per cent of Estonians are connected via ICT, making Estonia a premium hosting choice for online info-tech and e-commerce. Referred to as the Silicon Valley of the Baltic, Estonia demonstrates the power of collaborative innovation, bringing together companies, universities and citizen organisations to support an e-government that is highly trusted for transparency and efficiency (Anthes, 2015) and demonstrates the linkage between knowledge and moral civic power (Björklund, 2016). These examples may or may not draw close parallels to the ATLAS or CMS experiments and their visible impact on society but they illustrate the importance of sustained engagement between key stakeholders, and the maxim that luck often favours the well-prepared.

## 12.3.3 Enterprise Solutions

Enterprise solutions to social problems are a demonstrated reality in an increasing number of countries. The Sustainable Development Goals (SDGs) provide an anchor for social businesses. Impact investors are more prepared than ever before to back up social investments. Regulatory regimes are changing to accept social enterprises. However, the everyday outcomes of basic or fundamental research, despite being a significant contributor to scientific advancements worldwide, are less visible to the general public. Scientists can actively promote knowledge transfer by actively creating opportunities to present their research to counterparts, which is an acknowledged driver of national self-reliance.

Scientists have an inescapable role in inspiring innovation, and Japan is often cited as a constantly evolving haven for collaboration between the government, the business sector, and the research community. In 2018, the Japanese government launched the 'sandbox framework' with regulatory reform to boost hi-tech innovation, presently aimed at financial services, the healthcare industry, and mobility (JETRO, 2018; HBR, 2020).

China demonstrates another model of success in what is referred to as 'top-down innovation' (Xu, 2017), where under a national innovation strategy focused on breakthrough technologies, the government works with private investors to finance small and medium enterprises at a large scale, spearheading R&D through innovative organisational solutions. Despite criticism, the approach to innovation appears to have invigorated the innovation landscape in several enterprise clusters across the country, particularly in semi-conductors and bio-pharmaceuticals (Zhang et al., 2022).

By engaging with counterparts in government, business, and citizen organisations to showcase well-ordered research wherever possible, scientists strengthen the necessary link between science and society to establish a well-earned stake in shaping the future of enterprise-led development towards a better world.

## 12.4  Use of Big Science Ideas for Societal Applications

Not all Big Science projects are candidates for taking ideas from laboratories to industry. It is useful to examine some of the early efforts by individuals to steer some fundamental research outcomes to commercial applications. One such area is the medical applications of CERN's detector technologies as far back as the 1970s. At that time, knowledge was primarily transferred through the initiative of passionate individual researchers. Georges Charpak, for example, was responsible for developing the multiwire proportional chamber (MWPC) at CERN in 1968, and he was awarded the Nobel Prize in 1992 for his innovation. Multiwire chambers gave rise to further developments in the art of detectors, some of which are highly innovative. Most high-energy physics experiments make use of these methods, but their application has extended to widely differing fields such as biology, medicine, and industrial radiology (Charpak, 1992). The MWPC's ability to record millions of particle tracks per second opened a new era for particle physics. In medical imaging, its sensitivity promised to reduce radiation doses during imaging procedures, and in 1989, Charpak founded a company that developed an imaging technology for radiography that is currently deployed as an orthopaedic application.

CERN continued to build a culture of entrepreneurship. Systematic efforts in technology transfer can be traced back to 1988, when the CERN Industry and Technology Liaison Office was founded to stimulate interaction with industry and to assist in issues related to CERN's intellectual property (Nilsen and Anelli, 2016). Most Big Science organisations, including CERN and ESO, use technology transfer and knowledge diffusion activities, including the transfer of licensing of intellectual property rights, making software and hardware available under open licences, building industry interactions, and forming international collaborations, as transfer modalities. Such technology transfer opportunities also feed curiosity and the exploration of new paths for innovation and knowledge creation.

## 12.5  Effective Knowledge Transfer of Big Science Knowledge

In Big Science the basic notion is that all scientific results shall be made openly available to the public. Within this conviction, knowledge transfer mechanisms are not explicit. In recent times, knowledge transfer has frequently been advocated as a strategy to solve some of the most pressing issues, such as medical (Covid-19) and climate change of our times.

How is it possible to be more cost-aware when making the necessary equipment for research purposes? In general, industries are used to develop new technologies using prototypes and pilot plants using available research knowledge. Innovation requires a combination of technical and non-technical inputs. The coordination and communication among research and development teams across industry, universities, and research-collaborating institutes are necessary for the successful transfer of knowledge and continuous technological development.

At CERN, all the changes in future experiments allow for coping with the increase in size, strategic focus, and innovation opportunities of the projects and of the participating institutions and universities, which have been absorbed with time into existing work practices. For example, LHC beams are squeezed into very small beam sizes to maximise the rate of proton collisions as required for rare processes like Higgs production. In addition, the angular divergence of the beams at the interaction points was reduced, and these special settings allowed the ALFA and TOTEM experiments to measure proton–proton scattering angles down to the microradial level.

Any further changes required by the LHC and the High Luminosity LHC (HL-LHC) have been just a natural extension. The time from conception to realisation and functioning is now so large that scientists must bargain between their wish to use the most updated possible applications and the possibility of really doing it.

It is worth remembering here that CERN's accelerator construction and its upgrades are mostly accommodated within the CERN budget. The cost of the detectors and their upgrades as needed to carry out the research are mostly covered (more than 90%) by the collaborating institutes. Furthermore, in LHC experiments, almost 50% of the participants are not from CERN Member States, implying that the spillover of technological learning extends to international borders.

Big Science knowledge and ensuing technology transfers aim at broader societal impacts. However, this has not been the best case. The interests of many practicing physicists and engineers have specific research strategies in mind. Societal development is a related issue and often something hard to conceive.

Scientific performance is measured not only in terms of research outcomes but also in terms of social deliverables such as education and knowledge transfers. CERN and ESO as astrophysics experiments, the Human Genome Project, the Human Brain Project, and the European Molecular Biology Laboratory are shining examples of what they can deliver to society at large. Since its creation in 1954, CERN has had a long tradition of technology transfer, mainly through people and purchasing (Schmied, 1975; Bianchi-Streit et al., 1984; Autio et al., 2003), and collaboration agreements (Bressan et al., 2008; Florio et al., 2016).

The most revolutionary technology that has impacted our daily lives is the World Wide Web (WWW), also called the Web (Gillies and Cailliau, 2000; Berners-Lee and Fischetti, 2000; Berners-Lee and Cailliau, 1990). WWW was invented by Tim Berners-Lee in 1989. He drafted a one-page proposal (perhaps the shortest proposal at CERN) and presented it to the CERN management, outlining the general information management protocol about the accelerators and experiments at CERN. His solution was based on a distributed hypertext system. The outcome of this proposal, leading to the invention of WWW, made it possible to connect the entire world (Figure 12.3).

The need to get such scientific developments and knowledge transfer into a more structurally effective return to society led to the creation of technology transfer offices and technological parks within university campuses and near Big Science Centres.

**Figure 12.3** The inventor of the WWW, Tim Berners-Lee at the CERN Computer Centre
*Source:* © CERN

In addition, glass ceramic, laser guiding systems, ESO's RAMAN fibre amplifier technology, and software developments found commercial applications in cancer diagnostics, the telecommunications industry, and life and geophysical sciences. The scientific capability of making real-time observations, interactive data analysis, and automated processes in laboratories and institutes around the world is also having a large impact on society.

## 12.6  Knowledge Transfer and Knowledge Management

Several models have been created and developed to study the knowledge translation processes adopted by Big Science organisations such as CERN. These models of knowledge transfer help to understand and leverage this important process, which is at the core of innovation, entrepreneurship creation, and societal impact. CERN as an organisation, and similarly any other Big Science Centre, has its own epistemology, with its own tacit and explicit knowledge and creating entities (individuals, groups, and their organisations). The multicultural scientific and technological environment is also very important for individual and organisational knowledge creation.

Most knowledge transfer models are applied in companies that do not consider scientific knowledge acquisition and the scientific process. This is why a new knowledge management model for scientific organisations such as CERN has been created (Bressan 2004), incorporating Nonaka's four modes of knowledge conversion (socialisation, externalisation, combination, and internalisation) developed for business purposes (Nonaka and Takeuchi, 1995) with the knowledge acquisition model developed for didactic purposes (Kurki-Suonio and Kurki-Suonio, 1994) as shown in Figure 12.4.

The model of knowledge management in science is the continuous transferring, decoding, and utilising of existing knowledge to produce more knowledge. From individual perception, assessment, and analysis of the context and tools in which the five LHC experiments evolved, it has been possible to track the various aspects of knowledge acquisition. Social interaction, relationship quality, and network ties existing in the multicultural environment of LHC experiments have been shown to be associated with knowledge acquisition (Bressan et al., 2008), and contribute to innovation.

Big Science closely interacts with industry for the purpose of advancing research infrastructure and components such as magnets. Most innovative companies respond to both the push of research and development and the pull of the market. In some leading companies, attempts were made to change approaches to technological innovation. For example, using 'Lean thinking' business methodologies delivers more benefits to society by placing people first rather than technological systems (Womack and Jones, 1996).

In a knowledge-based economy, digitisation and a circular economy go hand in hand, and they can help shape new business models. Another concept is the use of Social Return on Investment (SRoI), which was used for the first time in 2000. In

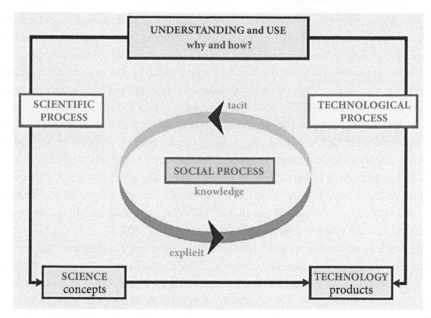

**Figure 12.4**  A knowledge management model
*Source:* © CERN

Standard Return on Investment (SROI), the social returns are based on cost versus benefit analysis to determine the value of projects using project evaluation and quantitative impact assessment. Such methods may assist the stakeholders in resource allocation, plan modification, social impact assessment, and the assessment of benefits that cannot be easily monetised. Such approaches may be used in organising Big Science projects to target social benefits.

## 12.7  Innovations: From Science to Society

Technology transfer, from invention to innovation and impact on society, is no longer a question resting solely on efficiency (Oliveira and Teixeira, 2010). Incidentally, technology transfer is primarily discussed from an economic perspective. However, the global context also needs to consider other important factors like ownership, culture, public policy, education, equity, and impact.

In other words, the success of a given technology transfer may rather lie in an equilibrium between the efficiency and resilience of the targeted outcome. (Abdurazzakov et al., 2020). Ciborowski and Skrodzka (2020) have shown that cooperation within economic ecosystems positively impacts innovation among countries.

Given the comparable aim of supporting social development, the same principle should be applied to technology transfer. This leads to proposing an extension of the Agency Theory (Mitnick, 2015), as a possible new sustainable economic and societal model. A Circular and Trustworthy model combines to transform new knowledge into invention and its dissemination to society as new technology products, processes, and services. In the original Agency Theory, there is one 'Principal', i.e. the donor of an invention, on the left, and one 'Agent', i.e. the user of an invention carrying out an action. The potentially unaligned relationship is handled by the setup of a contract between the Principal and the Agent (e.g. a software licence). According to this model, society is (re)introduced in the equation as a second Principal, on the right of Figure 12.5, to consider societal needs, incentives, and cost versus benefit aspects. For this virtuous circle to ignite, it is believed that both market and societal dimensions must be considered in the valuation of a given invention. Indeed, to maximise knowledge transfer to society, inventions must be forged in an accountable, transparent, and traceable manner so they can be turned into open, inclusive, virtuous, and circular technologies (Manset et al., 2017, 2023).

Motivation, and remuneration in particular, are clearly important but remain often the sole criteria of economic efficiency. The model therefore must be extended until the eventual beneficiary, i.e. society. In doing so, the model becomes virtuous and circular and thus opens the pathway to resilience in inventions and societies, which is key to progressive and long-term success in achievements. The beneficiary must therefore be an integral part of the construction process and convinced of the importance of the development.

The diffusion of innovations is largely influenced by the types of innovation, the communication channels available to diffuse new ideas, the time and degree of adoption, and the social systems. Big Science leads to clusters of innovation. Although there may be clusters adopting innovations, there are still barriers to innovation diffusion that must be overcome, such as a lack of local involvement (Rogers, 1962, 2003).

Today, with the consolidation of distributed ledger technology (DLT), emerging studies indicate mounting research interest in applying blockchain technology to knowledge-sharing frameworks (Zareravasan et al., 2020; World Bank Group, 2018) by reducing information asymmetry in collaborative networks (Schinle et al., 2020), and then projecting this catalytic potential to entrepreneurial innovation (Hashimy et al., 2021). Big Science makes it possible for multiple channels of innovation.

More particularly in the creation of peer-to-peer research and innovation networks, DLT can be seen to help in shaping the virtuous model[5] developed for this chapter and presented in Figure 12.5. This model explains the role of collaboration and economic relations in transferring knowledge. In fact, blockchains can serve the purpose of making inventions transparent, accountable, and traceable over time, while enabling their open, inclusive, and circular use in society. It can do so thanks to open information sharing not only in the ledger but also in the market and societal impact valuation.

---

[5] The authors of this model are David Manset, Beatrice Bressan, and Marilena Streit-Bianchi derived from Manset et. al., (2017).

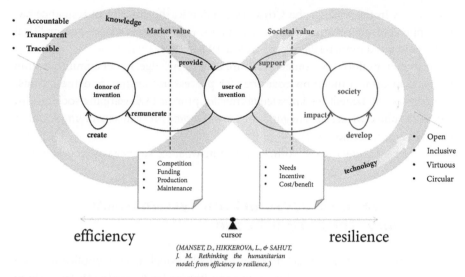

**Figure 12.5** The circular and trustable model

*Source:* David Manset, Beatrice Bressan and Marilena Streit-Bianchi (2023). *Developed from a Model Presented in 2015 D.* Manset Master Thesis 'Humanitaire éthique, de l'efficience à la résilience'

Reflecting on the humanitarian model from efficiency to resilience, the development of a new societal and economic model to reshape humanitarian aid with the beneficiary at the centre is necessary. This is because the process of building resilience requires the involvement of the beneficiaries. Applying new standards in promoting resilience is as necessary as considering the aid recipient in terms of the means to implement an action as well as in prioritising funding. As a consequence, the beneficiary must become a participant in the construction of a more suitable model (Manset et al., 2017). While the work addressed the humanitarian field, the underlying foundations came from former research work carried out in ecology. Indeed, to further understand the necessary relationship between efficiency and resilience, let us draw an analogy between society and nature. In their work on Ecological Complexity (Ulanowicz et al., 2009), it is explained that nature does not optimise the effectiveness of the natural ecosystem alone, but rather manages to strike an unequal balance between efficiency and resilience.

Some common basic rules and principles can be applied in this interconnected world, one of which is sustainability. Big Science may not always produce rapid worldwide benefits, and some technologies like the medical technology derived as a result of Big Science take years of continuous improvements and technology transfer efforts. Even the World Wide Web (WWW), one of the most globally impactful examples of a recent technology offered free by CERN, has taken considerable time to provide social and commercial benefits.

The term 'resilience', once used to describe people who had been through traumatic experiences, is now more commonly used to describe career resilience, which is defined as effective vocational functioning under challenging circumstances (Rochat et al., 2017). In the wake of the Covid-19 pandemic, the concept of resilience has been applied to the economy and development, with emphasis placed on political instances to build more resilient post-Covid-19 societies.

Collaborative efforts are needed to diffuse tacit knowledge. Better communication and participation in the knowledge transfer process are central, and these transfer processes will benefit the knowledge transfer process in developing societies. To achieve the objective at stake, negotiations to overcome problems involving social or political borders are necessary.

Examples of several cases in Annex 1 illustrate what big science has accomplished.

## 12.8 Innovation Ecosystems and Systematising Serendipity: The ATTRACT Case

The pathway of Big Science technologies to breakthrough market applications is a highly serendipitous one. It is difficult or even impossible to predict the fraction of fundamental knowledge that will end up leading to new businesses and products that will transform our society. One of the numerous causes of this phenomenon is that these technologies are primarily created by research communities with the sole objective of expanding the bounds of fundamental science beyond the scope of their original mandate: to make feasible the envisioned aims or pathways for industrial and even smaller markets; to make relevant to industrial and to even fewer market-envisioned goals or pathways. Despite this intrinsic difficulty, it would be desirable to rely, at least partly, on some sort of rule of thumb or heuristic approach that ultimately allows Research, Development and Innovation practitioners and policy-makers to develop strategies for streamlining, as much as possible, the 'Lab to the Fab' odyssey.

Annex 2 outlines the serendipitous process from basic science to market and the innovation ecosystems required for beating the odds from laboratory to market.

Innovation ecosystems constitute a potentially successful construction to somehow 'systematise' this serendipitous process precisely because they integrate different actors with complementary roles and motivations towards technology and innovation. Therefore, it is important to note that successful innovation ecosystems should have a carefully crafted but flexible structure capable of satisfying the opposing interests and goals of a diverse community of stakeholders.

Numerous studies have examined how serendipity, innovation, and science interact, and it has been found that there is a trade-off between productivity and serendipity (Murayama et al., 2015). Paradoxically, a new paradigm is emerging, called 'systematising serendipity', especially within the field of digital information retrieval (Wareham et al., 2022). Besides its novelty, there are some preliminary definitions in the current literature that suggest 'systematising serendipity' could be understood as improving the chances of making connections that lead to new discoveries.

The ATTRACT (breAkThrough innovaTion pRogrAmme for deteCtor infrAstructure eCosysTem) was an initiative to support the serendipitous process of knowledge transfer (Nordberg and Nessi, 2013).

The ATTRACT is a pilot initiative that aims to provide a new breakthrough innovation ecosystem based on the 'Open Science, Open Innovation, and Open to the World' philosophy. It is steered by a consortium comprising pan-European research infrastructures, European industrial sector organisations as well as business and innovation specialists, with the help of funding from the European Commission. The overarching goal of ATTRACT is to establish a European ecosystem with a wide scope in the field of breakthrough detection and imaging technologies. These would range from sensors and detectors to computing technologies for transforming data into information and ultimately knowledge. ATTRACT's special focus on detection and imaging responds to the following factors:

i   They are the backbone technologies that help European research infrastructures and their research communities push the limits of Basic Science; and
ii  They are the core of future industrial developments, applications, and businesses.

ATTRACT's operational goal is to increase the chances and accelerate the translation of Basic Science technologies into marketable products. In other words, generating the boundary conditions for systematising serendipity (see Annex 2). Though still in the early stages, various qualitative and quantitative studies are beginning to be developed regarding the ATTRACT ecosystem, allowing for the initial identification of some key factors for, as was mentioned in the previous section, beating the odds. Some of them are:

1. Public funding: The ATTRACT ecosystem is leveraging public funding sources. As reported by different scholars, public funding is key for helping nascent breakthrough technologies, many of them even at the conceptual level, reach the necessary maturity for raising the interest of private capital;
2. Phase approach towards technology maturity: The ATTRACT initiative considers that 'not all Valleys of Death' look the same (Figure 12.6); and
3. This is especially the case for breakthrough technologies conceived for Basic Science purposes. Unlike more incremental technologies, which only require one stage to reach a point of interest for private capital, these technologies

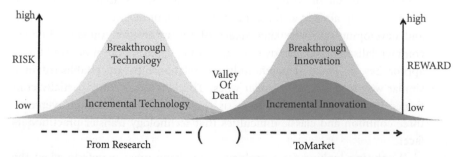

**Figure 12.6** Qualitative illustration of the different 'Valleys of Death'—incremental and breakthrough innovation

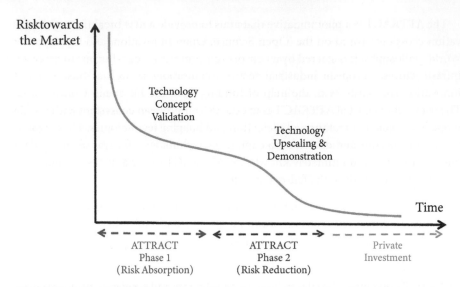

**Figure 12.7** Qualitative illustration of the two phases operationally implemented within the ATTRACT initiative

*Source:* https://www.openaccessgovernment.org/attract-programme/121400/ Last access 22/8/23

require two stages because of the high risk associated with their initial invest-ments before becoming a market product. First, there is the risk-absorption stage (Boisot et al., 2011), where ideas and concepts could reach a proto-type level. Second, a risk-mitigation or risk reduction stage, where the most promising concepts are further helped in raising towards a pre-market product (Figure 12.7).

4. Co-innovation: Co-innovation is understood within the ATTRACT initia-tive as a bridge between two communities (research and industry) with, in principle, different motivations and goals for undertaking R&D&I (cap-ital and/or resource-intensive) efforts· It entails the identification and col-laborative pursuit of win–win outcomes, both by research and industrial actors, starting already at the conceptual stages of technology development and developing until the later stages of the innovation value chain (e.g. commercialisation). Co-innovation, therefore, departs from more traditional approaches in which research–industry relationships are established in a similar way to customer–supplier ones. The hypothesis, only partially con-firmed by the preliminary studies, is that co-innovation would overcome three common and classic difficulties that technology transfer practitioners face:

   i 'A solution looking for a problem'. This issue often manifests when the research communities only establish relationships with the industrial ones at later stages of the innovation value chain.

ii 'A problem looking for a solution'. This issue could be considered the other side of the coin with respect to the previous one and occurs when industrial communities try reinventing the solution for a problem that might be well known and already solved in the academic realm.

iii The common development of 'know-how' between industry and research communities often does not occur in a purely supply–demand context. This could lead to the development of collaborative practices that mutually build trust among collaborators.

ATTRACT is still a novel innovation ecosystem in development. Nevertheless, useful data has already been extracted during its first phase (risk-absorption) from 170 funded breakthrough technology concepts, some at an early prototype stage (CERN, 2021d). Preliminary quantitative and qualitative results indicate the validity of the hypothesis outlined above (e.g. co-innovation). Also, the confirmation of the elements mentioned and others as key ones for increasing the chances and accelerating the translation of the Basic Science discoveries into the market. The ATTRACT initiative facilitates the integration of interdisciplinary teams of MSc-level students working side by side with professional researchers from academia and industry.

The organisations steering the ATTRACT initiative are embarking on its second phase (risk mitigation), leveraging the most promising opportunities emerging out of the first phase and again relying on public funding from the European Commission (CERN, 2021d).

A manifold and interdisciplinary socio-economic study will be realised not only to achieve full confirmation of the philosophy and hypotheses behind the ATTRACT initiative but also to demonstrate that the emerging paradigm of systematising serendipity is possible in practice. These initiatives continue to diffuse knowledge generated in Big Science experiments and initiatives.

## 12.9 Conclusions

Social development is intrinsically linked with enterprise development and wealth creation. Private enterprises play a key role in the progress of human civilisation. Big Science contributes systematically to the advancement of knowledge and innovation to solve complex issues and contribute to society's development. The examination of the relationship between science and complex social systems needs multiple perspectives from economic, political, and scientific considerations. The greater the complexity of Big Science projects, the harder it is to translate useful knowledge into tangible social outcomes. Complexity in the social system acts as a barrier to utilise knowledge freely. Very often, localised development propositions are necessary to address local social complexities.

A complex interconnection also exists between Big Science and enterprise development. Big Science initiatives are different from other forms of research ventures in the sense that they lead to varied opportunities through the production of enormous amounts of data, information, and knowledge that are relatively free to access and can be utilised in the public domain.

Many approaches are available to the translation of Big Science into useful products and processes. A culture of open science is needed for such translations, which involves traditional technology transfer as well as serendipitous processes. Garnering public trust in scientific research is important for the continuous progress of the scientific community in making giant leaps in socially beneficial innovations. Methods for converting Big Science into practical goods and procedures entail serendipity in both the research and its application. Such approaches also contribute to a new social order and an economic model that influence humanitarian aid with the beneficial development of the public contribution of Big Science. Together, complexity and serendipity act as catalysts to transfer Big Science knowledge by breaking down complex problems into simplified components.

The translation of Big Science knowledge into market applications might be a serendipitous process, but attempts have been made and approaches taken (e.g. ATTRACT) to systematise this process. The process should be founded on a well-ordered science that has the ability to ask the right questions and connects research with practice using multiple research processes, facilities, and collaborative inquiry. Reliable evidence and subjecting knowledge to rigorous scrutiny are required for the fundamental knowledge pursued in Big Science.

Knowledge translation processes in Big Science follow strategic and spontaneous paths, and the innovation ecosystems provide potential opportunities to establish reinforcing feedback loops between industry, universities, and government stakeholders.

# 13
# Future of Big Science Projects in Particle Physics

## Asian Perspectives

*Geoffrey Taylor and Shantha Liyanage*

## 13.1 Introduction

High energy physics colliders have dominated the Big Science landscape for decades and have contributed greatly to the understanding of the fundamental particles and forces in the Universe. The discovery of the long-sought-after Higgs boson at CERN's Large Hadron Collider (LHC) in 2012 is considered one of the most significant scientific achievements of recent decades. The discovery of the Higgs boson has completed the Standard Model (SM) of particle physics after several decades of painstaking investigation into the theory of elementary particle physics and experimental physics (CMS Collaboration, 2022).

The *LHC* is the world's largest and most powerful particle accelerator, showcasing European and international high energy physics capabilities. However, what comes next in the field? What is the future of high energy physics and astrophysics? What could the role of Asia be?

In this chapter, the authors describe Asia's participation in current high energy physics initiatives as well as potential roles anticipated in the future. It first provides a brief overview of Big Science initiatives for future experimental particle physics. Some comments are provided on the impact of the geopolitical environment within which these activities are growing and the challenges it imposes. The authors also provide some country-specific accounts of the major players in Asian particle physics, as well as the future opportunities and challenges they are contemplating.

## 13.2 Preparing the Future for Particle Physics

The LHC has reached a centre-of-mass energy of 13.6 TeV, exceeded the design luminosity, and produced a wealth of remarkable physics results. CERN is currently deep into a major upgrade of the LHC to significantly increase its luminosity and collision rates.

Geoffrey Taylor and Shantha Liyanage, *Future of Big Science Projects in Particle Physics*. In: *Big Science, Innovation, and Societal Contributions*. Edited by: Shantha Liyanage, Markus Nordberg, and Marilena Streit-Bianchi, Oxford University Press.
© Geoffrey Taylor and Shantha Liyanage (2024). DOI: 10.1093/oso/9780198881193.003.0014

Future strategies of the high energy physics community include further exploitation of the physics potential of the Large Hadron Collider (LHC) through its high-luminosity upgrade (HL-LHC). The HL-LHC is expected to operate into the 2040s and more details can be found in Chapters 3 and 5.

The 2020 Update of the European Strategy for Particle Physics (ESPPU 2020), was the result of two years of consultation both with the European and broader international physics communities. The EPPSU 2020 highlighted, 'Europe, together with its international partners, should investigate the technical and financial feasibility of a future hadron collider at CERN with a centre-of-mass energy of at least 100 TeV and with an electron-positron Higgs and electroweak factory as a possible first stage.' As stated in its high-priority future initiatives 'The timely realisation of the electron-positron International Linear Collider (ILC) in Japan would be compatible with this strategy, and, in that case, the European particle physics community would wish to collaborate.'

This post-LHC option being considered includes a linear $e^+$ $e^-$ collider as well as circular lepton or hadron colliders, as discussed in Chapter 5.

Four advanced proposals for an electron–positron Higgs factory are currently being pursued.[1] Two are linear, namely the International Linear Collider (ILC) and the CERN Compact Linear Collider (CLIC), where electrons and positrons are accelerated in opposite directions along a straight line and collide head-on at the centre.[2] The two circular colliders are the Future Circular Collider, e+e⁻ (FCCee) being studied at CERN and the Circular Electron Positron Collider (CEPC) in China. In circular colliders, the electrons and positrons circulate in opposite directions and come to collide at one or more locations around the collider (Yamamoto, 2021). Other less mature proposals include the Muon Collider[3] and the C3 linear e+e-collider.

The key merits of a linear collider are that beams can be polarised and that the collision energy can be extended to higher energies in future upgrades.

On the other hand, circular colliders can accommodate multiple collision points and higher luminosities especially at lower energies. Furthermore, the same circular tunnel may be used for proton–proton collisions in future developments as technology evolves.

There are several advantages in electron–positron collisions compared to proton–proton collisions in a Higgs factory. Since electrons and positrons are elementary particles, the Higgs boson can be produced from the fundamental process of annihilation in a clean environment. (Komamiya, 2001). At higher energy, pairs of top quarks can be created in a similarly clean way. Currently, the ILC is the most advanced of all the proposals, it is the least costly to build and has the lowest power requirements.

---

[1] In the 'Snowmass 2022' conference, additional concepts were considered for the further future.

[2] It should be noted that in both the 2013 and 2020 European Particle Physics Strategy Update, the ILC was identified as the first choice of Higgs factory for Europe and the world.

[3] For details see 'Muon collider hold a key to unravelling new physics', APS, Nov. 2021, Vol 30(10), https://www.aps.org/publications/apsnews/202111/muon.cfm.

The ILC is the result of decades of development. Since the 1990s, there has been a push to build a linear electron-positron collider capable of producing copious Higgs bosons. The ILC Global Design Effort (GDE) was supported internationally by the USA, Germany, and Japan, initially all contenders for hosting the proposed collider. By 2004, the International Technology Review Panel, instigated by the International Committee for Future Accelerators (ICFA), as part of the internationalisation of the project, recommended superconducting radio frequency (RF) cavities as the technology of choice for the ILC. The recommendation, endorsed by ICFA, was further developed by the ICFA-instigated GDE, resulting in the completion of the ILC Technical Design Report (TDR) in 2013, detailing a 500 GeV linear collider.

Most of the GDE work had been carried out before the discovery of the Higgs boson in 2012. In 2013, in view of the relatively low Higgs boson mass (125 GeV), Japanese physicists proposed a 250 GeV version of the ILC (Adolphsen et al., 2013). At this energy, the electron-positron collisions can produce a clean data set of Higgs bosons. The reduced size, enhanced by a joint US–Japan cost-reduction exercise, further makes the ILC250 the front-runner for the world's first Higgs factory. The benefit of the linear design is the ability to upgrade to higher energies by extending the length of the linear accelerator and by enhancements to the RF cavity performance, which add to its attraction.

The ILC250 thus offers an affordable yet highly capable option for the Higgs factory, which remains the global physics community's highest priority (ICFA, 2019, ESPPU, 2020). The ILC is seen by most as the shortest and least costly route to the Higgs factory. Nevertheless, it is clearly understood that it is the combined expertise and resources of the international community that will make the ILC a success.

Importantly, it remains the hope of the global high energy physics community that Japan will host the ILC, although obstacles remain for the Japanese government to formally accept this role.

CERN has called for a feasibility study for a next-generation frontier proton-proton machine, the Future Circular Collider (FCC). It would have a circumference of approximately 100km and, with major improvements in superconducting magnet technology that could be capable of ultimately producing 100TeV energy or higher proton–proton collisions—nearly an order of magnitude higher than the LHC. Further, in view of the cost and magnitude of the magnet development required, an electron–positron collider within the 100km tunnel (FCC-ee) is discussed as an intermediate step. This would itself be a very large undertaking.

Meanwhile, China has also been developing plans for a 100km-class collider, initially as an electron–position collider (CEPC) with a possible upgrade path to a future proton–proton collider (SppC), in a parallel plan to the FCC.

Whilst competition remains a key driver in all sciences, contemplating the duplication of such global facilities seems extravagant and even counter-productive for the field. For China to take the lead, it would require developments outside of as well as within science. Transparency in commitments and investments would be essential. Clear and proportional commitments to existing facilities and experiments will be required of funding agencies and their governments, in order for the international community to embrace and support major new initiatives. The benefit of large-scale cooperation and collaboration is clear. Bringing to bear the scientific resource base of the major Asian economies with that of Europe and the Americas, through formal agreements could make possible the next-generation of global-scale facilities. Asia could share the forefront of high energy physics; however, some major hurdles remain to be overcome.

With the swift ascent of Japan, China, India, and Korea on the technology front and in the larger economy during the past few decades, Asia and Asian science and technology have made some significant strides. How can this enthusiasm of diverse communities and cultures be translated into the political and economic power needed to harness the strategies of international high energy physics?

## 13.3   Enabling Factors for Collaboration in Asia

Asia's comparative strength lies in its human resources. Almost four billion people of the world's population live in Asia, with a cultural heritage extending over 5000 years. This makes Asia an economic and technological powerhouse. A recent UNESCO report described the Asia-Pacific as a vast, dynamic region, encompassing some three-fifths (59%) of humanity, with close to half of global economic output (45%) and strong expenditure on research and development (R&D, 42%). It is home to some of the world's most dynamic technological powerhouses (UNESCO, 2021). Interestingly, the report also highlighted the massive untapped resource of women physicists who can play an important role in future development. The number of women scientists majoring in physics has increased from about 15% in 1990 to 30% in 2010 in Asia (Kim, 2009).

Physicists across Asia are highly motivated to participate in international collaborations, although such opportunities are somewhat limited. Large countries in the Asia-Pacific, such as China, Japan, India, and South Korea; medium-sized countries like Australia and Taiwan; and smaller countries such as Vietnam and Thailand, participate in CERN programmes, most recently predominantly in the LHC. Participation is via national collaboration not via an Asia-specific programme. Many such countries also participate successfully at US laboratories. National laboratories in Asia, such as KEK in Japan and IHEP in China, host scientists from other Asian countries as well as from around the globe. Even the significant language and cultural differences found in Asia by western nationals have not deterred such collaboration. In fact, many experiments, especially in Japan but also in China and India, have significant numbers of participating international scientists.

However, Asian collaboration could neither emulate a version of the EU nor even of CERN. Experience shows that most Asian countries are often fiercely independent and aim to strengthen their national capabilities and international reputation. To produce internationally significant programmes, US and other Western nations have also driven their own ambitious Big Science programmes. International collaboration in major infrastructure facilities is now the norm for scientific communities.

While there is an imperative to produce significant national infrastructure, an important driver of national science programmes is to contribute in a significant way to high exposure, multinational facilities.

Whilst most countries have a variety of approaches to improving their core competencies in science and technology, a key driver is the development of national scientific capability and a rich intellectual property (IP) regime. These drivers are paramount in China and elsewhere in Asia, however, in fundamental science, 'Open Science'[4] and open collaboration are still dominant factors.

It is apparent in laboratories in Western nations that openness between researchers is an important determinant for improved outcomes for solving complex problems and feeding national productivity. For many years CERN had an open policy on IP rights that allowed scientists and member states to develop commercial applications. A relative openness remains.

The US government, through its various funding agencies, is much more concerned about leakage of IP, with considerable focus in recent years on IP developed in national laboratories being exploited in China and elsewhere. Europe too is concerned by the transfer of expertise and technology into Asian countries, particularly China, without due recognition or compensation for IP.

In very recent years, geopolitical considerations have become a serious barrier to large US contributions to Chinese projects, for example. Although political influence in scientific collaborations is anathema to most scientists, growing distrust at government levels will be a deterrent for scientists to collaborate. Recently, even major US scientific contributions to international scientific facilities have been discouraged if Chinese participation is taking place. In fact, recent research suggests great concerns over research transparency, lack of reciprocity in collaborations and consortia, and reporting of commitments and potential conflicts of interest related to these actions (Long, 2019). Trust is a fundamental ingredient in developing a healthy collaboration and a lack of trust and reliability can work against forming long-lasting collaborations that are essential in high energy physics experiments.

Such impediments, although clearly understandable, frustrate potential collaboration. Experience has shown that whilst scientists are more than happy to collaborate across all national borders, restrictions on such flow can easily become problematic bilaterally.

While international collaboration has been demonstrably successful and encouraged by scientists, future expansion of multinational facilities in Asia requires government-level agreements and trust to be further developed.

Trust, of course, cuts both ways. The selection procedure for the ITER (International Thermonuclear Experimental Reactor) appears to have damaged Japan's trust in taking the lead in important global facilities[5] (Nature, 2004). The world particle physics community is hoping Japan will host the future ILC. Whilst Japan carries the

---

[4] CERN's core values include making research open and accessible for everyone. For details see https://openscience.cern/node/22.
[5] See details: 'Time for Japan to shine?', Nature, 427, 763 (2004), https://www.nature.com/articles/427763a.

lead in this initiative, the Japanese government has not yet formally agreed to take on the host role.[6] The government remains adamant that clear indications are required from key nations of their willingness and capacity to provide a significant share of resources and expertise for the construction of the ILC. A catch-22 has arisen, with Europe and the US wanting a clear statement that Japan is willing to host the facility before committing substantial resources to its development.

In 2020, ICFA instigated the ILC International Development Team (IDT), to advance a programme to complete the final engineering development for the ILC, to develop ideas around the governance of a new international ILC laboratory, and to assist international discussion for the joint funding of the ILC.

In 2021, the IDT produced a detailed proposal, the ILC 'Prelab',[7] which would be responsible for carrying out these activities.

In a recent review of the ILC by the Japanese ministry, MEXT, requirements have been enunciated for the proposal to be taken further. To paraphrase, the key missing component is a broad agreement amongst major particle physics nations, that the next major collide beyond should be an e+e- collider, and that these nations are willing to fund the ILC Prelab as a global venture. (The IDT stresses that participation in the Prelab should not be considered a commitment to the full ILC.) The approval of the IDT proposal was deemed premature, with a formal agreement between funding agencies being a precondition for the Prelab.

This situation is not necessarily limited to Asia. In large projects in the US, such as LBNF/DUNE, there is also an expectation of contributions from Europe and Asia. Similarly, the LHC two decades ago had major contributions from outside Europe, especially from the US and Japan. In its current upgrade, the HL-LHC, international commitments from outside Europe have also been undertaken. However, whether it is the resources available or the confidence in assuming international resources will flow, large projects in the US and in Europe have proceeded before the completion of collaborative negotiations. In both projects, it should be noted that the facilities in question were commenced by existing laboratories, as part of their evolution. International demand for these facilities was, of course, important. However, there was no stipulation that a broad international agreement should be signed before work in the new facilities could commence. The host laboratories, in effect, acted as guarantors for the projects. It is understandable that for Japan, the scale of the ILC project, both capital and operational, demands a clear agreement for shared responsibility.

## 13.4  Towards a Unified Asian Focus?

Is it possible that the ILC or CEPC can be the foundations of a change for truly global-scale pan-Asian cooperation in high energy particle physics?

---

[6] See details in Normile, D '"Japanese government punts on decision to decision to host international linear collider', *Science*, March 2019, https://www.science.org/content/article/japanese-government-punts-decision-host-international-linear-collider.
[7] International Linear Collider International Development Team, 'Proposal for the ILC Preparatory Laboratory (Pre-Lab)', arXiv:2106.00602, June 2021.

In an attempt to bring a unified Asian focus to potential future high energy accelerator facilities, the Asian Committee for Future Accelerators (ACFA) was established at POSTECH, South Korea, in 1996. Its inspiration was taken from ECFA (the European Committee for Future Accelerators), largely set up to provide a forum for national CERN user groups and government agencies. ACFA has the clear goal of developing a unified view of future major accelerator facilities in Asia. With the discussion of a future Asian linear collider, ACFA aims to strengthen regional collaboration on accelerator-based science in Asia, promote next-generation accelerator scientists and encourage future accelerator projects in Asia (Kurokawa, 2005).

Currently, there are 13 members in ACFA including Australia, Bangladesh, China, India, Indonesia, Japan, South Korea, Malaysia, Pakistan, Singapore, Taiwan, Thailand, and Vietnam. The diverse levels of development and technical capabilities, along with an equally diverse range of political systems and economic status, have resulted in ACFA being little more than a forum to discuss national programmes. It has so far failed to focus attention on the potential for very large, international facilities in Asia. Without an equivalent of the European Community for formalising pan-Asia facilities, ACFA has not provided unified support for a high energy collider at the forefront of international facilities. Whilst national facilities of significance have been successfully developed and operated for decades, the step to host a multinational or international-scale laboratory in Asia remains elusive. The leading nations in Asian high energy physics, as in many other fields, are Japan and China, with significant contributions to the field also from India and Korea. They are important members of international bodies which promote and support future collider development work. In particular, participation in ICFA (International Committee for Future Accelerators) and FALC (Funding Agencies for Large Colliders) is of direct relevance to the future of the field.

As noted, an electron–positron collider has been highlighted in European and US planning exercises as the next-generation facility for the field. Japan and China are both, largely independently, pursuing a Higgs factory for precision studies of the Higgs boson; Japan through the ILC, China through the CEPC. Schematic overviews of CEPC and ILC250 are offered in Figures 13.1 and 13.2.

Asia's science and technology development has undergone a renaissance with the rapid rise of Japan, China, India, and South Korea as economic forces and developers

**Figure 13.1** Layout of the Circular Electron Positron Collider (CEPC) Project
*Source:* IHEP

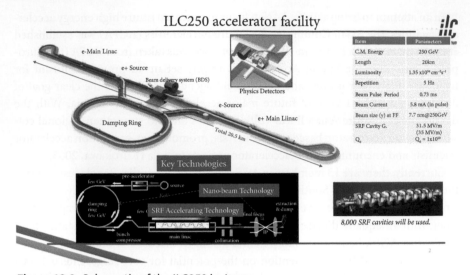

**Figure 13.2** Schematic of the ILC250 in Japan

*Source:* Shin MICHIZONO International Development Team (IDT) WG2/ KEK

of technology. As noted in Chapter 12, Asian development is often attributed to a successful model of technology and knowledge transfer for economic growth and prosperity.

New Big Science initiatives in high energy physics and astrophysics on the horizon, such as the ILC, the Square Kilometre Array, the Long Baseline Neutrino Facility; the Electron-Ion Collider; are some collaborations in which Asian participation is expected (Banks, 2020).

The Covid-19 pandemic took the world by surprise and raised new challenges for human societies, including the way we operate in large-scale research initiatives. One of the most noticeable impacts was on the research operations of Big Science organisations like CERN and ESO. These Big Science organisations had to postpone, cancel, significantly reduce their international collaborations, logistical support, and procurements of services.

In fact, Covid-19 resulted in a rise in research collaboration between some countries, whilst in others, including China and the USA, it has declined (Maher and Noorden, 2021). Nevertheless, the quest to resolve hitherto unsolved natural phenomena such as the origins of the universe, dark matter, and dark energy continues unabated. It requires effort not just from Europe and the USA but from the entire world. Big Science organisations have a key role to play in global scientific leadership scattered around many nations with comparative advantages such as optimised geographic locations of facilities for better research outcomes. In fact, Big Science communities have been remarkably effective in evolving complex processes and mechanisms to enable international collaborations (Robinson, 2021).

CERN's unique organisational characteristics and research processes are examples of how complex research can be conducted. Money alone is not sufficient to attract scientists and engage them in the type of collaboration that is needed for breakthrough research and responding to human challenges. The key ingredients are innovative leadership, a strong culture of collaboration and the ability to use consensual governance models to fund developing Big Science initiatives.

## 13.5 Major Players in Asia

The most active countries in experimental particle physics in Asia include Japan, China, Korea, and India.

## 13.5.1 Particle Physics in Japan

The most advanced collaboration between Europe, the USA, and Japan is channelled through the Japanese High Energy Accelerator Research Organisation (KEK). Japan made major contributions to the magnet programme of the LHC, continued with the HL-LHC and made major contributions to the ATLAS/LHC experiment. In 2014, an agreement was signed between KEK and CERN that cemented this relationship further to increase the collaborative effort on accelerator R&D and construction projects of mutual interest. These interests extend to key projects such as the LHC and its luminosity upgrade, the LHC injectors, linear collider studies, and the associated accelerator test facilities (ATF) and ATF2 (CERN, 2014). In 2018, joint US–Japan research on reducing construction costs for the ILC commenced.

Japan has been consistent in the development of its national facilities that enjoy international participation. After the KEK constructed the proton synchrotron and the Tristan e+e- collider (both of which were successful technically, but did not make major particle physics discoveries), Tristan was subsequently converted into the asymmetric e+e- KEKB 'B-factory' in direct competition with PEP-II/BaBar (SLAC). KEKB and the associated Belle experiment were highly collaborative with broad international participation. Both were successful technically and in their discoveries in particle physics, resulting in the 2008 Nobel Prize in Physics going to Nambu, Kobayashi, and Maskawa, to the great national pride of Japan and especially to Japanese science. The competition between the KEK and SLAC was keen and constructive. SLAC terminated the programme in 2008 while KEK continued data taking for a couple more years and went on to be upgraded to the highly innovative SuperKEKB/Belle II development. Data taking commenced in 2018. Many international collaborators from BaBar joined Belle II, bringing membership to over 900 collaborators from many nations, dominated by Asia but also with strong participation from Europe, the USA, Canada, and Australia. Various successful accelerator developments, including highly technical and important 'nano-beams', are also central to the ILC design.

With KEK's highly successful B-physics programme, Japan has clearly become an attractive destination for international collaboration on a large scale.

In neutrino physics, Japan has also experienced great success, resulting in clear national pride (the neutrino was dubbed Japan's national particle). Masatoshi Koshiba (died 12 November 2020) shared the 2002 Nobel Prize for physics for measuring neutrinos from a supernova with the ground-breaking Kamiokande detector, which was later expanded to SuperKamiokande, which resulted in the 2015 Nobel Prize being shared by Takaaki Kajita with Arthur McDonald (a former student of Koshiba) for the discovery of neutrino oscillations, which shows that 'neutrinos have mass'. Super-K is currently being upgraded to Hyper-K, along with the major upgrade to the J-Parc proton accelerator required to provide the increased flux of neutrinos for future studies of neutrino mass and CP-Violation in the neutrino sector. (Kajita's work measured the properties of neutrinos produced in the atmosphere.) The progress from K2K to T2K and the future HyperKamiokande involves a very important development, with a very large resource commitment, of accelerator-produced neutrino beams requiring ever-increasingly large detectors. T2K and HyperKamiokande have both attracted significant international collaboration.

The accelerator developments of the early as well as contemporary Japanese machines resulted in a highly capable and highly regarded expert community. This skills-base has been highly significant to the extent that there is an international expectation of Japanese participation in all major global facilities. Japan is an attractive destination for international scientists, both in collaboration via their home institutes and in interaction with staff members at Japanese institutions.

Japanese industry has been highly engaged in Big Science both in Japan and in equipment provided by Japan as in-kind contributions or purchased worldwide for experiments at CERN, the USA, and elsewhere. Japanese industrial technology capability has become very important for the commercial supply of components for international scientific infrastructure for use in experiments outside of Japan. An example is the Hamamatsu Company, which is known for its ubiquitous supply of customised, high-precision silicon particle detectors for major experiments all over the world.

Japan has provided major in-kind components, both large (for example, super-conducting magnets) and small (such as silicon detectors). The Japanese physics community made considerable contributions to international industrial and commercial benefits and earned a high reputation for enhanced industrial capability, highly valued among the international scientific community.

Following the discovery of the Higgs boson at CERN in 2012, the worldwide high energy physics community designated an e+e- collider 'Higgs Factory' as the highest priority for the next major machine in the field. Among other countries, Japan plays a very important role in these developments, and is considered by the worldwide particle physics community as the key candidate for hosting the multibillion-dollar ILC facility.

Obstacles remain for the Japanese government to make a definitive statement about whether they can host the ILC. These include both technical and resource-sharing questions. The developments carried out under the GDE and since have resulted in a mature technical design. However, a complete engineering design is required before construction can commence. This would necessarily include civil works planning and hence detailed site information. Having at least some partner nations agree to become members of the future ILC laboratory (or even just of the ILC Pre-Lab initially) with resource sharing agreements in place is an essential step for progress. Even providing the resources to complete the remaining technical developments will require agreements between key parties.

Whilst there is very clear and very strong support for the ILC as a Higgs Factory worldwide and for its establishment in Japan, strong political (both domestic and international) support will be essential for its success. Concern has been expressed along the lines: 'CERN discovered the Higgs boson and now the field just wants to improve on the measurements.' The quest for a deeper understanding of the fundamental nature of the Universe drives searches for unexpected phenomena. The detailed understanding of the Higgs boson is considered by many in the field to be the most promising path to pursue in the coming years. The so-called Higgs portal to new physics will have to be explained to populations, business and academic leaders, and governments alike, if the ILC is to become a reality. In Japan, an increased effort to broaden the knowledge of and interest in the ILC Higgs factory is underway. The entire international community also needs to pursue such communication activities.

Japan has undergone considerable setbacks due to the loss of the Fukushima reactors in 2011. The Kamioka Liquid Scintillator Antineutrino Detector (KamLAND) is an electron antineutrino detector at the *Kamioka Observatory*. It is an underground neutrino detection facility in Hida, Gifu, Japan. The KamLAND detector uses Japanese nuclear reactors as its neutrino source and the loss of the Fukushima reactor and ensuing radiation have impacted the experiments. Reardon (2011) suggests that the biggest problem for high energy physics researchers is the continuing need for power, as the Japanese nuclear power plants are still mostly shutdown. This serves as a reminder to the field of the need to look for sustainable options in the future.

## 13.5.2  Particle Physics in P R China

In comparison with Japan's perceived nervousness, China seems to be in a mood to expand its impact and exposure in wide-ranging technological and scientific projects, from space technology to advanced light sources, and in a range of commercial and military technologies. National pride is a clear driver, although the very largest projects are expected to generate technical expertise and commercial benefit.

Collaboration, in the broad sense, with the expectation of mutual benefit, as a clear driver, takes a secondary position to the primacy of national benefit. Western nations, as well as Japan, have been highly cooperative in helping to develop China's capabilities. There is concern that this highly collaborative approach is not a mutually balanced one with China.

China's growth as an economic power has been seen as an incentive for the international community to welcome the nation as a full member of the highly collaborative high energy physics field. The resources that could be added to the worldwide programme would be considerable, as would the very large intellectual capacity of China.

There is a long history of US–China cooperation in particle physics. Following Mao's era, Deng acknowledged that the world was getting increasingly interconnected, and China needed to open her doors. He adopted a Chinese market economy and, through openness successfully achieved unprecedented economic success.

Chinese physicists have trained in the USA for many years. Some remained, and some returned to China, following the end of the Cultural Revolution. The impact of the re-establishment of US–China diplomatic relations led to scientific interaction led by key physicists including T.D. Lee, C.N. Yang, and C.S. Wu and also Sam Ting for L3 at CERN. Of particular note is the friendship (both personal and professional) between Wolfgang 'Pief' Panofsky (Director of SLAC) and Lee, leading to a design for the Beijing Electron-Positron Collider (BEPC). This followed a proposal in the mid-1970s to build a 50Gev proton synchrotron, which was strongly discouraged by both Panofsky and Bob Wilson (Fermilab). They argued that such a machine and the associated experiments would be too expensive, at an energy level below those at Fermilab and CERN, and would need a level of technology and magnitude of effort that would result in China being unable to make a significant contribution to such a project ('like trying to jump on a train moving at high speed' (Panofsky, 2007)).

Colliding beam e+e- physics was newly developing. BEPC would bring China into the field at an early phase. BEPC became a highly visible and important national project in China. Deng Xiaoping personally wielded the shovel at the groundbreaking ceremony for the project. An upgrade of the BEPC to a tau-charm factory has been contemplated for years.

Panofsky's efforts to make BEPC a success cemented a strong relationship between the countries. Fundamental physics collaboration between SLAC and IHEP flourished, leading to the diplomatically significant United States—China Agreement on Cooperation in Science and Technology, signed by Jimmy Carter and Deng Xiaoping. Under this agreement the Joint Committee on Cooperation in High Energy Physics was formed and met annually until 2018.

Daya Bay near Hong Kong was one of a trio of experiments located in close proximity to nuclear reactors, along with the Reactor Experiment for Neutrino Oscillation (RENO) in South Korea and Double Chooz in France, which were responsible for seminal measurements of a critical parameter in the Pontecorvo–Maki–Nakagawa–Sakata matrix, which quantifies aspects of neutrino mixing (CERN Courier, 2021). In October 2007, neutrino physicists' broke ground 55 km north-east of Hong Kong to build the Daya Bay Reactor Neutrino Experiment. This facility consisted of eight 20-tonne liquid-scintillator detectors sited within 2 km of the Daya Bay nuclear plant to measure neutrino oscillations using antineutrinos produced by the reactor. After years of data taking, the Daya Bay has paved the way for the much larger and highly anticipated Jiangmen Underground Neutrino Observatory (JUNO) to tackle the neutrino mass hierarchy—the question of whether the third neutrino mass eigenstate is the most or least massive of the three—and to search for evidence of a neutrino–antineutrino asymmetry.

Most recently, Chinese physicists have also responded to the discovery of the Higgs Boson at CERN. Aiming to build upon the world 's familiarity with CERN and the notoriety it has obtained from the LHC, with its huge international collaborations, they are proposing the CEPC and SppC to follow. If indeed the Chinese government decides to move in this direction, reaching complete success will require major involvement by the international community. Expanding contributions to international laboratories would be seen very positively by the worldwide community.

In fact, for China to become a global facility, it will need to increase its provision of resources for the development of the many technical aspects equivalent to other international Big Science endeavours. The complaint by Western nations of an unbalanced flow of skills, communication, resources, and benefits will need to be addressed (Jia and Liu, 2014). China will need to become considerably more generous towards the established players in the field, to ensure the uptake of international physicists in a future Chinese facility.

It is interesting to note some parallels between these early developments of the Chinese particle physics programme and the current discussion around the proposal of CEPC and SppC (CEPC-SPPC Study Group, 2015). This study group was formed in Beijing in September 2013 to investigate the feasibility of a high energy Circular Electron Positron Collider (CEPC) as a Higgs and/or Z factory and a subsequent Super Proton–Proton Collider (SPPC).

The reliance on collaboration remains. The transfer of skills and technology that were required for the BEPC design, construction, and operation has resulted in a very important national capability. However, the design of the CEPC project has been heavily borrowed from the development of the proposed CERN FCC-ee machine (Innovationnewsnetwork, 2021). The huge step to the 100TeV proton–proton collider to follow both FCC-ee and CEPC (if either or both eventuate), will almost certainly demand global collaboration, wherever it is built.

The Xi Jinping era has seen a remarkable change from the relatively quiet opening of China pursued by Deng. It is clear from statements, strategies, and actions, that China is now on an expressly overt drive to establish itself as a world leader in a broad range of activities, including commerce, military strength, international politics, and scientific and technological prowess.

In recent times, major developments in a range of forefront sciences have been encouraged. The requirements for a major proposed funding programme required that the proposed facilities be world-leading and that there be a strong desire for international collaboration for any such facility in China. Through significant international recognition of China's top facilities and a strong desire to participate, Chinese national pride would be appeased.

The Belt and Road Initiative was described in a communiqué signed by many world leaders at the Second Belt and Road Forum in Beijing in 2019 as the modern version of the ancient Silk Road to strengthen world connectivity, 'promoting peace and cooperation, openness, inclusiveness, equality, mutual learning and mutual benefit'. However, the Belt and Road initiative has been viewed in the West as a means of replacing the post-war US hegemony. A parallel programme, 'The Thousand Talents', is a policy directed at 'bringing back' ethnic Chinese with their expertise and knowledge created in the West. These policies are quite unique and are difficult to reconcile with the usual notion of collaboration in global science projects.

The government's policies have resulted in scepticism and distrust of the unequal balance of resources and technological skills and ideas. As a result, a dramatic reduction in Sino–US scientific collaboration is experienced, impacting mutual cooperation and collaboration with scientists. In Europe and elsewhere, equity worries have raised concerns about continuing with open relations without any strings attached. The geopolitics of the current era are not conducive to major scientific collaboration. The generous approaches to international scientific collaboration are no longer universal.

In particle physics, the major next-generation facilities will certainly require global collaboration and cooperation. Within the scientific community, openness to collaboration is still highly valued. The current climate in international relations clearly precludes open science and collaborative approaches. Previous Big Science collaboration experiences, however, show that scientists typically follow the directions of their funding agencies and governments, which are usually non-negotiable. Scientists will generally fall in line if diplomatic intergovernmental relations preclude collaboration, as enforced by funding threats.

## 13.5.3  Particle Physics in South Korea and India

South Korea is also a significant player in high energy physics. Beyond particle physics, their investments in South Korea include the Pohang Light Source, a Rare Isotope Accelerator Project in Daejeon, the High Flux Advanced Neutron Application Reactor (HANARO), and the Reactor Experiment for Neutrino Oscillations (RENO), which was funded to the tune of US$9 million to collect data until 2009 (Bhat and Taylor, 2020).

South Korea had strong connections to international scientific infrastructure development programmes in Japan (KEK), the USA, and Europe, in addition to its very successful national programmes. For example, the RENO neutrino project has gained international attention for its precision neutrino oscillation measurements. However, this was a national-scale operation confined to Korean institutions.

India is an interesting case in that it is a very large country in terms of population and commercial capacity, but it lacks scientific facilities considered to be world-leading scientific infrastructure. Homi Bhabha, a theoretical physicist, laid the foundation for the revival of Big Science programmes (in relative terms) at the Tata Institute of Fundamental Research in 1945 to support nuclear and particle physics research (Wadia, 2009). In fact, India has had a strong nuclear science programme for decades, with its desire to develop nuclear capability fuelled by border disputes with Pakistan and, to a lesser extent, China. Over the years, particle physics has seen significant growth in expertise that has led to a significant programme of particle physics being carried out in Europe, the US, Russia, China, and Japan. The results of these initiatives led to closer collaboration between India and CERN culminating in a significant programme in LHC research. Subsequently, India recently took up Associate Membership at CERN. India also hosts many scientists from developing countries in a range of fields. A key characteristic of Indian science is the desire to generate homegrown technology from the ground up. This strategy has led to the rather slow development of national science and technology facilities, which are still predominantly nuclear rather than high energy in nature.

India had collaborations with several particle physics organisations, including agreements with CERN for cold testing on magnets using their human resources, which began in 2001 and lasted for three years. Indian scientists were involved in the cold testing of 1706 superconducting magnets for the LHC. This work continues with magnet R&D at Fermilab in the USA. Recently, India has launched a project to collaborate with LIGO and build an advanced gravitational wave detector in Maharashtra, India, to be completed by 2030. This will be made to the exact specifications of the twin Laser Interferometer Gravitational-wave Observatories (LIGO), in Louisiana and Washington in the US.

The Indian Neutrino Detector, an underground facility for cosmic neutrino studies, has been a casualty of a very extended development and construction timeline. The future for Indian particle physics lies in collaborating with other nations at CERN, in Japan and the USA, and perhaps in China and Russia if their future plans materialise. Big Science undertaking is not simply about finding funds but also about assembling scientists and laboratories capable of collaborating at international levels.

## 13.6 Conclusions

Asia is not a latecomer in particle physics. It has been involved in the field for many years. Recently, in addition to growing domestic programmes, some nations in Asia have been involved in front-line high energy physics research both in the US and in Europe. Japan and China, and to a lesser extent, South Korea and India, are now positioning themselves in the preparations for the next steps in Big Science projects in high energy physics. The two dominant Asian nations, Japan and China, whilst also having developed a strong national programme, are not yet seizing the opportunities to take leading roles in global-scale collaboration for the next generation of facilities for two quite different reasons: political differences and differences in research priorities.

Japan, for its lack of confidence in international partners willing to commit sufficient resources and expertise to the ILC project, seems unwilling to unilaterally declare an official interest in hosting the ILC. And China, perceived by many nations as an unwilling, inward-looking partner, has a long way to go before it can be considered a genuine collaborating nation. China has not yet demonstrated its willingness to host global-scale collaborative facilities with Western-style openness.

This discussion is not disconnected from the increasing role Asia is playing in global science, technology, and economics. With roughly half of the global economic resources and with growing global challenges, there is also a strong desire for Asian participation in a broad range of activities. With potentially huge economic and human resources available in Asia, the world's scientific and technical community eagerly encourages major contributions to the future benefit of humankind. As noted earlier in the book, humanity is facing significant challenges, including poverty, safe drinking water, housing, climate change, combatting epidemics and poverty, food shortages, education, and sanitation, to name few of these challenges.

Asian involvement in particle physics is riddled with challenges. Most of these are deeply rooted in the breadth and diversity of cultures and political systems. Many examples of excellent collaboration can be found. However, at a very broad scale, deep engagement is required for developing future global-scale facilities. Unfortunately, the commitment to global-scale collaboration is tentative at best, with national interests, a lack of trust between nations and even different levels of concern for basic human rights posing serious impediments for future global science.

There is growing concern that nationalism and economic expansionism are resulting in wariness and caution about the openness inherent in Big Science. Science is an international currency. Collaborative and participatory approaches must be fostered in order to increase Asian participation in Big Science projects.

If there are any lessons to learn from the past about the power of collaboration, Asia can profit from looking closely at how other Big Science facilities, such as CERN and ESO have fostered, developed, and sustained open international scientific collaboration for the greater benefit of science globally.

# 14

# The Social and Educational Responsibility of Big Science

*Steven Goldfarb, Christine Kourkoumelis, Viktorija Skvarciany, Christine Thong, and Shantha Liyanage*

## 14.1 Introduction

Over many centuries, humanity has benefited from the development of one of the most complex and precise scientific instruments ever conceived. It can resolve physical structures on the order of tens of microns and pinpoint celestial bodies millions of light years away. As humans evolved, they used this device to improve and expand their understanding of nature and to provide the data they needed to develop complex models of the universe.

This instrument, the human eye, when coupled with a continually developing brain and an ever-growing body of data, formalised and recorded over time, has helped our species to survive and evolve.

In more recent times—the past few millennia—the knowledge and methods accumulated and taught to subsequent generations helped them to develop new instruments capable of reaching greater and greater precision, looking both outwards to the very depths of the cosmos, and inwards to the elementary building blocks of our universe.

The previous chapters described the powerful nature and complexity of the devices that have evolved from this process, as well as the international collaborations necessary to build, operate, and consolidate the information provided by these tools. It is a never-ending process driven by the human need to understand the world we live in, to satisfy innate curiosity, and to provide the tools future generations will require to survive.

Here, we discuss problems that society and science are currently facing, the skills that will be needed to address these problems in the future, and the ways that Big Science can improve pedagogy, curricula, and teacher professional development to make learning relevant, interesting, and engaging. A discussion will follow on how critical learning processes and deep thinking in Big Science can help to understand creative and inquiry-based learning to prepare teachers and students take up the challenges of the next generation. The role of education in Big Science is also an

Steven Goldfarb et al., *The Social and Educational Responsibility of Big Science*. In: *Big Science, Innovation, and Societal Contributions*. Edited by: Shantha Liyanage, Markus Nordberg, and Marilena Streit-Bianchi, Oxford University Press. © Steven Goldfarb et al., (2024). DOI: 10.1093/oso/9780198881193.003.0015 important means to engage the public and drive public support for Big Science initiatives.

This chapter examines educational and learning processes and systems that are closely associated with Big Science organisations. These learning structures and processes are often developed within large-scale scientific organisations such as CERN, LIGO, ESO with a view to the creation and dissemination of new knowledge. Training among the scientists and engineers, who are involved in these projects, is also an important facet of these learning and educational activities.

Interestingly, while these ties are often developed to facilitate the production of new experiments and to recruit new talent for the field, their effects have a much deeper impact on society, establishing a basis for public support for science and influencing the methods utilised in the development of public policy. In a world in which imprecise information is generated and propagated at an ever-increasing rate, the value of developing and supporting the educational methods described in this chapter, designed to instil an early and profound appreciation of science, is both evident and vital.

Big Science makes significant inroads into education and science communication, and graduate students will be able to become acquainted with and demonstrate direct participation in ground breaking original research, which will not only motivate students to pursue original research but also teach and collaborate in complex problem-solving processes.

In essence, Big Science and learning are both essential and integral components of scientific progress and addressing complex societal challenges in the future.

## 14.2  Knowledge and Information Flow in Big Science

Knowledge and information flow are the lifeblood of all scientific organisations. Information can flow from peer to peer and is often multidirectional. As discussed throughout this book, Big Science experiments, such as ALICE, ATLAS, CMS, and LHCb at CERN, as well as astrophysics facilities, such as LIGO in the USA and Virgo in Italy, have made efforts to standardise, simplify, and optimise the flow of data across various networks of researchers within and outside Big Science experiments. Scientific publications are the direct output of explicit knowledge transfer. Big Science experiments use them as the primary source to disseminate this new knowledge for the public good (Bozeman and Youtie, 2017). There are several other channels and professional networks that facilitate the transfer of tacit and experiential knowledge. Knowledge generated through the scientific processes of these experiments is typically very specific and focused on targeted scientific aims. The interpretation and comprehension of these data and scientific knowledge require an additional understanding of experiential knowledge and technology. For instance, the interpretation of ATLAS collision data requires a thorough understanding of the detector, its components, its encoded signals, and condition data.

Over time, Big Science organisations have developed and refined processes to enhance knowledge and information flow to various parts of the organisation. Big Science organisations are also big at producing a vast amount of data and information and sharing it with their collaborating partners. The flow of knowledge in Big Science organisations needs to be centred not around individual scientists or scientific processes but on the cornerstone of a communication-centred approach. Some scientific papers produced by ATLAS, CMS, and other experiments have thousands of authors, each of whom is a scientist who actively contributed to the design and operation of the experiment, as well as data analysis. Naturally, individual scientists and organisational routines are part and parcel of this communication-centred approach.

A report by the OECD (2010) on large international research infrastructures categorised current and future facilities accordingly:

a. Experiments such as ITER ('The Way' in Latin), JET, the CERN accelerators (e.g. the Large Hadron—LHC) and its detectors (e.g. CMS and ATLAS), Pierre Auger Observatory in Argentina, ILC initiative, and SuperB in Japan;

b. User facilities for a small number of simultaneous users, such as ALMA in Chile, big optical telescopes, Square Kilometre Array telescope (SKA), and the Extremely Large Telescope (ELT) of ESO; and

c. Common facilities for many simultaneous users, such as the European Synchrotron Radiation Facility (ESRF) in France and The Institut Laue-Langevin (ILL) in Grenoble, European XFEL GmbH and Facility for Iron and Antiproton (FAIR) in Darmstadt, Germany, and the European Spallation Source (ESS) in Lund Sweden.

Big Science provides an opportunity for the open science regime and offers prospects for technology transfer and dissemination. The difference is that these experiments operate at the frontiers of science, hence attracting students from all over the world. Scientists working in these experiments are a select few, often those who have achieved a high level of scientific competency; hence, they act as Master Craftsmen and mentors. Students are eager to work under such talented scientists on these unique Big Infrastructure projects.

Big Science conveys to the public the positive aspects of investing in fundamental science by undertaking outreach and communication activities. For example, the European Particle Physics Communication Network (EPPCN) and the International Particle Physics Outreach Group (IPPOG) work to increase public contribution and visibility through the transfer of knowledge to public users. Other Big Science programmes in Astrophysics such as the European Organisation for Astronomical Research in the Southern Hemisphere (ESO), the Very Large Telescope (VLT) and the Atacama Large Millimetre/submillimetre Array (ALMA), among others, have also developed effective educational outreach activities.

Given the importance of data assimilation, curation, and analysis, it is necessary to describe some of the basic terminology used. Data generally refers to the basic and most common denominator that refers to facts and figures in a numerical sense which can be codified. Both numeracy and literacy provide valuable data. When data are contextualised, categorised, and organised, they become information that explains a particular fact that we need to know and rely on. Knowledge is the culmination of data and information that provide useful insights, meaning knowhow, understanding, and intuition that one can do something with the information provided. Big Science provides an opportunity to open up new science regimes and to provide opportunities for innovation, technology transfer, and dissemination.

As noted in the previous chapters (Chapters 5 and 8), the knowledge exchange across different disciplines involves not only particle physics, nuclear physics, astrophysics, and life sciences communities, but also scientists and professionals from information technology, electronics, nano technology, material sciences fields, etc. The synthesis of such knowledge often leads to the emergence of new applications and contributes to rich knowledge reservoirs.

Data and information produced in Big Science experiments are archived for future use and terabytes of such data can be made available for public use. Such valuable data and information are mined to develop new theories, to verify concepts and our understanding as well as to reshape or reformulate theories and experimental knowledge giving rise to whole new fields of inquiry. Big Science knowledge becomes a public asset when it is opened to investigation by the broader research community in a planned and systematic manner.

Big Science collaborations such as CERN and Advanced LIGO (Laser Interferometer Gravitational Wave Observatory) have worked tirelessly to develop and refine data characterisation techniques, data quality protocols, and extensive analysis processes. Machine learning has proved to be a powerful tool for the analysis of massive quantities of complex data in astronomy and related fields of study (Cavaglià et al., 2020). In a data-rich environment, students can learn numerous new techniques like Machine Learning (ML) which involve the science of design, development, and applications of computer algorithms that 'learn' to perform specific tasks. These tasks automatically improve their performance through the use of adaptive techniques and procedures (Guest et al., 2018). ML is an example of the power of learning from Big Science events.

Furthermore, Deep Leaning (DL) has emerged as a powerful tool to process data at scale, with similar sensitivity to traditional algorithms, but at a fraction of their computational cost, and is used in gravitational-wave astrophysics (Wei et al., 2021).

Data management plans are in place to ensure a systematic approach to the preservation and access of data (Anderson and Williams, 2017). Such approaches allow raw data generated in experiments such as ATLAS, CMS, and LIGO to be used and reused for current and future scientific outcomes.

Besides the use of data for scientific discoveries, the charter of most Big Science organisations extends to educating the public and helping students appreciate leading advances in science. In 2014, the LIGO Open Science Centre (LOSC) has built a web presence (Vallisneri et al., 2015), which now has terabytes of significant content from the S5 run, all reviewed and approved by the LIGO Scientific Collaboration (LSC). The key to the Open Science movement is that it allows not only LSC members but also non-members to access this valuable data.

Similar efforts have been made at CERN and ANSTO in Australia, with public education programmes aimed at disseminating scientific knowledge to students and teachers. In addition, the LIGO Gravitational Wave Open Science Centre (LOSC) provides a wealth of information to students and teachers, making available data from gravitational wave observatories as well as tutorials and software tools.

The culture of an organisation is the key to understanding the flow and dissemination of data and information. All Big Science experiments have distinct concepts, values, behavioural patterns, and acceptable norms for organisations. For example, LIGO's Open Science policy allows data to be stored, shared, and used by audiences other than those responsible for collecting such data. Such considerations are specific to the culture of Astrophysics and Gravitational Wave communities and the way they operate and relate to each other in scientific inquiries. As with all public investments, a degree of accountability and justification of big spending is required. There are cost-benefit considerations and justifications for public investment in Big Science (Florio, 2019).

CERN, as an organisation, has a specific scientific culture that is unique to the particle physics community (Knorr Cetina, 1999). The organisational culture may be driven by specific characteristics of the epistemic culture of individuals, groups, and the scientific community to which they belong. When scientists and technologists meet regularly at the CERN cafeteria, they converge to exchange ideas, concepts, and conjectures freely and without intellectual property considerations. The CERN cafeteria is a knowledge-learning hub where informal discussions become vital in deriving new theories and looking at data and information from a different perspective. These are instances of knowledge 'bonding' or relationship building that often happen in such a community of practice. As Lave and Wenger (1991) suggested, such a community of practice contains 'a set of relations among persons, activities, and the world, over time and in relation with other tangential and overlapping communities of practice'. Without bonding or developing an epistemic culture, the free flow of data, information, and scientific knowledge will not progress. Often, subtle details and insights are postulated and formulated in informal settings. This is because serendipity is often needed to steer to the next level of earth-shattering ideas by connecting and synthesising tacit knowledge, insights, and imagination. Learning plays a central role in Big Science organisations.

These learning interactions allow the participants to tease out individual biases and examine what works and what is right about the facts and events observed. Very often, there can be disagreements and an agreement to disagree or try different approaches to arrive at common solutions. This can lead to the birth of anomalies, and such anomalies might, in turn, lead to breakthrough science (Kuhn, 1962/2012). Indeed, some of the outstanding designs for ATLAS and CMS have come from pieces of paper scribbled at the CERN cafeteria and later finalised in a more formal setting and discussion (Jenni, 2017). The flow of knowledge and information is socialised and supported in a more amicable atmosphere.

Informal settings include casual conversation in the corridor, at a lift, during a walk to the next building, or at a café. Formal conversations take place in organised seminars, internal communications, and published papers, where findings are tabulated, illustrated, and delivered, thus allowing for cross-examination and rigorous review of findings in open forum. In large complex science organisations such as CERN and diverse high energy physics and Astrophysics communities, learning takes place continuously, and the flow of information and knowledge takes place in both informal and formal settings.

These formal and informal methods of communication provide the organisation with an opportunity to advance the free flow of information and knowledge. Learning in complex scientific organisations relies heavily on efficient learning processes that are organised as well as informal. The rubrics of an organisation to effectively function require an efficient flow of information which is part of the responsibility of the scientific community; without it, the organisation ceases to function, and scientific outputs like seminars, refereed papers, and reports slow down. Not all data, information, and knowledge are useful and necessary for all, and adding value to such communications result in discoveries from those who are embedded in the scientific processes. For example, the value and use of CMS and ATLAS data are intrinsic to those who are thickly involved in those experiments and understand the nuances of meaning and context when the right questions are being asked. As discussed in detail in our previous book (Boisot et al., 2011), information must address and provide answers to the questions that need solving; information and knowledge should be accurate, current, and state of the art, and they must also have utility value.

All complex organisations have limitations due to the inefficiency of information and knowledge flows. This can be due to undue bureaucracy or knowledge-hoarding. Even in knowledge-driven organisations such as CERN, knowledge flow is not entirely value free. Scientists are responsible for the cognitive value of research results and the reliability of technologies (Kitcher, 2001; Lekka-Kowalik, 2010). There are many organisational mechanisms that can be put in place to reduce lethargy in information and knowledge flow. For example, LIGO's Data Plans and processes of technical committees are set up to address specific issues, and the minutes of these committees can be shared to disseminate knowledge and information to all participants. In this way, the organisation can increase the flow and reduce information and knowledge hoarding.

Within the sphere of learning, investigation into the reliability of knowledge is important. How do scientists determine what is reliable and what is not? Information and consequent knowledge must have sense and making sense is critical. That is, it is necessary to provide some perspective, model, viewpoint, framework, and diagram so the recipients can formulate their own mental model and make sense of what is explained in a complex concept or idea.

Scientists participating in Big Science have an important role to play in shaping the future skills and knowledge of young scientists. The clarity of the latest knowledge and understanding and its effective diffusion are helpful for researchers, teachers, and students alike. The correct interpretation of the meaning can be considered to have a signal-to-noise ratio, with a high ratio indicating the high quality of the knowledge. This relies on independent, reliable verification. In the case of any scientific undertaking, information must address specific problems, be relevant to the investigation, and remain in line with prior findings.

Teachers and students can be selective in absorbing knowledge that is pertinent to their curriculum and pedagogy. As Biesta (2017) puts it, the distinction between poiesis and praxis helps us to see that teachers do not just need knowledge about how to do things (techne) but also, and most importantly, need practical wisdom (phronesis) in order to judge what needs to be done. Thus, Big Science experiments provide big opportunities to explore new knowledge and technology.

## 14.3  The Critical Role of Informal Education

As young children, we are fascinated by the beautiful blue sky. We share that fascination with those around us, who confirm they also see a blue sky, and teach us the name of the colour. Before long—perhaps immediately—we wonder why. Why is the sky blue is explained in his book on insights into medical discovery (Comroe, 1978)? After years of formal education, we might learn the mathematical equations needed to understand the relatively complex model of sunlight scattering, and we can calculate that the intensity of the light passing through the atmosphere is proportional to the fourth power of the frequency of its waves (deGrasse Tyson, 2010). It is a fallacy to think we need to learn to become scientists. We have been researchers since the day we were born. As soon as our eyes open, we look around, take in our environment, and try to make sense of it. What changes from that day forward is our understanding of the current knowledge base, the proposed models to describe it and make predictions, and the methodologies employed to build these models from the data.

Unfortunately, somewhere along the way, between kindergarten and the undergraduate electromagnetism course, many have lost the thread connecting the initial thrill of discovery and the formal education required to develop a deep understanding of its meaning. Great teachers recognise this and do what they can to bring that thrill back to the classroom. Perhaps they can relate the lesson to their own current research or, at least, to current science headlines. But this is not always an easy task and often the latest headlines involve seemingly complex topics unfamiliar to the teacher or the students.

This is where informal science education can make an impact. Scientists involved in current research know that much of their work is anchored in very basic concepts. Those who are active in public engagement recognise this and develop methods to convey the fundamental aspects of recent advancements in language that is accessible to the general public: dark matter and conservation of momentum, the Higgs boson and a person moving through a crowded room, gravitational waves and billiard balls on a sheet, viral infection and dominoes. By working together with formal educators, these scientists can bring the excitement of current research to the classroom and use these concepts to catalyse the learning process.

The methods employed by active researchers often differ from current formal education in a fundamental way. For example, researchers do not start from so-called 'first principles', deducing expected results from a basic rule, such as $F = ma$ or $E = mc^2$, as is often the case in the classroom. Rather, researchers start by gathering data, comparing it to other data then, through induction, seek to define the basic rules. Laboratory classes, equipped with inclined planes and pendula, help to convey these concepts, but their Sisyphus-like nature often lacks the excitement of tying these concepts to current research and discoveries. On the other hand, if a student is shown how conservation of momentum can be used to search for dark matter in high energy proton collisions, that excitement can be reignited.

Those of us involved in large-scale fundamental scientific research are fortunate. We play a role in one of the most fascinating fields of science, one that lies at the very core of human understanding of the universe. Furthermore, the hardware, electronics, and computing challenges of our field are forever pushing limits and, in many cases, their solutions result in important, tangible improvements to our daily lives.

Equally important, the expertise needed to address these challenges goes beyond the capabilities of any one institution or nation. It requires a concerted worldwide effort, involving international teams of researchers, engineers, and technicians working together, each bringing their own cultural backgrounds and points of view to the table. We thus have a golden opportunity to both teach the scientific process as well as promote the values of international collaboration around the globe.

Such lessons are brought to many high school, undergraduate, and graduate students, not via teachers in classrooms, but rather through their experiences visiting, participating in informal hands-on programmes, and/or working in internships. And the opportunities are growing fast.

As discussed in Chapter 9 of this book, several countries have invested hundreds of millions of dollars in synchrotron radiation sources and neutron sources. Australia, for example, developed large infrastructures including the Australian Synchrotron (AS), the Australian Centre for Neutron Scattering (ACNS), the National Deuteration Facility (NDF), Parkes Radio Telescope, and Australian Nuclear Science and Technology Organisation (ANSTO) facility. These facilities provide varying opportunities for students to study structures and dynamics in the physical world. They open up opportunities to learn from the real-life experience of working scientists in these facilities and work as research apprentices.

Big Science also provides invaluable infrastructure facilities for Open Learning. The learning of complex scientific processes is made possible through the Open Learning and Innovative Environment Frameworks (Segedy et al., 2015). Large Science Infrastructure facilities provide an innovative environment for technology-based authentic learning. Some examples include visits to the underground detector caverns, such as those of ATLAS and CMS, and interacting with scientists in these facilities, allowing students unique access to informal education activities.

Open Learning Environments allow students to think deeply and critically, to evaluate their learning tasks, and to realise their creative talents. Big Science projects provide opportunities for students to learn entirely new disciplines and acquire future employable skills such as data mining and modelling. Large Infrastructure facilities such as CERN and ESO, thus provide students with the possibility to widen their employment opportunities in emerging industries. For example. CERN graduates working in data modelling and networking have ended up working for prestigious financial companies. Similarly ESO outreach activities bring international network of scientists and science communicators from all ESO member states.

One should note that, in these cases, informal education comes from learning rather than teaching. That distinction is important (Biesta, 2017) and this often takes place in Big Science infrastructure, where student-centred learning is profound and multidisciplinary opportunities are abundant.

## 14.4  Global Challenges Facing Science Education

Although the benefits of informal science education and public engagement are evident, it is often the case that people in positions of power or seeking such positions will attempt to undermine these efforts or even attack the fundamentals of science directly in order to attain personal gain. Ironically, tools developed by our own community to support our work by improving communication within the community are easily used to disseminate intentionally false or misleading statements in direct contradiction with science.

Unfortunately, large audiences, whether due to naivety, ignorance, or simply fervent belief in the sources of the misinformation, are susceptible to these campaigns (Olan et al., 2022). This is not new to humankind by any means, but current communication tools and methods, including social media and some professional privately owned media platforms, are able to disseminate false and dangerous information across the globe to vast audiences essentially instantaneously. Countering the effects of these activities is not only important, but given the social, health, and ecological issues facing our planet, it has become urgent.

Interestingly, the most important lesson we bring to the classroom strikes at the core of the problem. That is, as scientists, we must make careful and thorough measurements and we only publish results when the evidence has been well-tested and shown to be significant. It is key that educators and researchers alike also stress the importance of expecting the same from authorities in all parts of society. The students need to hone their own scientific and technological skills at deciphering the messages coming from their own heroes in public.

The methodologies and procedures developed in large international scientific collaborations for the publication of results provide effective decision-making models for students. Simply tracking the life of a single measurement from its roots as an idea to publication provides an enlightening story of scepticism, testing, and persistence. The typical publication gauntlet includes local teams, analysis subgroups and groups, team presentations, publication committee edits, revisions and re-writes, peer reviews, and, in many cases, failure to publish. Learning about these challenges not only exposes students to the process of collaboration, but it also provides a stark contrast to many public sources of information, thus improving students' analytical and methodological ability to discern the value of what they are being presented to the scientific community.

## 14.5  Education Strategies for Large Research Infrastructures

Large Research Infrastructures (LRI), such as those belonging to the EIRO forum— the Forum for the European Intergovernmental Research Organisations including, for example, CERN, European Synchrotron Radiation Facility (ESFR), European Space Agency (ESA), and European Molecular Biology Laboratory (EMBL)—have developed and implemented a variety of innovative approaches to science education. Their activities aim to develop and strengthen ties between their scientific research and civil society. Effective approaches include direct involvement between researchers and teachers, students and citizens in their local communities and beyond. In this manner, they bridge the gap between scientific research and civil society by using open formal and informal school education as mediators and by showing and making accessible to local communities the positive impact of the scientific discoveries and the applications derived from them. This is not only a social obligation to inform society of their findings, but also a means to build and maintain support from those who fund the research.

The International Particle Physics Masterclasses (IPPOG, 2021), described in Section 14.6 below, started in 2005 and are another example of creating effective synergies between schools, universities, and research centres. A further example of informal education is the programme of virtual visits to the LHC experiments at CERN. This enables the students to 'visit' experiments online, giving them a chance to talk to researchers from their classroom, thus sparing travel money and time. Of course, field trips to the LRI, when possible, provide much better insight and experience into scientific working life. CERN annually receives around 400,000 requests for on-site visits, with about 60% coming from schools.

The laboratory can only accommodate about one-third of its visit requests. This was the main reason for the ambitious 'Science Gateway' project, which was launched by the Director General of CERN in 2021. CERN Science Gateway is a new education and outreach centre, which was inaugurated in October 2023—for education, training, and outreach targeting the general public of all ages. It is hoped to be able to accommodate all requests for visits to CERN, and will include a 900-seat auditorium, hands-on experiments, and several in-depth exhibitions on the science and technology of particle physics. The general public over the age of five is included and CERN inspires and encourages young people to pursue science, technology, engineering, and mathematics (STEM) education.

The above examples, as well as a number of Outreach projects supported by the European Commission Horizon 2020 programme, aim to bridge the gap between scientific research and civil society by making accessible both the positive impact of the discoveries and the direct transfer of the cutting-edge technology that made the discoveries possible.

CERN has implemented several educational initiatives and programmes for students, teachers, doctoral students, and junior researchers. Through these programmes, the trainees learn valuable lessons in science and international collaboration, and when they return to their home countries, they become ambassadors of this open scientific culture.

CERN also plays a key role in training the trainer. In particular, CERN's teacher programmes—often taught in the teachers' native languages—provide invaluable training in the form of apprenticeships. The trained teachers then initiate school projects with their students based on the subjects chosen, according to the students' preferences, and in consultation with the affiliated research institutions. The schools open their doors to society, enterprises, and local stakeholders in order to co-design and disseminate the student projects.

Through the training available in Big Science organisations and the transfer of both education and technology, educational policy-makers are able to articulate learning outcomes as they develop programmes where students learn in an interactive manner, including hands-on activities with real data, student-developed information apps, web-fests, and science-inspired art. A further benefit is that the formal education curricula become more oriented towards inquiry-based science education and more focused on modern discoveries and their explanations. The students are encouraged to pursue deeper learning instead of memorising material and to develop skills in collaborative problem solving. These skills are necessary in a competitive working market where they will seek employment regardless of the field chosen. It is widely known that the fields of mechanical precision, electronics, medical industry, banking, stock exchanges, etc. greatly appreciate the skills acquired by Masters and PhD students in particle physics. Of course, the ones who seek science-oriented careers find themselves with extra advantages, having participated in activities directly related to scientific methodology and contracts with Big Science organisations.

## 14.6  National and International Science Education Networks

Researchers engaged in informal science education have been diligently bringing these lessons into secondary classrooms around the world for many years now. Although scientists often do this of their own accord, today most of the activities are coordinated through educational programmes supported by dedicated universities, national networks, laboratories, and experimental collaborations.

Some of the larger national networks currently engaged actively in particle physics education and outreach include Netzwerk Teilchenwelt in Germany, QuarkNet in the United States, the public engagement components of the Science and Technology and Facilities Council (STFC) in the United Kingdom, Institut national de physique nucléaire et de physique des particules (IN2P3), in France and Istituto Nazionale di Fisica Nucleare (INFN) in Italy. These networks foster the development of relations between local schools and researchers and provide common platforms and tools for the development of national programmes.

The European Union (EU) through funding of outreach programmes (in Life Long Learning, FP7, Horizon 2020, and Erasmus+) supported the creation of e-infrastructures for engaging science in the classroom based on inquiry-based learning. Programmes such as Learning with CERN's ATLAS (Long, 2011), Discover the COSMOS, the PATHWAY to Enquiry Learning Teaching, the collection of labs hosted by the GO-LAB initiative, the Inspiring Science Education, the OSOS which supports a large number of European schools to implement open schooling approaches, the FRONTIERS Project, bringing Nobel Prize physics to the classrooms, and finally the most recent Citizen Science project REINFORCE are noteworthy. All of the above programmes brought together a network of educational communities, science centres, museums, and research centres in activities to diffuse the IBSE across Europe. At the same time, they created a very large repository of tools which are at the disposal of teachers. Each of the above networks has trained more than 1,000 teachers in workshops and implemented their resources for many thousands of students across Europe.

The International Particle Physics Outreach Group (IPPOG) is an international collaboration of scientists, science educators, and communication specialists working across the globe in informal science education and outreach for particle physics. It was founded in 1997 by then-CERN's Director General Chris Llewellyn Smith, with the goal of coordinating the efforts of these various independent entities to use the limited resources more effectively to reach classrooms and the public across the globe.

Although the members of what was originally called EPOG (European Particle Physics Outreach Group) came from CERN's member states, the group expanded and took on a broader scope. IPPOG became an international collaboration in 2016, adopting a Memorandum of Understanding, and, at the time of writing, hosts 40 members, including 33 countries, six experiment collaborations, one international laboratory (CERN) and two national laboratories (DESY and GSI) as associate members.

A primary global activity of IPPOG is the International Particle Physics Masterclass programme introduced above. The programme pairs scientists involved in current particle physics research with students, either in the home institute of the scientist or the classroom of the students, to engage in the excitement of fundamental research. Students learn about the physics goals and the functioning of an experiment, then get a chance to be physicists for the day by doing their own analysis of real data from the detector. At the end of the day, the students meet with other students via videoconference to discuss their results, thus getting a first-hand experience of international collaboration. In the 2019 edition of the International Masterclasses (before the Covid-19 pandemic), 14,000 students from 54 counties and 225 institutions participated.

While the focus of these Masterclasses is particle physics and much of the activity utilises data and tools from major experiments, the lessons taught are much broader in scope. Students' inherent interest in science is re-awoken through exposure to cutting-edge research activities and close engagement with the scientists involved in that research. They not only about science but also about the importance of scientific methodology and its human understanding of nature. They also learn first-hand the excitement and value of international collaboration, and how it contributes to scientific advancement through human diversity and worldwide cooperation.

Another international programme supported by IPPOG is the Global Cosmics initiative, which provides an organisational umbrella for the variety of experiments and data-sharing activities based on cosmic particle detectors located in classrooms around the globe. Events like International Cosmic Day organised by DESY, Netzwerk Teilchenwelt, IPPOG, QuarkNet and Fermilab, and International Muon Week organised by QuarkNet, expand the reach to locations where collider physicists might not have the opportunity to bring Masterclasses.

Through the development and coordination of these major programmes, IPPOG provides tools for scientists and educators around the globe to complement formal educational curricula with the excitement of cutting-edge research. In some cases, national education programmes are experimenting with the introduction of more current material, adding Masterclass-type activities to their official curricula. In these cases, IPPOG and local researchers can serve as a resource for material and tools, partner with teachers, and connect (physically or remotely) with classrooms.

Outside the classroom, recent initiatives have focused on bringing scientific material into non-scientific environments, such as art, culture, and music festivals. In many cases, festival organisers have sought out entertaining scientific activities to broaden their cultural base and extend their audience reach. In other cases, scientists have proactively sought out the organisers and convinced them of the added value to their events. The common result goes beyond science education to develop cross-cultural comparisons that not only engage audiences effectively, but also deepen our own understanding of the societal implications of fundamental research.

## 14.7  Coopetition: The Collaborative Aspects of Big Science

Communicating the value of international collaboration is a key objective of the education and outreach efforts of the particle physics community. In terms of importance for the wellbeing of future generations on this planet, it should be seen in equal light to the efforts outlined above for teaching the scientific process and instilling appreciation for fundamental research. There is no doubt that, as populations grow and challenges become more and more global in nature, humankind will need to embrace more cooperative worldwide strategies.

Large-scale international particle physics collaborations provide realistic and proven models for the development of such environments. The technological and financial scope of current facilities in experimental particle physics necessitate the need for their construction and operation by multiple institutional and national bodies. This has led, over the past few decades, to the establishment of large, diverse multinational collaborations hosting thousands of physicists from around the world.

Demonstrating the effectiveness of these collaborations, despite the diversity of national and cultural backgrounds would already be a success story, but it would miss the point. The fact is that the very diversity of these groups in a scientific environment is what gives them their strength. Each project launched by a collaboration, be it hardware development, computing, or data analysis, benefits from the broad range of ideas born by the diversity. These are extremely important lessons to bring to both the classroom and to the public in general, especially during times of political nationalism, anti-globalisation, and populism.

Beyond the advantage of diversity, the experiments also provide lessons regarding competitive and cooperative relations within and between collaborations. The need for independent confirmation of scientific measurements often necessitates the construction of at least two experiments exploring similar domains of the field. In high energy physics, this translates into the construction of detectors installed on the same or comparable accelerators exploring similar energy ranges.

The Large Hadron Collider (LHC) at CERN, as an example, hosts four major experiments situated at locations on the accelerator where the particle beams are brought into collision. Two of the experiments, ATLAS and CMS, are considered to be general-purpose detectors, designed to measure a very broad range of physics, from precise measurements of the Standard Model to searches for new and exotic phenomena (see Chapters 1 and 2 for details). These experiments have similar capability and host major international collaborations, each with around 5000 members contributing to the development, running, and maintenance of the detectors, as well as the extraction and analysis of data and the publication of results.

A typical (albeit simplified) work cycle for typical analysis necessitates a competitive stage, during which the independent experiments collect and analyse new data, simultaneously developing and improving their detection methods and analysis techniques, and then reporting their new and/or improved results to each other during major conferences. Each conference launches a cooperative stage, during which the experiences gained and methodologies created by the researchers are shared in detail, along with the new results. From this moment on, the scientists learn about and often integrate each other's methods into their own, making improvements to future measurements and thus improving the field's knowledge base and expertise.

One of the projects organised through CERN's IdeaSquare utilised these lessons for the training of management teams from professional, non-scientific backgrounds. In events organised with the ESADE School of Business, management executives were split into two independent groups, given Masterclasses by scientists from the competing ATLAS and CMS experiments, and then asked to give presentations to each other in the style of a typical physics conference. Their methods and results were shared and then put together to combine results for better outcomes. Conversely, another educational activity hosted by IdeaSquare is the Crowd4SDG citizen science programme, which encourages non-scientists to use entrepreneurship and innovation to tackle science problems. These lessons can provide an important basis for the development of future cooperative trends between large-scale science and the private sector (see also Chapter 6).

## 14.8  Diversity and Inclusion in Big Science as a Social Responsibility

Big Science provides opportunities for better integration of extensive international, cultural, and social connections. In such a knowledge environment, a spectrum of ideas, technical knowledge, methodologies, and different perspectives strengthens a collaboration's ability to imagine and implement new solutions. This includes a wide range of activities, spanning detector hardware, electronics design, magnets and superconducting materials, crystals, data, and software development. Such a variety of work, makes it possible to foster a broad spectrum of imagination, ingenuity, and thinking skills.

Given this highly innovative and knowledge-intensive work, high energy physics is still somewhat confined to a narrow group of intellects. Low socio-economic groups, minorities, under-privileged communities, women, and non-binary genders, are arguably under-represented. After all, when more than half of the world's population is under-represented, (Makarova et al., 2019), then the field is not inclusive, and it is depriving itself of key contributions. Efforts to address diversity in high energy physics and astrophysics have received more attention during the last decade. Dedicated offices and boards within the major laboratories and experiments are identifying inequalities and helping to define new policies aimed at increasing inclusiveness and accessibility for everyone. CERN and ESO already have focused on employing more female physicists. This is, however, not only confined to high energy physics but also common to most sciences and physics in particular.

Concerning the physical sciences, such as high energy physics, men have traditionally dominated the field. As an example, of the 218 Physics Nobel Laureates between 1901 and 2021, only four were women (Nobel Foundation, 2019). And, despite efforts over the past few decades to increase the participation of women in the field, the major LHC collaborations each contain fewer than 30% of women on their author lists.

The diagram in Figure 14.1 for example, presents an update to the core values, as defined by the Office of Diversity and Inclusion (CERN, 2021c).

Still, the problem is deeply imbedded throughout the field. In most countries, the fraction of female physics students is relatively low and decreases further as one

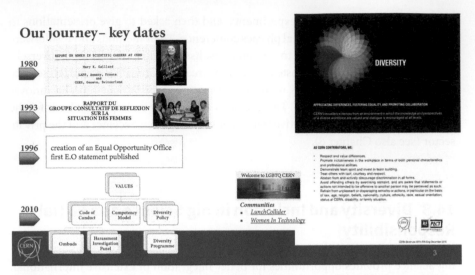

**Figure 14.1**  Core values supporting diversity and gender equality initiatives at CERN
*Source:* © CERN

moves up the academic hierarchy. The reasons have been studied at length by international organisations such as the International Union of Pure and Applied Physics (IUPAP), the Institute of Applied Physics (IAP), and the United Nations (UN). Several initiatives have been launched in order to attract young females to physics such as the UN's 'International Day of Women and Girls in Science'. On this day, for example, IPPOG organises a set of International Particle Physics Masterclasses dedicated to girls and young women and featuring a diversity of volunteer women researchers as mentors (see Figure 14.2).

Despite such efforts, there are a variety of factors that continue to limit the retention of women in the field. As a historically male-dominated field, certain practices present unreasonable roadblocks for young women. These include a highly competitive job market during the early years of professional development, continuous pressure to publish and the need to present at conferences held in distant locations and often over weekends. These constraints can be difficult for anyone, but they particularly affect women with young families. They are also often unnecessary and only serve to maintain the status quo. Efforts in education and outreach, working with diversity and inclusion groups, serve to expose these problems, let young women know they are welcome in the field, and given them a voice in their collaborations.

Concerning formal education, science teachers need to encourage girls to continue their studies of mathematics and science by increasing their self-confidence, utilising cooperative and collaborative work paradigms instead of competitive ones, and giving them positive images of careers in mathematics and science. The required balance between working hours and personal life should not act as a barrier to pursuing successful careers. And a roster of outstanding female scientists can provide role models for young women.

**Figure 14.2**  A group of business leaders being introduced to the concept of 'coopetition' through an IPPOG Particle Physics Masterclass, CERN
*Source:* © CERN

Furthermore, it is imperative to provide these students with well-trained teachers and mentors from within their familiar environment to spark the girls' interest, creativity, and participation.

Efforts to increase the inclusion of the scientific community consisting of sense-disabled people (especially visually disabled) and senior citizens should also be mentioned. Several dedicated projects (e.g. REINFORCE), develop special sonification protocols and tools for data analysis by disadvantaged citizens. The ambition is to extend the senses through scientific inference. The objective is to design user-centred software to produce audio-visual outputs from 1D and 2D data which integrates multidisciplinary and interdisciplinary astronomers, engineers, computer specialists, software designers, educators, disability specialists, bioengineers, neurobiologists, and sociologists, both blind and sighted, and addresses the topic of accessibility to scientific data particularly in the astronomy field.

The senior citizens programme aims to both counter the increase in the isolation of many members of this social grouping, by involving them in a wider research community—where they can work from home—and provide a useful and interesting pastime. With organised transportation to and from the close-by research site, they are encouraged to attend monthly meetings between themselves and the scientists involved. This way, their critical thinking will be promoted and diffused.

**Figure 14.3** A virtual visit to ATLAS from a classroom in Ghana
*Source:* © CERN

## 14.9  Science Enriching Education in Other Disciplines

Big Science facilities create educational initiatives in multidisciplinary areas by increasing connections with other disciplines, such as CERN accelerator development, design, social sciences, business, engineering, and IPPOG. IPPOG has a variety of tools to reach even the most remote parts of the world and conducts virtual Masterclasses as shown in Figure 14.3.

Big Science initiatives are derivative technologies that inspire the exploration and design of new products, experiences, built environments, and systems for the betterment of society. Methods from creative practices guide the exploration and are enhanced as a result of working with Big Science. New curricula and learning activities for applied innovation are shaped by the complexity of Big Science, which encompasses many new disciplines and learning environments.

Oxman proposes a similar train of thought by proposing Kreb's Cycle of Creativity, (Oxman, 2016), which combines nature and culture, production and perception, through four domains representing creative exploration: Art, Science, Engineering, and Design. This is a metaphor for 'creative energy' that generates, consumes, and regenerates across domains. There has also been a connection between design and business, with design thinking becoming a tool for understanding complex relationships among consumers, producers, and business stakeholders. Technology transfer offices have also been developed to use business and entrepreneurship to apply technology derived from science.

Deep technology is essential for innovation, as it has the potential to transform markets and solve complex environmental and health-related challenges. However, deep technology can often fail to translate from lab to market, leading to the 'Valley of Death' and the development of new products early in technology development. Design and entrepreneurial skills for applied innovation have the capacity to identify new products early in technology development and help traverse this valley of death. As discussed previously in the case of the social sciences, understanding how to work with Big Data will also be helpful in decision -making when designing new products, systems, experiences, and built environments.

In responding to the United Nations' Sustainable Development Goals (SDGs) that are being explored, future vision canvases and implementation roadmap techniques have been developed to equip students with skills and techniques relevant to technology development, social innovation, and developing design applications for the future. To address this, CBI student teams often have CERN scientists as coaches who provide a CERN perspective and technical viewpoint. The CBI programme has seen new innovation activities develop by working with Big Science, with new ways to approach radical innovation by 'thinking in orders of magnitude'. These small and tangible approaches spark curiosity, a sense of agency, and provide inspiration. While students may not study in the domains of Big Science, they are being inspired by concepts and their development process.

## 14.10 Conclusions

Big Science organisations needs to diffuse knowledge to the public in a variety of creative ways if society is to benefit from such knowledge. Education and learning are crucial knowledge transfer processes and education is one of the Big Science's immediate and recognised societal impacts. Big Science organisations often adopt open science policies to disseminate such knowledge to public and private sector organisations.

CERN, ESO, and other large-scale science programmes have strong commitments to education and outreach programmes. Learning and dissemination of scientific knowledge and data are integral parts of their mission. These organisations adopt educational and knowledge transfer policies to inspire young people with outreach programmes and provide training opportunities by offering a range of hands on cutting-edge technology engagements, fellowships, doctoral studies, and industry placements.

Learning is a powerful knowledge transfer process where students and scientists in Big Science organisations become the agents of change. Not only peer-to-peer learning, but numerous undergraduate students benefit from CERN's summer programmes, educational research engagements and internships. These programmes act as conduits to transfer knowledge from laboratories to the public. Working together with industry and also with financial assistance from governments, various Big Science educational initiatives, both formal and informal learning programmes, are offered through scholarships, industry attachments, and other direct and indirect educational activities. These initiatives help to inspire the next generation of scientists, engineers, and innovators.

Learning is also a powerful tool for connecting and sharing knowledge in human civilisations. It is a process bridging different disciplines and allowing science to flourish since the growth of Big Science programmes in the 1970s. Learning is also an integral part of the Big Science community, where collaboration facilitates the sharing of research facilities, exchange of data, and contributing to building human curiosity and capability.

Learning-by-doing plays a key role in Big Science experiments as complex machines like the LHC get built for the first time without any previous model to follow. In this sense, learning is also a process that is internal to Big Science organisations. This includes a significant number of postgraduate students. Big Science projects provide unique learning opportunities for a variety of disciplines.

IPPOG, IdeaSquare, and ATTRACT programmes are among the signature Big Science project initiatives that connect scientists with students. Big Science connects with big data, communication, and the publication of scientific outcomes. All these activities are closely connected to educational programmes and initiatives that enhance curiosity, promote science education, and facilitate the dissemination of knowledge to the public.

While making sense of such knowledge is key to informing the public, it is also important to cultivate a culture of continuous learning and knowledge sharing. Understanding the connections between knowledge and how to engage students of all types is essential for explaining complex topics to the general public, such as dark energy and dark matter.

Learning in Big Science settings requires the development of different learning models and effective innovative pedagogies such as Masterclasses, multitask project-based learning, boot camps, challenge-based innovation, and others interactive learning programmes. Learning complex concepts requires simplification. The explanation requires making sense of what people observe, then interpreting the knowledge in terms of scientific theories, experiments, and their outcomes. Learning about Big Science is a challenging process.

Both learners and teachers have to use innovative metaphors, visual aids, design artefacts, real-world examples, practical examples and engagement. Taking on board the opportunities and challenges offered by digital education and making use of available tools and programs will also be beneficial (Streit-Bianchi et al., 2023).

Benefits can be derived from outreach activities as they allow the natural progression of supporting a knowledge culture that links with Big Science organisations and in particular, particle physics and astrophysics communities. Big Science projects involve complex learning processes linked to new technology, new techniques, and complex ideas.

Learning is indeed an intrinsic process in Big Science. Scientists and researchers have to engage in complex learning that is required for continual learning and innovation. It is an art, a channel of connection, a social experiment, and a process that adds value to scientific discovery in a multitude of ways.

# 15
# The Future of Big Science and Social Impacts

*Shantha Liyanage, Markus Nordberg, and Marilena Streit-Bianchi*

## 15.1 Introduction

The examples and practitioners' views shared throughout the previous chapters offer four key takeaways. First and foremost, the focus on fundamental knowledge is the main generative mechanism in Big Science. To qualify as Big Science, the scope and intent of these projects by the scientific community need to be focused on significant fundamental science issues and problems. It could be about testing a specific grand unification theory, hypothesis, or experimental verification of the Higgs boson, understanding gravitational waves, or dark matter and dark energy. Contemplating even larger, more technically challenging attempts, like the Future Circular Collider (FCC) at CERN, serves as an indispensable example of the build-up of cumulative scientific knowledge, technological capability, and the synthesis of human ingenuity.

How large collaborations are successfully put together is a culmination of many factors and is an arduous process that depends on the drive, determination, and commitment of key individuals. It can be regarded as a churning process that requires a continuous cycle of systematic formulation of project ideas, engaging with different people and organisations, and refining theories and experiments. In addition, such ideas need to obtain political and organisational support. Sometimes, the state of technology can be a barrier. For example, for both ATLAS and CMS, developing radiation-tolerant electronics remained a major hurdle towards functioning signal processing systems for a long time (Brianti and Jenni, 2017).

The second key takeaway is the path dependence of Big Science. Big Science ideas can be described as an extension of both theoretical and instrumental dependence (Peacock, 2009). Many examples discussed in this book reveal a particular pattern, organisation, and design that is noteworthy in the development of experimental physics research. The case of ATLAS and CMS at CERN, for instance, suggests that instrumental or experimental dependence is necessary for the falsification or confirmation of a theory—in this case, the Standard Model of particle physics, which is a theory concerning electromagnetic, weak, and strong nuclear interactions developed throughout the latter half of the twentieth century. This raises further questions:

Shantha Liyanage et al., *The Future of Big Science and Social Impacts*. In: *Big Science, Innovation, and Societal Contributions*. Edited by: Shantha Liyanage, Markus Nordberg, and Marilena Streit-Bianchi, Oxford University Press. © Shantha Liyanage et al., (2024). DOI: 10.1093/oso/9780198881193.003.0016

Where do big ideas come from? What is the process of selecting those ideas? and how to fund and take those ideas forward?

Third, as this book highlights, Big Science differs from more traditional science in view of its complexity in organisations and management. In a resource-intensive environment, Big Science projects appear to run with responsible leadership and management processes requiring systematic, clear reporting, and documentation. These mechanisms are indispensable to ensure scientific rigour and accountability to international partners, to circumvent any potential pitfalls and to deal with uncertainty. Careful planning, resource allocation, teamwork, and dynamic leadership processes are key success criteria for successful Big Science operations.

Big Science also involves careful management processes to draw on the right types and levels of collaboration that allow building interdisciplinary teams and groups to last over decades. One of the main challenges is the effective management of communication and coordination among team members spread across various research organisations in different geographic locations. ATLAS, CMS, and LIGO had thousands of scientists working from different laboratories across the globe.

The capability of dealing with diverse epistemic cultures requires significant personal skills. Above all, managing complexity in these experiments can be difficult due to the sophistication of the technologies. For example, the ATLAS inner tracking system (ITS) is a highly complex detector system that requires effective management to ensure efficient operation and maintenance. ITS is also a modular system that needs careful engineering integration and installation, requiring expert knowledge to deal with various components, testing, continuous monitoring and reliable operations. Such insights ignite more important discussions and raise the following questions: How can we manage knowledge generation and translation processes?; What has learning or networking got to do with Big Science operations?; and How do we determine what works and what does not?

Fourth, Big Sciences give rise to new disciplines and novel knowledge systems. In fact, a discipline is a body of knowledge that is practiced by a group of scientists who are disciplined to adhere to standards for the creation and dissemination of knowledge in a particular field. As complexity grows, Big Science experiments in high energy physics, advanced telescope projects, astrophysics experiments, and molecular biology experiments use some of the most complex and deep scientific and technological techniques, knowledge, and skills.

Big Science initiatives are designed to solve grand challenges and find answers to very complex problems. Naturally, this requires bringing together diverse groups of talented scholars from varied disciplines converging from different nations all around the globe. Big Science undertakings thus unite countries and scientific communities irrespective of their individual political or national, geographic interests.

CERN, for example, has developed competencies over 70 years, building effective international collaborations even among fierce adversaries. Similarly ESO established in 1962 operated over 60 years. Building effective collaboration is the most difficult challenge in Big Science operations. Many protocols need to be put in place for diverse groups to work together. Time will reveal how the global community will unite to tackle future Big Science challenges. There is potential for countries in Asia, Europe, Africa, Central America, or North America to work together despite cultural and geopolitical differences.

When working on Big Science projects, trust becomes an essential ingredient for effective international collaborations. Collaborative efforts at each stage of the development of complex experiments need transparency, good communication, and open discussions. In this modern world, the prospects for establishing and sustaining cooperative Big Science initiatives have been negatively impacted by ongoing conflicts, a lack of trust, commitment, transparency, and accountability. To maintain support for Big Science initiatives, science diplomacy is essential to building success in collaborative scientific efforts.

Considering all chapter contributions, we present the following conclusions to explain the Big Science processes and develop the Collaborative Innovation Framework, a general purpose framework to illustrate factors that contribute to knowledge generation, development, and diffusion (COIF).

## 15.2  Chapter Reviews and Findings

### 15.2.1  Connecting the Dots: Big Science, Breakthrough Innovation and Society

We have come a long way in appreciating the historical roots of Big Science; learning with respect, how it takes decades to build a world-leading facility, building advanced technical infrastructure, and forming outstanding scientific teams. The ultimate success of all these entities relies on unity amongst dynamic networks of universities and supportive industrial bases across nations around the globe, along with a substantial degree of political support.

The Introduction (Chapter 1) provided an overview of the Big Science concepts covered in this book. It addressed the following questions: Why is Big Science important? How does it contribute to novel scientific knowledge and how does it impact public goods and social benefits? Based on a sample of practitioner accounts of Big Science endeavours, we identified missing links between integrating Big Science methods that are dispersed among technological tools, industrial opportunities, educational possibilities, and broader societal considerations.

The following are some of the most important lessons from Chapter 1:

1) The historical roots behind Big Science indicate that it takes several decades to build a world-leading facility with supporting technical infrastructure and future projects will require an even longer timescale;
2) Although a Big Science facility is typically centrally located, its success relies on building a dynamic network of (international) universities for carrying out the projects, data sharing, analysis, and a supportive industrial base; and

3) Big Science's contributions to society seem to be primarily serendipitous, meaning that their full societal ramifications have been rarely correctly predicted in advance.

## 15.2.2 Isn't it the Difficult Journeys that Lead to Beautiful Destinations?

As Richard Feynman, Nobel Laureate and a well-known physicist, said: 'It doesn't matter how beautiful your theory is, it doesn't matter how smart you are; if it doesn't agree with experiment, it's wrong.'

The iconic success stories of LHC's ATLAS and CMS experiments provide vivid perspectives on how Big Science projects come about: how they evolve from pushing the frontiers of science, driven by ambitious scientific goals and producing advanced technologies.

Chapter 2 describes how these experiments work in tandem with one another while also using various technologies and methods on their own. The genesis of ATLAS and CMS confirms that big ideas and concepts come from individuals (not necessarily laboratories where they work) and epistemic culture is the magnet that creates teams and groups who want to work together. The collaborations build around such individuals and teams that complement each other intellectually and socially. Collaborations naturally build around them. It is, however, naive to believe that all good ideas will receive funding and that the ability to attract funding support will determine which idea is a winner. There is always healthy competition among scientists, laboratories, and technologies (e.g. circular and linear colliders). Competing groups are working in a wide range of fields such as astrophysics, dark matter, and dark energy.

As explained in Chapter 2, the capital-intensity and complexity of these projects require long time frames for research inquiries and investigations. These experiments have a specific scientific and technological scope, complex design, and advanced engineering know-how that require the collective expertise of some of the best brains in the field of high energy physics.

Some of the key lessons from Chapter 2 are noted below:

1) The CMS and ATLAS experiments contributed to the development of new technologies and techniques in detector design, microelectronics, data processing, and computing;
2) Big Science projects are inherently complex undertakings and no individual can expect to solve all issues;
3) The level of precision and accuracy required is immense, with very high energy, radiation, and the intensity and speed of collision rates;
4) Lean engineering of the overall concepts of the ATLAS and CMS detectors was of paramount importance. In that sense, one can say that each one of them was its own prototype; it has a pre-determined specific task and implied life

cycle, not designed to be identically copied or replicated but to allow possible upgrades; and

5) Developing and scaling up Big Science experiments require innovative thinking, human and financial resources, time, multiple iterations and building strategic partnerships. They also require collaborating with leading experts in science and engineering to address specific scientific scope with due consideration to limitations (e.g. current and evolving science policies, geopolitics, and economic and funding cycles).

## 15.2.3 A Handful of Wisdom from a Success Story

To provide the reader with an understanding of the depth and scope of Big Science projects, the book has zoomed in, on the illustrative example of the LHC machine, which marks the successful culmination of over 80 years of continuous and tireless development of new technology in particle accelerators. Naturally, the LHC played a pivotal role in the discovery of the Higgs boson. It was a massive collaborative effort and a true reflection of what humans can accomplish when they collaborate. The Higgs boson discovery is not an isolated single event. It was the culmination of decades of both theoretical and experimental research.

Even with comprehensive mitigation mechanisms in place, Big Science projects are risky. However, the rewards can be significant: the discovery of the Higgs boson and the associated Nobel Prizes are examples of the latter. The joint discovery of the Higgs boson by ATLAS and CMS in 2012, a missing piece of particle physics' Standard Model (SM), is a significant recent achievement.

The following lessons can be derived from Chapter 3, which deals with the construction of the world's largest machine—the Large Hadron Collider:

1) Big Science projects must be willing to take calculated risks in order to realise their scientific goals. When incidents happen, problems must be addressed quickly, openly, and collaboratively as a team and new approaches may not always succeed. Everyone collaborates and works together to solve the problem, avoid cost escalation, mitigate risks, and manage them effectively;

2) Responsible governance is to work out all possibilities and remedies to protect researchers, organisations, and public safety in a high energy and radiation environment;

3) Governance structures need to consider the project's overall life-cycle stage and the complexity of each part of the connected system, for example, upgrade projects may require separate management structures from those involved in daily operations;

4) Organisational structures are streamlined to facilitate project work packages with overall directors taking consultative approaches while ensuring a rapid and effective decision-making process; and

5) Due to the complexity and advanced level of selected technologies, Big Science projects can protect themselves against massive, unforeseen effects through partitioning and securing them in 'blocks' or 'sectors' that will contain any unanticipated damages.

## 15.2.4  Versatile Big Science

Chapter 4 describes the various disciplines and associated technologies of accelerators and detectors that have enormous economic and societal benefits. Distinctive examples were drawn in this regard from medical, biomedical, energy generation, energy transmission, space, computing, and other industrial fields.

The following related lessons can be derived from Chapter 4:

1) A strong relationship exists between Big Science and innovation. Partnership with industry benefits both collaborating parties with innovative practices that combine creativity and intellectual diversity;
2) Human aspects such as multi-ethnic and multinational environments, as well as cultural differences between research organisations and industry, should not be overlooked in Big Science projects;
3) Big Science projects are sources of ground-breaking technologies. Partnerships evolve over time from idea refinement to the final stages of industrialisation, including the R&D and prototype phases;
4) Development of new technologies from Big Science requires new approaches to innovation: pushing the frontiers of superconducting electrical transmissions for the HL-LHC suggests potential benefits and applications for society; and
5) Practical applications that go far beyond the accelerator fields, for example, range from climate change to cultural preservation.

## 15.2.5  Here and Now Determines the Future

Chapter 5 leapfrogs into the future to illustrate how CERN's past experience can shape future high energy research frontiers. Aside from particle physics, other fields such as astrophysics and cosmology must collaborate in order to fully understand the big, open questions about the nature and behaviour of the universe. Scientists in general collaborate to develop new and more efficient scientific tools that pave the path for major discoveries. Progress in science certainly calls for unity among the different communities, including not only particle physicists, information technology professionals, and other professionals, but also a multitude of other stakeholders, including industry personnel.

The lessons that can be derived from Chapter 5 are:

1) Future Big Science initiatives like the FCC, coordinated by CERN, can potentially keep pushing the frontiers of science well into the twenty-first century;

2) Effective collaborations need to be open, transparent, and diverse. A collective understanding of Industry and academia from an early stage in the life cycle of long-term projects is necessary to facilitate technology development and rapid diffusion;

3) New types of organisational management approaches are necessary to ensure continuity, as well as new approaches to reward and motivate the young generation of researchers in long-term experiments and projects;

4) In designing the next generation of Big Science projects, the evaluation of the socio-economic impact should be integrated from the early phases of the project life cycle; and

5) Continuous review, evaluation, and monitoring of projects will maximise returns from such large public investments.

Big Science projects typically attract large numbers of leading researchers, engineers, technicians, and students from thousands of universities, and research laboratories all around the world. There is a strong chance that these intellectual powerhouses will want to work more closely with businesses and governments to form alliances that will help them tackle other complex or urgent problems.

The painstaking process of extrapolating from past experiences to future situations can be useful and rewarding because no individual—certainly not only the project leaders—can anticipate all the risks and rewards of complex experiments.

## 15.2.6  Creative Constructs: Big Science, Learning Cycles, and Design

A number of chapters covered designs, leadership, medical technologies, and examples from astrophysics to show how innovation works in Big Science organisations and experiments.

### Simplicity and Significance

All complex research, like detector-based technologies, can be simplified with creative designs, simplifications, and innovation. To this end, the concept of Social Learning (SLC) presented in Chapter 6 provides valuable insights into how design thinking can be used to solve Big Science complexity and how, if more widely used, it can support innovation processes involving both particle physics and astrophysics experiments.

SLC simplifies and structures the innovation process. These learning cycle approaches extend beyond the academic domain and, when combined with an interdisciplinary approach, can even bring together distinctly different domains such as science and the humanities, as well as levels and domains of expertise.

Such an open innovation approach can assist in understanding the current operations of Big Science and in planning for future research projects, for instance, in the search for dark matter and dark energy, where the use of open data sources and the sharing of information are becoming increasingly crucial.

The following related lessons can be derived from Chapter 6:

1. Design concepts offer new and innovative ways of codifying the process of how knowledge generated in Big Science projects could be disseminated for the benefit of society (e.g. experiments carried out at IdeaSquare at CERN are provided as examples);

2. Design artefacts, demonstrate the importance of incorporating the (end) user experience in the design process as early as possible and call for a multidisciplinary approach;

3. Design practices can codify, abstract, and generate new meaning for Big Science knowledge by synthesising human, technical, and economic considerations into tangible design artefacts. Social Learning Cycles (SLCs) can be expanded beyond the scope of Big Science projects (e.g. ATLAS to why they were originally applied) to the level of knowledge and technology transfer impacting applied sciences beyond the hosting organisations;

4. Designs have taken on a major role in Big Science in the visualisations and image reconstruction of events in LHC experiments and astrophysics; and

5. Beyond detector and accelerator technology, design concepts, engineering prototypes, and artefacts have contributed to innovations in various components and devices.

## 15.2.7 Driving the Vision to Reality

Human interactions and leadership, which have long been central interests in Big Science operations, were covered extensively in this book. Chapter 7 discusses leadership from the perspective of complexity and its application to Big Science projects. Leadership demands versatility in new skills. Big Science leadership requires skills that go beyond those of a typical leader and the usual project management skills that call for a combination of complex knowledge and abilities.

Big Science ethos reminds us of the need to place emphasis on transparency, empathy, ethical behaviour, as well as building trust. A common emphasis on collaborative leadership was then observed, on the scientific and technical credibility of the elected spokesperson as well as the responsibility of the leader towards the scientific community.

The following lessons can be derived from Chapter 7:

1) Leaders of Big Science need to be inspirational, credible, and competent;

2) Leaders across Big Science experiments can have diverse and different leadership approaches due to the size of the collaboration, geographical location, and disciplinary orientation;

3) Leaders have to deal with complex project structures, diversity of technology, and budgetary constraints;

4) Leadership traits include ethical, authentic, and shared leadership, stakeholder management, listening to employees, valuing diversity, building trust, empathy, and having diverse leadership culture to foster innovation; and

5) Leaders need to pay attention to gender issues: the participation of females in high energy physics is still relatively low. Efforts to increase diversity in leadership roles and to increase female participation will bring unique perspectives, creativity, and insights into scientific endeavours.

## 15.2.8 Never Believe the Sky Is the Limit

The marvel of astrophysics discoveries is another important pillar of Big Science. Using significant examples, Chapter 8 demonstrated the critical role of technological innovations in astronomical discoveries.

The complexity, cost, and audacity of these investigations require large collaborations modelled on high energy physics collaborations, regarded as pioneers. The astrophysics community displays a unique epistemic culture, in which data and analysis sharing has become the norm, with open access and open communication.

The authors describe the spectacular discoveries ranging from the search for Earth-like planets—the existence of life in nearby solar systems—to measuring subnuclear scale displacements that capture the wisps of passing gravitational waves produced in cataclysmic black hole collisions or attempting to identify the elusive nature of dark matter through the presence of neutron stars.

These discoveries take decades in parallel with the development of new technologies, techniques, and leading research. Hence, sustained long-term commitments from all stakeholders, particularly governments and funding agencies, are necessary for breakthrough innovation in Big Science.

The following lessons can be derived from Chapter 8:

1) Pushing audacious ideas results in ambitious Big Science astrophysics projects that rely heavily on observational science;

2) Technology development in the field of astrophysics contributes to fundamental research through extensive data collection using big telescopes, satellites, and radio astronomy to study the universe;

3) The path to discovery can extend over decades due to technological change and the use of new techniques such as multi-wavelength;

4) Astrophysics involve multiple stakeholders around the world and has complex data sharing and analysis functions. Sustained long-term commitment to all stakeholders, particularly governments and funding agencies, is necessary for long-term success; and

5) Astrophysics and particle physics research communities have complementary epistemic goals and cultures with common interests in the study of the universe.

## 15.2.9 Breakthroughs in Medical Technology

Chapter 9 traces the development of detector technologies in medical applications, which have been in use since the early days of Big Science and have thus contributed and will continue to contribute to societal well-being. Using examples of medical technology such as radiation therapy and cyclotron-based proton therapy, the authors focused on the technological trajectories and potential infrastructure contributions to miscellaneous fields of medicine using specific combined treatments, such as Neutron Enhanced Captured Therapy where Big Science may well play a role.

The following lessons can be derived from Chapter 9:

1) LINACs and medical detectors have revolutionised medical diagnostics, radiation therapy, and cancer treatment. Due to high capital costs, accelerator, and detection technologies have a long adoption curve;

2) Big Science has contributed to the development of imaging technologies used in medical diagnosis such as CT, MRI, and PET scans that have not only assisted in detecting accuracy in (e.g. beam-based methods and algorithms) the LHC, ATLAS and CMS and other experimental technology components for any defects but also contributed to significant advances in medical applications;

3) Once a new technology is adopted, there is a long tail of continuous improvements that follow and are used in various interconnected disciplines;

4) Technologies that are not clinically applicable may still be useful in addressing changing needs in both established and developing science and technology sectors; and

5) Advances in accelerator technology are making medical technology more affordable and accessible, lowering capital costs, and thus making these medical treatment types more accessible to a larger number of people.

## 15.2.10 Multiple Perspectives: Big Science and Society

Several chapters addressed the organisational and social construction of knowledge and all emphasise the importance of embedding learning in Big Science projects to translate knowledge into usable forms and learning experiences for future generations.

Chapter 10 explored the multifaceted and entangled relationships between Big Science and society, offering varied perspectives on its complexity and richness. The authors note that this avoids the use of single-lens, closed, and rigid valuation frameworks. The challenge therefore remains to capture both ontological and epistemological aspects of the scientific activity, as well as the idiosyncrasies of the individual actors and communities taking part in it.

A potential starting point could be to consider why and how scientific activities emerge out of fundamental characteristics of human nature, such as curiosity, imagination and serendipity and how they are capable of generating collaborative communities and, ultimately, collective and individual value. The widespread Covid-19 pandemic has already demonstrated that the knowledge produced by Big Science is not a luxury but rather a necessity for addressing both current and upcoming problems facing our planet.

The following related lessons can be derived from Chapter 10:

1) Big Science knowledge production is complex and multifaceted; it is an interdisciplinary human enterprise;
2) Human complexity in networking and relations is path-dependent and evolutionary;
3) The outcomes of Big Science need to be judged from various lenses without prejudice;
4) Big Science experiments are open to scrutiny and constructive criticism for their simplicity and effectiveness; and
5) Curiosity, imagination, and serendipity are intertwining forces for solving complex problems and Big Science knowledge and technology are global public goods.

## 15.2.11 Facing Big Data Challenges

Chapter 11 shows how data modelling, artificial intelligence, and data mining have hugely contributed to data analysis in Big Science projects. As the next-generation Big Science instruments are going to be loaded with ever-massive data generators, the quest for collection, storage, and analyses of data are challenging tasks. Currently, the impact of newly developed algorithms, rare signal discrimination, and methodologies to reconstruct images in high energy physics and astrophysics is visible.

Even more intriguing questions have been asked about how to share, manage, and use massive amounts of data generated in particle collisions and astronomical observations. Open access will remain a requirement for publicly funded research data, regardless of the scientific domain or geographical boundaries. The data life cycle is increasingly becoming valuable for science while making inroads to contribute to social, educational, and economic development. Artificial intelligence and quantum technologies are starting to have an impact on research fields.

The following related lessons can be derived from Chapter 11:

1) Big Science acts as a stimulus for science–industry interactions and has an impact on society as a co-developer, lead user, and a source of inspiration;
2) Big Science and big data techniques are intertwined, and techniques such as machine learning, grid computing, data mining and modelling, predictive analysis, and artificial intelligence will enable new technological discoveries and innovation;
3) Concepts such as openlab at CERN and Open target and Industry programme at EBI-EMBL can be used to test the organisational transition of multidisciplinary science–industry interactions;

4) Big Science can foster interactions between scientific progress and technological needs, as well as between technological solutions and new scientific pathways; and

5) Big Science can serve as a testbed for the transition of data gathering, analysis, programming, and algorithm developments from scientific applications towards broader societal impact. It may also be able to aid in a change of mindset.

## 15.2.12  Big Science's Call for Entrepreneurs for the Common Good

Although not strategically targeted, the impacts of Big Science connect with centralised economies and have the potential to generate large-scale prosperity in society through the development of enterprises. The authors of Chapter 12 have recognised that these enterprises need social equity mechanisms based on Big Science collaboration and values for cultural transformation.

The following related lessons can be derived from Chapter 12:

1) The positive impact of fundamental science in Big Science is connected with centralised economies that have brought large-scale prosperity through free enterprise, which is used as a social equity mechanism for transformation;

2) Big Science can transform research into social good and give it a direction, for example, in achieving universal developmental aspirations by using reliable, circular models and contributing to the achievement of the Sustainable Development Goals of the United Nations;

3) By collaborating with different stakeholders (such as in ATTRACT), Big Science has the potential to address social issues;

4) Innovation serendipitously ignites by using knowledge management tools with social capital, thereby overcoming obstacles to achieving quality of life; and

5) Serendipity can initiate the process of transforming fundamental science into breakthrough commercial innovation, and it may be possible to 'systemise' serendipity (e.g. CBI process at IdeaSquare at CERN).

## 15.2.13  An Outlook on Asia's Positioning in Big Science

The authors in Chapter 13 look ahead to the possible leadership role Asia may assume in the future in Big Science projects in particle physics. Asia has been strongly involved in front-line particle physics research for some time, like in the US and Europe. Despite some Asian countries making large investments in Big Science projects, they are still lagging behind their US and European counterparts in attaining scientific excellence in high energy physics and/or astrophysics.

The following related lessons can be derived from Chapter 13:

1) Although several Asian countries are actively involved in major Big Science projects in particle physics, there does not so far appear to be a strong common foundation or consensus to take on leadership in Big Science;

2) Japan and China are both signalling intent, but their approaches and motivations appear to be very different;

3) Concrete initiatives from Asian countries are necessary to spearhead Big Science initiatives (e.g. LIGO-India) collaboration;

4) Collaborative Big Science initiatives in Asia are necessary to combine growing talents in the region while pursuing fundamental scientific ambition and keeping technological and economic growth in perspective.

## 15.2.14 The Future of Big Science Stands on Shared Wisdom

The majority of Big Science organisations are focused on producing fundamental scientific knowledge. However, these projects have had significant direct and indirect spill over effects on the public good in terms of knowledge dissemination and learning.

Learning is recognised in Chapter 14 as a crucial element that influences peer-to-peer learning, academic learning, and the success of many postgraduate students through initiatives like the CERN summer programmes and other initiatives. Outreach activities can provide significant benefits and contribute to the development of a strong epistemic culture in specific areas of science.

The following related lessons can be derived from Chapter 14:

1) Large-scale international collaborations have developed structures and processes that facilitate the flow of information and knowledge. A mixture of competition and cooperation, driven by shared curiosity in diverse mindsets, helps to optimise this flow and can be an example for others;

2) Education and public engagement are critical for garnering support for large-scale scientific projects (e.g. the FCC);

3) Educational and outreach programmes facilitate two-way interactions between scientists and the general public, fostering future research; and

4) Big Science can thrive with public support and engagement, particularly in the recruitment of young scientists to take over future high energy physics development.

A strong commitment to education and learning is a key feature of Big Science. The development of the next generation of scientists, engineers, and technicians is crucial, but it is not the only factor driving the internal work that Big Science organisations must do. Moreover, there is a need to instil faith and interest in fundamental research, as well as to teach and share the methodologies that go with it, to help future generations develop interest and engagement in science.

## 15.3  Towards an Analytical Framework

The findings outlined in these chapters suggest that Big Science should be viewed as a complex system that interacts with its own components (Robertson and Caldart, 2008; Palmieri and Jensen 2020). Big Science displays elements of complexity, serendipity, open nature, networking, social design, and creative thinking that connect to society. Big Science complexity is highly structured and process-driven, as demonstrated in Big Science experiments. There are many subsystems within the complex structures, like the Inner Detector, Calorimeter, Muon Spectrometer of the ATLAS and CMS detectors. Some of these complexities increase as energy levels and the sophistication of experiments increase.

Multidisciplinary teams and groups must work together in order to solve some of the most complex operational system problems while staying within budget. Big Science creates vast knowledge networks and an innovation ecosystem that is characterised by knowledge-based and knowledge-driven open innovation.

The findings also support the view that the rationale for public expenditure and political support for large-scale science infrastructure underpins Big Science's benefits and outcomes (Wagner et al., 2015; Gastrow and Oppelt, 2018; Hallonsten, 2021).

Big Science collaborative processes seem to facilitate the seamless transfer of fundamental knowledge to the technology development of initiatives. Useful outcomes trigger as a result of cumulated knowledge on working with complex systems and subsystems. The diffusion of Big Science knowledge into useful outcomes is never a linear process.

The fundamental characteristics of research prevent too much simplification of meaning and translating knowledge across the social learning cycle (SLC). It is therefore necessary to untangle the underlying epistemological and ontological positions as to what scientific knowledge is feasible to transfer, what is not, and under what conditions. The question then remains: how to effectively integrate such processes within scientific collaborations. That is, how to explain the complex processes involved in the collaborative work of epistemic groups such as scientists, social scientists, and business managers and how useful ideas get translated from Big Science facilities such as CERN and ESO.

Based on the insights gained from the previous chapters, some of the key issues to consider here include:

a) How much does Big Science contribute to the common pool of public resources?

b) How does it promote greater collaboration?

c) How does it produce primarily non-excludable and non-rivalrous pure public goods, allowing anyone to use them without restriction regardless of whether they contributed to their creation? and

d) How does it create, to a certain level, intellectual property that can be developed in conjunction with public goods?

It is clear from the examples given in the previous chapters how theoretical and experimental knowledge came to be a crucial component of Big Science organisations and how these organisations developed over time. As seen, the organisation and management of this body of knowledge have complex dynamics, and not all knowledge is easily transferrable. Moreover, most of the valuable knowledge remains the tacit knowledge of the scientists and researchers who created it in the first place. They may be unavailable or unwilling to actively participate in knowledge diffusion, or they may lack the time or desire to serve as transfer agents or consultants in related technology transfer projects. It is necessary to translate tacit knowledge into useful organisational knowledge, but these processes can be quite inefficient or challenging.

In a systematic approach to the above questions and based on the findings from the previous chapters, we propose a simplistic framework called the Collaborative Innovation Framework (COIF) in order to show the connection between knowledge creation, development, and diffusion. Further work is naturally required to test the applicability of this framework. The proposed framework, captures the essential components of Big Science knowledge processes and the dynamics of fundamental

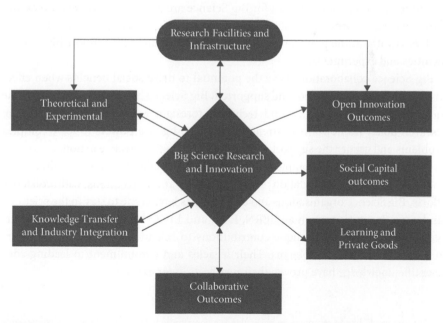

**Figure 15.1** Collaborative Innovation Framework (COIF) for Big Science
*Source:* Created by author S. Liyanage

knowledge diffusion as illustrated in Figure 15.1. The framework exhibits the key components and dynamic relationships.[1]

The basic premise in Figure 15.1 is that in Big Science, fundamental knowledge drives open innovation, which in turn paves the way for new applications resulting from, for example, detector and magnet systems, which in turn create new fields of knowledge. The COIF model demonstrates three types of relationships: the process leading to the domain of knowledge; the knowledge validation process and the development of knowledge; and finally, the constituents of the knowledge conversions.

## 15.4  Concluding Remarks

The fundamental tenet of Big Science and society, is the need for collaboration, collegiality, openness and sharing benefits of knowledge. The ability of human beings to live in harmony has the potential to drive science, technology, and social change through Big Science collaborations. While Big Science collaborations have the potential to drive social change, there are scientific and technological challenges that require concerted human efforts. In this book, we have outlined, with examples, many challenges facing the progress of Big Science. Such progress is determined by complexity, serendipity, design, and knowledge diffusion processes, together with the human desire to converge intellectual power and knowledge.

Public investment and support for Big Science are essential to solve complex and growing complex problems that are worth solving. Scientists alone cannot resolve all problems without the political and social support to fund and support Big Science facilities and experiments.

Big Science collaborations have the potential to drive social benefits when effectively coordinated, managed, and supported. Big Science has the power to overcome most barriers with scientific and technical foresight and with carefully chosen research policy frameworks to strengthen its investigative powers to solve complex problems and garner the support of all nations through collective action.

Given the nature and complexity of issues such as climate change, health, social and environmental degradation that require advanced solutions with collective efforts, Big Science organisations and experiments have come to stay in the scientific landscape. Institutions such as CERN, ESO, and LIGO have shown time and time again their ability make unique contributions to scientific understanding require to solve such complex problems. Their legacies and commitment to leading-edge scientific knowledge have proven their justification for existence.

---

[1] Authors would like to recognise, in particular, the contributions from the following persons: Anita Kocsis, Tim Boyle, Christine Thong, and Panagiotis Charitos.

Since their emergence in the 1960s, Big Science projects have undergone considerable development and evolutionary changes. Growing complexity with new technologies, sophistication of interdisciplinary collaborations, data-driven research, and open science initiatives have changed the nature of Big Science, leading to social renewal, human progress and social transformation. These initiatives have increase capacity to bring together interdisciplinary groups from different countries and cultures to work towards a common goal.

By nature, Big Science is a creative movement with visionary undertakings. Scientists set bold research agendas and go about designing and constructing advanced technologies with well-defined processes. Such processes, as described in this book, call for the coordinated teamwork from diverse disciplines. These organisations often engage in solving novel, fundamental and complex problems that require the application of cutting-edge scientific and technological knowledge. Hence, Big Science is a different league of its own governed by different knowledge synthesis, a philosophy of collaboration, dialogue and open discussions. Big Science also evolve through sharing of scientific and technological infrastructures, the constant search for breakthrough knowledge and the mobilisation of significant financial and human resources. It is, however, naïve to think such collaborations are easy to put together and are free from competition.

Potentially, most Big Science initiatives gives rise to useful innovation. The past decades of operations of Big Science demonstrated that scientific knowledge, methodologies, and findings together with technical instrumentation designed for purely basic research purposes have eventually ended up in elegant solutions that are fundamental to practical applications in our daily lives as well as to medical, environmental and economic developments. Numerous applications discussed in this book provide some examples. Big Science has evolved significantly over time, both in terms of the complexity of experiments undertaken and the way in which they are conducted using open science and open innovation to promote social benefits. The development of new technology for Big Science experiments can have broader applications beyond the intended use and can transform human society.

There are many future challenges for Big Science organisations. Its nature and interdisciplinarity are expected to change dramatically with grand undertakings like FCC, Linear Collider, and Dark Matter searchers. Rapid advances in technology and data-driven analytics will facilitate greater participation of multidisciplinary groups to come up with innovative solutions to the world's complex problems and challenges in climate, energy, and health.

Operations and maintenance of Big Science are quite challenging and difficult tasks. Working in these organisations can be difficult for younger and upcoming scientists to demonstrate their creative talents. Some publications have more than 1000 scientists as authors and some young people may be among the thousands of those authors contributing to a single scientific publication. Individualism and intellectual freedom can be marginalised when working among experienced and highly accomplished researchers. A continuous search for pool of talent with precision and accuracy is required for a dynamic evolution of Big Science experiments.

Moreover, there are limits and restrictions on the types of research problems that are possible to investigate. Not all ideas will become part of Big Science investigations.— In other word, Big Science is an integral part of social construction.

Big Science are also subjected to some restrictions and political pressures. There can be organisational restrictions on valuable beam time and telescope time. Politics and funding can influence the research scope and agenda. Very often, Big Science organisations have specific long-term strategies for high energy physics or astronomical research that are constantly reviewed and modified in consultation with scientists (for example the European Strategy for Particle Physics[2]).

Life cycles, capital investment sizes, and the sizes of the participating scientific communities all seem to be growing in Big Science projects. The LHC also resembles modularity in design and how various interactive components can be assembled independently like a jigsaw puzzle and bring them all together to produce the desired innovative and collaborative solutions and outcomes.

We noted that the guiding principles of Big Science initiatives are constantly evolving. Those principles promote 'open science' and 'open data' concepts that encourage transparency and make data available for public use. These principles also promote ethical collaboration and governance that uphold morality, diversity, and ethical considerations in science diplomacy.

The opportunities for Big Science to flourish are immense. We support the view that Big Science undertakings will continue to be a global phenomenon and that the most effective multidisciplinary and collaborative way to solve humanity's complex problems by combining human intelligence and resoluteness. Building a world-class scientific instrument such as the LHC, which can create extreme conditions similar to those immediately after the Big Bang and then analysing the results with extraordinary precision, is a daunting challenge for scientists and the LHC has proven that such endeavours are possible with human collaboration.

Many scientific issues that are fundamental in nature, such as climate change and the origin of the universe, are too big and are complex problems to solve by an individual, single country, one scientific institution or a nation.

Besides its contribution to scientific fundamental knowledge, Big Science can be more human-centric and the driver of humanistic-based economic principles. In a recent book, Professor Stephen Hill (Hill et al., 2022) outlined the power of human beings to assert fundamental values and build harmony across different cultures. The backdrop of unrelenting destruction caused by ongoing wars (e.g. between Russia and Ukraine in 2022) and other manufactured human conflicts and miseries serve as a stark reminder that the humanity has to be vigilant about the importance of building the spirit of sharing and collaborating for the good of society. Given the

---

[2] See details in CERN documents: https://europeanstrategy.cern/european-strategy-for-particle-physics.

tumultuous current geopolitical trends, humans are urged to view human collaboration as a powerful tool to solve social problems—a tool similar to hunter-gatherers' splint stones for collective good.

We invite our fellow scientists and policymakers to contemplate either launching or participating in new Big Science undertakings to benefit from the key messages and potential lessons outlined in this book. Sharing the thrill and wonder of scientific discovery, we wish our readers a journey of learning filled with enriching and inspiring insights.

# Annexes
## Chapter 12

### Annex 1: Illustrative Examples of Big Science Accomplishments

---

**Medipix**

---

Medipix is an example of a successful transfer of CERN technology that spans over 20 years. It all started in the mid-1990s with the development of the first Medipix chip, a pixel detector read-out chip derived from the Omega3 chip, used in the $\Omega3$ tracker in the CERN West Area. The Medipix2 collaboration was established in 1999, followed by Medipix3 in 2005, and Medipix4 in 2016. The first chip showed its advantages and the possibility of use in medical and other fields. Michael Campbell, spokesperson for the Medipix Collaborations, believes in the importance of transferring technologies developed for CERN's needs to society. In 2000, a collaborator from Nikhef started to discuss the use of Medipix2 with colleagues at Philips Analytical (then PANalytical and afterwards Malvern-Panalytical) in Almelo, The Netherlands. Philips Analytical was a market leader in X-ray diffraction and fluorescence spectroscopy equipment used in the materials industry. After some years, they successfully introduced the Medipix2 chip into its instrumentation. This success paved the way for the Medipix3 chip's development.

Since 2012, thanks to the efforts of collaborating members at the Institute of Experimental and Applied Physics in Prague and the University of Houston, Texas, five Timepix chips have been plugged into the USB ports of the laptops on the International Space Station and have immediately started delivering data on the radiation environment to the Earth-based team in Houston. In 2014, the same team incorporated two Timepix chips into a battery-operated system that flew on the NASA Orion rocket. NASA subsequently base-lined this technology for dosimetry applications on future manned missions. Four spinoff companies in medical imaging detectors have been created from Medipix3 collaborations: ADVACAM s.r.o. (CZ), Amsterdam Scientific Instruments (NL), Mars-Bio Imaging (NZ), and X-Spectrum (DE). Noteworthy is the potential for early lung inflammation detection by MARS-spectral imaging using the Medipix3 chip. Furthermore, spectral CT by MARS which brings colour to X rays, is an important breakthrough in imaging.

The Medipix4 Collaboration was launched in 2016 with the aim of designing pixel read-out chips that, for the first time are fully prepared for Through Silicon Via (TSV) processing and may be tiled on all four sides, enabling large areas to be covered seamlessly. Moreover, TSVs permit the development of new read-out architectures by avoiding the need to send all the data to one side of the chip for read-out. Two new chip developments are foreseen: Medipix4, which will target spectroscopic X-ray imaging at rates compatible with medical CT scans, and Timepix4, which provides particle identification and tracking with higher spatial and timing precision. The new read-out chips might have applications in quantum computing.

## ELYSIA, LIFTT and newcleo

An interesting and very innovative way of doing business is illustrated by the business journey of Stefano Buono, an Italian scientist, CERN alumnus, and business developer.

Stefano Buono is a strong believer in the importance of transferring technological innovation to society and is a co-founder of ELYSIA Capital, his single-family office whose mission is to 'turn sustainable projects into great enterprises by investing in their evolution as a means to improve social wellbeing'. The aim is to invest in ideas and sustainable projects that make a positive contribution to society and smaller communities, as any social growth will contribute in an efficient way to economic growth. Improving the quality of life, investing in education and spreading skills across borders and cultures while keeping in mind the goal to building a sustainable world are the framework of supported actions. Just to quote some of them: (a) PLANET building smart city designs and smart, affordable social houses, creating sustainable multipurpose spaces to improve the quality of life through digital technologies, services, and social innovation; (b) FARMALISTO providing online individual health care services and products in Colombia, Mexico, and Peru; and ASIA strategic holding that supports the educational and other development initiatives in Myanmar and Vietnam. He is also the Chairman of LIFTT, *Giving Ideas the Highest Value*, a newly created private company founded by the Links foundation, whose funding institutions are the Torino Politecnico and the private foundation Compagnia di San Paolo, Italy. It is an investment company operating in the field of KEC (Knowledge Exchange and Commercialisation). It pursues the goal of stimulating and supporting technology transfer from research institutions and of generating revenues and a positive impact through the economic exploitation of research and innovation outcomes. In September 2021 as CEO he launched the company newcleo futurable energy to provide with a new approach safe and clean nuclear energy. The aim being to develop a latest generation of nuclear reactors providing a path to combatting both climate change and existing nuclear waste.

---

### Solar Hydrothermal Advanced Reactor Project (SHARP) for Biological Waste Management

---

The Solar Hydrothermal Advanced Reactor Project (SHARP) for Biological Waste Management was founded by Beatrice Bressan and Yakoov Garb in late 2013, with a visit to BGU Sde Boker campus in early 2014. On this visit, the potential of the solar panels was discussed, and some early characteristics of the combined system and its usefulness were developed. Some months later, considerable refinement of the project, as well as a decision for four panels to be sent from CERN to BGU, allowed the development of a pilot facility for the coming years.

The pilot project addresses this challenge: the use of a high-efficiency flat-plate solar thermal collector as a heat source directly driving a hydrothermal biological waste treatment reactor to produce sterile bio-coal and useable nutrients from animal waste, all at a scale that allows installation on a fixed or even mobile platform and operation in rural developing country settings. The goal is to explore whether a reasonably priced, robust device could allow a feasible way to treat biological waste using only solar energy, thus greatly expanding the range of contexts in which these wastes could be more effectively managed and utilised.

The use of the BGU/CERN solar hydrothermal reactor, a high-efficiency flat-plate solar thermal collector, as a heat source directly driving a hydrothermal biological waste treatment reactor can be suitable to produce sterile bio-coal and useable nutrients from animal waste, in a configuration suitable for robust off-grid operation in rural developing country contexts as well as niches in developing countries where off-grid in situ treatment of biological waste is desirable.

---

## Annex 2: Serendipity—From Concept to Market

---

### From Basic Science to Market: A Serendipitous Process

---

Let us consider that a future breakthrough technology, developed within the realm of Basic Science, would end up generating market applications at a future time, $t + nt$ ($n$ being an integer number such that $n > 1$), with probability $P$.[1] We assume that a similar case has happened in the past, at time $t$. Let us also contemplate a scenario of total serendipity. Even if a Basic Science discovery resulted in a market value at time $t$, in the past, it does not preclude that another one could follow the same pathway at a later time $t + nt$. In other words, an extreme memoryless 'from Lab to Fab' process. In mathematical form, an exponential memoryless probability process representing such a case can be written as:

$$P(t + nt|t) = P(t + nt) P(t) = e^{-\lambda(t+nt+t)} = e^{-\lambda t(2+n)} = e^{-\alpha t} \tag{1}$$

Where $\lambda$ is a constant not really relevant for our discussion and $a = \lambda (2+n)$ is for simplification as *discussed below*. Expression (1) is known, statistically, as the survival function of an exponential distribution. What expression (1) essentially indicates is that the probability, $P$, that a Basic Science technology ends up in the market, at time $T > t$, is extremely small. How small depends on the parameter, a. Mathematically, this is expressed by considering the Equation (1) above as:

$$P(T > t) = e^{-\alpha t} \tag{2}$$

Let us now introduce a new variable, X, to capture the number of created technologies or events.

$$X = x_0 e^t \tag{3}$$

By simple substitution, it is possible to transform the exponential distribution considered originally in (2) into:

$$P(X > x) = P\left(x_0 e^t > x\right) = P\left(e^t > x/x_0\right) = P[t > Ln(x/x_0)] = \left(\frac{x_0}{x}\right)^\alpha \tag{4}$$

Thus, $X$, has a Pareto distribution with parameters $x_0$ and $a$.

$$P(X > x) = \left(\frac{x_0}{x}\right)^\alpha \tag{5}$$

The relationship between the exponential and Pareto probability distributions is well known in the field of statistics.[2] The interesting fact in our context is that $(x_0/x)$ represents now the fraction of those Basic Science technologies ending up in market applications. It is worth noticing, as is also well known, that the similarity between the exponential and Pareto distributions increases when moving towards the so-called long-tails of both.[3] This would support the plausibility that only a very small fraction of the technologies developed for boosting Basic Science discoveries, due to the serendipity of the process, would indeed end up in commercial market applications and business.

[1] See for example, S. Ross (2019), *Introduction to Probability Models*, 12th edition, Elsevier.
[2] An interesting account of pedagogical reference is found in M. Hardy (2010), Pareto's Law. *Math Intelligencer* 32, 38–43.
[3] Cf. Ref 2.

# Beating the Odds from Lab to Market: Innovation Ecosystems

As referred to above, the Pareto distribution (Equation (5) above) reflects the fraction of Basic Science technologies ending up in market applications in the future.

Let us now consider the importance of index $\alpha$. Since that fraction $(x_0/x)$ must be between $0$ and $1$ (both inclusive), $\alpha$ must be positive. Additionally, in order for the total number of technologies to be finite at a given time, $\alpha$ must also be greater than $1$. The interesting consequence of these two conditions is that the larger the Pareto index is, the smaller the proportion of Basic Science technologies that would end up in the market. An interesting way to visualise it is by considering the odds as a measure for providing the likelihood of a particular outcome. They are calculated as the ratio of the number of events producing such an outcome with respect to the number that do not. In mathematical terms for our context, there is a close relationship between the odds and the $\alpha$ index expressed as:[1]

$$a = Ln\frac{x_0}{x - x_0} \qquad (6)$$

Table A.1  Illustration of some cases in our context relating the proportion of Basic Science discoveries ending up in market applications and the value of the $a$ exponent.

| Fraction | A |
|----------|------|
| 90–10 | 1.05 |
| 80–20 | 1.16 |
| 67–33 | 1.58 |

As an illustration, Table A.1 offers some examples, including the so-called 80–20 rule (Table A.1).[2] As can be inferred from it, in the context considered here, it would be more desirable to have a proportion of 67–33 rather than 80–20 regarding the fraction of Basic Science technologies ending up in the market. This is, in our opinion, where Innovation Ecosystems could bring a significant added value. In other words, the issue is how Innovation Ecosystems could help increase the value of the Pareto index, $\alpha$.

---

[1] N.L. Johnson, A.W. Kemp, and S. Kotz (2005), *Univariate Discrete Distributions*, 3rd Edition, Wiley.
[2] Reed, W.J. (2001), The Pareto, Zipf and other power laws, *Economics Letters*, 74 (1), 15–19.

# References

Aad, G., et al. (2015). 'Combined Measurement of the Higgs Boson Mass in pp Collisions at √ s = 7 and 8 TeV with the ATLAS and CMS Experiments'. *Physical Review Letters* 114: 191803, published 14 May 2015.

Aasi, J., Abbott, B.P., Abbott, R., Abbott, T.D., Abernathy, M.R., Ackley, K., Adams, C., et al. (2015). 'Advanced LIGO'. *Classical Quant. Grav.* 32(7). https://doi.org/10.1088/0264-9381/32/7/074001.

Abba, A. et al. (2021). 'The novel Mechanical Ventilator Milano for the COVID-19 pandemic'. *Phys Fluids* (1994) Mar., 33(3): 037122. doi: 10.1063/5.0044445.

Abbott, B.P., Abbott, R., Abbott, T.D., Abernathy, M.R., Ackley, K., Adams, C., et al. (2016). 'Observation of gravitational waves from a binary black hole merger', *Phys. Rev. Lett.* 116(6). https://doi.org/10.1103/PhysRevLett.116.061102.

Abbott, B.P., Abbott, B.P., Abbott, R., Abbott, T.D., Abernathy, M.R., Ackley, K., Adams, C., et al. (2017). 'GW170817: Observation of gravitational waves from a binary neutron star in spiral'. *Phys. Rev. Lett.* 119(16). https://doi.org/10.1103/PhysRevLett.119.161101.

Abbott, B.P., Abbott, R., Abbott, T.D., Abernathy, M.R., Acernese, F., Ackley, K., Adams, C. et al. (2017b). 'Gravitational waves and gamma-rays from a binary neutron star merger: GW170817 and GRB 170817A'. *Astrophys. J. Lett.* 848(2): L13. https://doi.org/10.3847/2041-8213/aa920c.

Abbott, R., Abbott, T.D., Abraham, S., Acernese, F., Ackley, K., Adams, C., et al. (2020). 'GW190521: A binary black hole merger with a total mass of 150 M☉'. *Phys. Rev. Lett.* 125(10). https://doi.org/10.1103/PhysRevLett.125.101102.

Abdurazzakov, O., Illés, C., Jafarov, N., and Aliyev, K. (2020). 'The impact of technology transfer on innovation'. *Polish Journal of Management Studies* 21: 9–23. https://doi.org/10.17512/pjms.2020.21.2.01.

Acernese, F., Agathos, M., Agatsuma, K., Aisa, D., Allemandou, N., Allocca, A., et al. (2014). 'Advanced Virgo: a second generation interferometric gravitational wave detector'. *Class. Quant. Grav.* 32(2). https://doi.org/10.1088/0264-9381/32/2/024001.

ACS. (1999). *The Discovery and Development of Penicillin 1928–1945.* London: A. Fleming Laboratory Museum.

Adolphsen, C., Barone, M., Barish, B., et al. (2013). *The International Linear Collider Technical Design Report—Volume 3.II: Accelarator Baseline Design.* ILC-REPORT-2013-040. arXiv:1306.6328v1.

Aghanim, N., Akrami, Y., Arojjo, F., et al. (2018). 'Planck 2018 results'. *Astron. & Astrophys.* 641 A1. https://doi.org/10.1051/0004-6361/201833880.

Aghanim, N., Akrami, Y., Ashdown, M., et al. (2020). 'Planck 2018 results. VI Cosmological Parameters'. *Astron. & Astrophys.* 641 A6. https://doi.org/10.1051/0004-6361/201833910.

AGLAE. (2021). 'A high-tech laboratory'. https://c2rmf.fr/analyser/un-laboratoire-de-haute-technologie-pour-les-collections-des-musees/aglae.

Akiyama, K., the Event Horizon Telescope Collaboration et al. (2019). 'First M87 Event Horizon Telescope Results. VI the shadow and mass of the central black hole'. *ApJL,*875L6. DOI: 10.3847/2041-8213/ab1141.

Aksaker, A., Yerli, S.K., Erdoğan, M.A., Kurt, Z., Kaba, K., Bayazit, M., and Yesilyaprak, C. (2020). 'Global site selection for astronomy'. *Monthly Notices of the Royal Astronomical Society.* 493(1): 1204–1216. https://doi.org/10.1093/mnras/staa201.

Alabama University (2008), 'History of radiation oncology'. University of Alabama at Birmingham, Comprehensive Cancer Center, https://www.uab.edu/medicine/radonc/about/history access on 9-12-2022.

Albrecht, J., Alves, A.A., Amadio G., et al. (2017). 'A Roadmap for HEP Software and Computing R&D for the 2020s'. arXiv:1712.06982 [physics.comp-ph] and The HEP Software Foundation.

Albrecht, J., Alves, A.A., et al. (2019). 'A Roadmap for HEP Software and Computing R&D for the 2020s'. *Comput. Softw. Big. Sci.* 3(7). https://doi.org/10.1007/s41781-018-0018-8.

ALICE Collaboration et al. (2008). 'The ALICE experiment at the CERN LHC'. *Journal of Instrumentation* 3: 1–245. https://iopscience.iop.org/article/10.1088/1748-0221/3/08/S08002/pdf.

Alphabeta (Australia part of Accenture) (2020). *Australia's Deep Tech Opportunity: Insights from the Cicada Innovations Journey.* A report undertaken on behalf of Cicada Innovations.

Amaldi, U. (1976). 'A possible scheme to obtain e–e– and e+e– collisions at energies of hundreds of GeV'. *Phys. Lett.,* 61B: 313–315. https://doi.org/10.1016/0370-2693(76)90157-X.

Amaldi U., and Kraft, G. (2005). 'Radiotherapy with beams of carbon ions'. *Reports on Progress in Physics* 68 (8): 1861–1882. https://doi.org/10.1088/0034-4885/68/8/R04.

Anadon, L., Chan, G., Y. Bin-Nun, A., and Narayanamurti, V. (2016). 'The pressing energy innovation challenge of the US National Laboratories'. *Nature Energy* 1, 16117. https://doi.org/10.1038/nenergy.2016.117.

Anderson, S., and Williams, R. (2017). 'LIGO Data Management Plan', June 2017. http://dcc.ligo.org/LIGO-M1000066/public.

ANSTO. (2017). 'Important climate study'. https://www.ansto.gov.au/news/important-climate-study.

ANSTO. (2020). https://www.ansto.gov.au/.

Anthes, G. (2015). 'Estonia: A model for e-government'. *Communications of the ACM* 58 (6), June: 18–20. https://doi.org/10.1145/2754951.

Antikainen, M., Mäkipää, M., and Ahonen, M. (2010). 'Motivating and supporting collaboration in open innovation'. *European Journal of Innovation Management* 13(1): 100–119. https://doi.org/10.1108/14601061011013258.

Apffel-Marglin F., and Marglin, S.A. (eds). (2015). *Dominating Knowledge: Development, Culture, and Resistance.* London: Oxford University Press. https://doi.org/10.1093/acprof:oso/9780198286943.001.0001.

Archer, B. (1979). 'Design as a discipline'. *Design Studies* 1(1): 17–20. https://doi.org/10.1016/0142-694X(79)90023-1.

Arena M.J., and Uhl-Bien M. (2016). 'Complexity leadership theory: Shifting from human capital to social capital'. *People and Strategy* 39(2). https://sagewaysconsulting.com/wp-content/uploads/2017/03/ComplexityLeadershipTheory_HRPS_39.2_Arena_Uhl_Bien.pdf.

Aronova, E. (2014). 'Big Science and "Big Science Studies" in the United States and the Soviet Union during the Cold War', in N. Oreskes and J. Krige (eds), *Science and Technology in the Global Cold War,* 393–430.MIT Press Scholarship Online https://doi.org/10.7551/mitpress/9780262027953.003.0013.

ARPANSA (2021). *Human Factors, Australian Radiation Protection and Nuclear Safety Organisation.* Australian Government, Canberra. https://www.arpansa.gov.au/regulation-and-licensing/safety-security-transport/holistic-safety/human-factors.

Arthur, W.B. (2009). *The Nature of Technology: What It Is and How It Evolves.* New York: Simon & Schuster.

Aso, Y., Michimura, Y., Somiya, K., Ando, M., Miyakawa, O., Sekiguchi, T., et al. (2013). 'Interferometer design of the KAGRA gravitational wave detector'. *Phys. Rev. D* 88(4). https://doi.org/10.1103/PhysRevD.88.043007.

Ass, B.M., Avolio, B.J., and Atwater, L. (1996). 'The transformational and transactional leadership of men and women'. *Applied Psychology* 45(1). https://doi.org/10.1111/j.1464-0597.1996.tb00847.x.

ATLAS Collaboration (1992). 'ATLAS: letter of intent for a general-purpose pp experiment at the large hadron collider at CERN', Letter of Intent, CERN-LHCC-92-003, http://cds.cern.ch/record/291061. Access on 11-12-2022.

ATLAS Collaboration (1994). 'ATLAS: technical proposal for a general-purpose pp experiment at the Large Hadron Collider at CERN', Technical Proposal, CERN-LHCC-1994-038 (1994); JINST 3 (2008) S08004, https://cds.cern.ch/record/290968?ln=en, access on 11-12-2022.

ATLAS Collaboration (2012). 'Observation of a new particle in the search for the Standard Model Higgs boson with the ATLAS detector at the LHC'. *Physics Letters* B 716(1): 1–29. https://doi.org/10.1016/j.physletb.2012.08.020.

ATLAS Collaboration (2019). *ATLAS, A 25-year Insider Story of the LHC Experiment.* Advanced Series. New Jersey: World Scientific. https://doi.org/10.1142/11030.

ATLAS Collaboration (2019b). 'ATLAS releases first result using full LHC Run 2 dataset'. https://home.cern/news/news/physics/atlas-releases-first-result-using-full-lhc-run-2-dataset.

ATLAS Collaboration (2020). https://atlas.cern/.

ATTRACT. (2020). https://attract-eu.com/.

Australian Academy of Science (2020). 'What are the impacts of Climate Change?'. https://www.science.org.au/learning/general-audience/science-climate-change/7-what-are-impacts-of-climate-change.

Autio, E. (2014). 'Innovation from Big Science: enhancing Big Science Impact Agenda'. Department for Business Innovation and Skills. The National Archives, Kew, London. https://www.gov.uk/government/organizations/department-for-business-innovation-skills.

Autio, E., Hameri, A.P., and Nordberg, M. (1996). 'A framework of motivations for industry-big science collaboration: A case study'. *Journal of Engineering and Technology Management* 13(3–4): 301–314. https://doi.org/10.1016/S0923-4748(96)01011-9.

Autio, E., Bianchi-Streit, M., and Hameri, A.P. (2003). *'Technology Transfer and Technological Learning through CERN's Procurement Activity'.* CERN Yellow Reports: Monographs, p. 78. Geneva: CERN. http://cds.cern.ch/record/680242.

Autio, E., Hameri, A.P, and Bianchi-Streit, M. (2003b). 'Can companies benefit from Big Science?'. *CERN Courier*, December. https://cerncourier.com/a/can-companies-benefit-from-big-science/.

Autio, E., Hameri, A.P., and Vuola, O. (2004). 'A framework of industrial knowledge spillovers in big-science centers'. *Research Policy* 33(1): 107–126. https://doi.org/10.1016/S0048-7333(03)00105-7.

Autio, E.A., Streit-Bianchi, M., Hameri, A. P., and Boisot, M. (2011), 'Learning and innovation in procurement: The case of ATLAS-type projects', in Boisot, M., Nordberg M., Yami, S., and Nicquevert, B. (eds), *Collisions and Collaborations: The Organisation of Learning in the ATLAS Experiment at the LHC*, 135–159. Oxford University Press. https://doi.org/10.1093/acprof:oso/9780199567928.003.0008.

Aymar, R. (2014). 'The origins of LEP and LHC, subnuclear physics: Past, present and future' 97–107. http://www.casinapioiv.va/content/dam/accademia/pdf/sv119/sv119-aymar.pdf.

Bach, L., and Lambert, G. (1992). 'Evaluation of the economic effects of large R&D programmes: the case of the European space programme'. *Research Evaluation* 2(1): 17–26. https://doi.org/10.1093/rev/2.1.17.

Baker, I., Maxey, C., Hipwood, L., Isgar, V., Weller, H., Herrington, M., and Barnes, K. (2019). 'Linear-mode avalanche photodiode arrays in HgCdTe at Leonardo, UK: the current status'. *Proc. SPIE 10980 Image Sensing Technologies: Materials, Devices, Systems, and Applications* VI. https://doi.org/10.1117/12.2519830.

Baldwin, C. and von Hippel, E. (2011). 'Modeling a paradigm shift: From producer innovation to user and open collaborative innovation'. *Organization Science* 22(6): 1369–1683. https://doi.org/10.1287/orsc.1100.0618.

Ballarino, A. (2002). 'HTS current leads for the LHC magnet powering system'. *Physica C: Superconductivity* 372–376(3): 1413–1418. https://doi.org/10.1016/S0921-4534(02)01042-0.

Ballarino, A. (2014). 'Development of superconducting links for the LHC machine'. *Superconductor Science and Technology* 27(4). https://doi.org/10.1088/0953-2048/27/4/044024.

Ballarino, A. (2019). 'Accelerator technology: Now and the future', in EUCAS 2019, Glasgow, UK. https://snf.ieeecsc.org/sites/ieeecsc.org/files/documents/snf/abstracts/RP109Ballarinoplenary.pdf.

Ballarino, A., Wagner, U., and Ijspeert, A. (1996). 'Potential of high-temperature super conductor current leads for LHC cryogenics', in *16th International Cryogenic Engineering Conference and International Cryogenic Materials Conference, Kitakyushu, Japan, 20–24 May*: 1139–1142. https://cds.cern.ch/record/326938?ln=sv.

Ballarino, A. et al. (2016). 'The Best Paths project on MgB2 superconducting cables for very high power transmission'. *IEEE Trans. on Appl. Supercon.* 26 (3): 1–6. https://doi.org/10.1109/TASC.2016.2545116.

Ballarino, A., et al. (2019). 'The CERN FCC conductor development program: A worldwide effort for the future generation of high-field magnets'. *IEEE Trans. on Appl. Supercon.* 29(5): 1–9. https://doi.org/10.1109/TASC.2019.2896469.

Bammer, G. (2008). 'Enhancing research collaborations: Three key management challenges'. *Research Policy* 37(5): 875–887. https://doi.org/10.1016/j.respol.2008.03.004.

Bancino, R., and Zevalkink, C. (2007). 'Soft skills: The new curriculum for hard-core technical professionals'. *Techniques: Connecting Educations and Careers* (J1) 82(5): 20–22. https://eric.ed.gov/?id=EJ764824.

Banks, M. (2020). 'The 10 most important future big-science facilities in physics'. *PhysicsWorld.* https://physicsworld.com/a/the-10-most-important-future-big-science-facilities-in-physics/.

Barth, R.F., Zhang, Z., and Liu, T. (2018). 'A realistic appraisal of boron neutron capture therapy as a cancer treatment modality'. *Cancer Communications* 38. https://doi.org/10.1186/s40880-018-0280-5.

Barzi, E., and Zlobin, A.V. (2019). 'Nb3Sn wires and cables for high-field accelerator magnets', in Schoerling D. and Zlobin A. (eds), *Nb3Sn Accelerator Magnets*, 23–51. Springer. https://doi.org/10.1007/978-3-030-16118-7_2.

Baskerville, R.L., Kaul, M., and Storey, V.C. (2018). 'Aesthetics in design science research'. *European Journal of Information Systems* 27(2): 140–153. https://doi.org/10.1080/0960085X.2017.1395545.

Bass, B.M. (1985). *Leadership and Performance beyond Expectations.* New York: Basic Books.

Bass, B.M., Avolio, B.J., and Atwater, L. (1996). 'The transformational and transactional leadership of men and women'. *Applied Psychology* 45(1). https://doi.org/10.1111/j.1464-0597.1996.tb00847.x.

Baudis, L. (2021), 'Cryogenic dark matter searches'. *EPN* 52/3, DOI: https://doi.org/10.1051/epn/2021304, https://www.europhysicsnews.org/articles/epn/pdf/2021/03/epn2021523p22.pdf.

Bauer, P., Ballarino, A., Devred, A., Ding, K., et al. (2020). 'Development of HTS Current Leads for the ITER Project'. *IOP Conference Series: Material. Science Engineering*. ICMC 756. https://doi.org/10.1088/1757-899X/756/1/012032.

Beck, S., Bergenholtz, C., Bogers, M., et al. (2020). 'The Open Innovation in Science research field: a collaborative conceptualisation approach'. *Industry and Innovation* 1–50. https://doi.org/10.1080/13662716.2020.1792274.

Beck, H.P., and Charitos, P. (eds). (2021). *The Economics of Big Science: Essays by Leading Scientists and Policymakers.* Cham: Springer. https://doi.org/10.1007/978-3-030-52391-6.

Bell, K.A., Brown, R.M., et al. (2004). 'Vacuum phototriodes for the CMS electromagnetic calorimeter endcap', in *Proceedings, 13th Conference on Computing Applications in Nuclear and Plasma Sciences, RT2003, Montreal, Canada, May 18–23*, 2003: 2284-2287. https://doi.org/10.1109/TNS.2004.836053.

Benavent-Pérez, M., Gorraiz, J., Gumpenberger, C., and de Moya-Anegón, F. (2012). 'The different flavors of research collaboration: a case study of their influence on university excellence in four world regions'. *Scientometrics* 93(1): 41–58. https://doi.org/10.1007/s11192-012-0638-4.

Benedikt, M., and Zimmermann, F. (2016). 'Towards future circular colliders'. *Journal of the Korean Physical Society* 69 (6): 893–902. https://doi.org/10.3938/jkps.69.893.

Benedikt, M., Blondel, A., Janot, P., Mangano, M., and Zimmermann, F. (2020). 'Future circular colliders succeeding the LHC'. *Nature Physics* 16: 402–407. https://doi.org/10.1038/s41567-020-0856-2.

Berkley, S. (2020). 'Covid-19 needs a big science approach'. *Science* 367(6485): 1407. https://doi.org/10.1126/science.abb8654.

Bernasek, A. (2003). 'Banking on social change: Grameen Bank lending to women'. *International Journal of Politics, Culture, and Society* 16(3): 369–385. http://www.jstor.org/stable/20020172.

Berners-Lee, T. (1989). 'Information management: A proposal'. https://www.w3.org/History/1989/proposal.html

Berners-Lee, T. and Cailliau, R. (1990). 'Worldwide Web: Proposal for a HyperText Project', 12 November 1990. https://www.w3.org/Proposal.html.

Berners-Lee, T., and Fischetti, M. (2000). *Weaving the Web: The Original Design and Ultimate Destiny of the World Wide Web by its Inventor.* New York: HarperBusiness.

Berry, S.D., Bourke, P., and Wang, J.B. (2011). 'qwViz: Visualisation of quantum walks on graphs'. *Computer Physics Communications* 182(10): 2295–2302. https://doi.org/10.1016/j.cpc.2011.06.002.

Best Paths. (2018). New Superconducting cable sets records for power transmission, http://www.bestpaths-project.eu/en/news/new-superconducting-cable-sets-records-for-power-transmission.

Bhat, P.C., and Taylor, G.N. (2020). 'Particle physics at accelerators in the United States and Asia'. *Nat. Phys.* 16: 815. https://doi.org/10.1038/s41567-020-0911-z.

Bianchi-Streit M., Blackburne N.F., Budde R., Reitz H., Sagnell B., Schmied H., and Schorr B. (1984). 'Economic Utility Resulting from CERN Contracts (Second Study)'. CERN Yellow Report CERN 84/14 and 1988. Quantification of CERN's Economic Spin-off. *Czech J. Phys.* B38: 23–29.

Bicarregui, J., Gray, N., Henderson, R., Jones, R., Lambert, S., and Matthews, B. (2013). 'Data management and preservation planning for Big Science'. *The International Journal for Digital Curation* 8(1): 29–41. https://doi.org/10.2218/ijdc.v8i1.247.

Biesta, G. (2015). 'What is education for? On good education, teacher judgement, and educational professionalism'. *European Journal of Education* 50(1): 75–87. https://doi.org/10.1111/ejed.12109.

Biesta, G. (2017). *The Beautiful Risk of Education.* Routledge.

Biscari, C. and Rivkin, L. (2019). 'Accelerator science and technology', in *Open Symposium—Update of the European Strategy for Particle Physics.* 13–16 May 2019, Granada, Spain. https://indico.cern.ch/event/808335/contributions/3380835.

Bishop L. (2015). 'The right to science: Ensuring that everyone benefits from scientific and technological progress'. *European Journal of Human Rights* 4: 411–430. Indiana University, Robert H. McKinney School of Law Research Paper No. 2015–46. https://ssrn.com/abstract=2690549.

Björklund, F. (2016). 'e-Government and moral citizenship: The case of Estonia'. *Citizenship Studies* 20(6–7): 914–931. https://doi.org/10.1080/13621025.2016.1213222.

Björklund, T.A., Keipi, T., Celik, S., and Ekman, K. (2019). 'Learning across silos: Design Factories as hubs for co-creation'. *European Journal of Education* 54(4): 552–565. https://doi.org/10.1111/ejed.12372.

Blanchard, Y., Galati, G., and van Genderen, P. (2013). 'The Cavity Magnetron: Not just a British invention'. *IEEE Antennas and Propagation Magazine* 55(5): 244–254. https://doi.org/10.1109/MAP.2013.6735528.

Bliemel, M., and van der Bijl-Brouwer, M. (2018). 'Transdisciplinary innovation'. *Technology Innovation Management Review* 8(8): 3–6. http://https://doi.org/10.22215/timreview/1173.

Blewett, J.P. (1979). 'Applications of linear accelerators', in R.L. Witkover (ed.), *Proceedings of the 1979 Linear Accelerator Conference, Montauk, New York, USA*: 1–4. Brookhaven National Laboratory Associated Universities, Inc.

Blondel, A., Gluza, J., Jadach, S., et al. (2019). 'Standard model theory for the FCC-ee Tera-Z stage'. *CERN Yellow Reports: Monographs* 3. https://doi.org/10.23731/CYRM-2019-003.

Bogers, M., Bekkers R., and Granstrand O. (2012). 'Intellectual property and licensing strategies in open collaborative innovation' In de Pablos Heredero, C. and López, D. (eds), *Open Innovation in Firms and Public Administrations: Technologies for Value Creation*. IGI Global Publishing. https://doi.org/10.4018/978-1-61350-341-6.ch003.

Borges, G.S., and Santos, B.F. (2021). 'COVID-19 Vaccine as a Common Good'. Journal of Global Health (1). https://jogh.org/covid-19-vaccine-as-a-common-good.

Boisot, M.H. (1995). *Information Space: A Framework for Learning in Organizations, Institutions and Culture*. Routledge.

Boisot, M.H. (1998). *Knowledge Assets: Securing Competitive Advantage in the Information Economy*. Oxford University Press.

Boisot, M.H. (2013). *Information Space (RLE: Organizations) e-book*. Routledge. https://doi.org/10.4324/9780203385456.

Boisot, M.H., and Canals, A. (2004). 'Data, information and knowledge: Have we got it right?', *Journal of Evolutionary Economics* 14(1): 43–67. https://doi.org/10.1007/s00191-003-0181-9.

Boisot, M., Nordberg M., Yami, S., and Nicquevert, B. (eds). (2011). *Collisions and Collaborations: The Organization of Learning in the ATLAS Experiment at the LHC*. Oxford: Oxford University Press. https://doi.org/10.1093/acprof:oso/9780199567928.001.0001.

Boisot, M., and Nordberg, M. (2011). 'A conceptual framework: The I-Space'. In Collisions and Collaborations: the Organisation of Learning in the ATLAS Experiment at the LHC, 28–54. Boisot, M., Nordberg M., Yami, S., Nicquevert, B. (eds), Oxford University Press

Bomford, C.K., and Kunkler, I.H. (eds). (2002). *Textbook of Radiation Therapy* (6th edn): 311. Churchill Livingstone.

Bonney, R., Shirk, J.L., Phillips, T.B., Wiggins, A., Ballard, H.L., Miller-Rushing, A.J., and Parrish, J.K. (2014). 'Next steps for citizen science'. *Science* 343(6178): 1436–1437. DOI: 10.1126/science.1251554.

Bortfeld, T., and Jeraj, R. (2011). 'The physical basis and future of radiation therapy'. *The British Journal of Radiology* 84: 485–498. https://doi.org/10.1259/bjr/86221320.

Bos, N. et al. (2007). 'From shared databases to communities of practice: A taxonomy of collaboratories'. *Journal of Computer-Mediated Communication* 12(2). Article 2: 652–672. https://doi.org/10.1111/j.1083-6101.2007.00343.x.

Bosma, A. (2017). 'Vera Rubin and the dark matter problem'. *Nature* 543: 179. https://doi.org/10.1038/543179d.

Bourguignon, D. (2014). 'Turning waste into a resource—moving towards a circular economy'. *EPRS, European Parliament*. https://www.europarl.europa.eu/RegData/etudes/BRIE/2014/545704/EPRS_BRI(2014)545704_REV1_EN.pdf.

Bourguignon, D. (2016). 'Closing the loop: New circular economy package'. *EPRS, European Parliamentary Research Service*. https://www.europarl.europa.eu/RegData/etudes/BRIE/2016/573899/EPRS_BRI(2016)573899_EN.pdf.

Boyle, W.S., and Smith, G.E. (1970). 'Charge-coupled semiconductor devices'. *Bell Syst. Tech. J.* 49: 587–593. https://doi.org/10.1002/j.1538-7305.1970.tb01790.x.

Bozeman, B., and Youtie, J. (2017). 'Socio-economic impacts and public value of government-funded research: Lessons from four US National Science Foundation initiative'. Research Policy, 46(8): 1387–1398.

Breakthrough Prize Foundation. (2021). https://breakthroughprize.org.

Breitwieser, L., and Hesam, A. (2020). 'BioDynaMo: an agent-based simulation platform for scalable computational biology research'. *bioRxiv* 2020.06.08.139949. https://doi.org/10.1101/2020.06.08.139949.

Bressan, B. (2004). 'A study of the research and development benefits to society resulting from an international research centre CERN'. PhD dissertation. HU-P-D112 Helsinki: Helsinki University. https://cds.cern.ch/record/1362734/files/astudyof.pdf.

Bressan, B. (2009). 'Knowledge transfer: from creation to innovation'. *CERN Courier*, April. https://cerncourier.com/a/knowledge-transfer-from-creation-to-innovation/.

Bressan B. (2010). *Knowledge Management in an International Research Centre*. Lambert Academic Publishing.

Bressan, B. (2014). *From Physics to Daily Life 1 & 2*. Weinheim: Wiley-Blackwell. https://doi.org/10.1002/9783527687077.

Bressan, B., Kurki-Suonio, K., Lavonen, J., Nordberg, M., Saarikko, H., and Streit-Bianchi, M. (2008). 'Knowledge creation and management in the five LHC experiments at CERN: implications for technology innovation and transfer', in ATL-OREACH-PUB-2009-001. 1–53. http://cds.cern.ch/record/1095892/files/ATL-OREACH-PUB-2009-001.pdf.

Brianti, G., and Jenni, P (2017). 'The Large Hadron Collider (LHC): The energy frontier', in *Advanced Series on Directions in High Energy Physics Technology Meets Research*, 263–326. https://doi.org/10.1142/9789814749145_0008.

Bronowski, J. (1988). 'Knowledge or certainty', in *ETC: A Review of General Semantics* 45(1), Spring: 43–51. http://www.jstor.org/stable/42579414.

Bronson, R.C., Kumanyika, S.K., Kreutzer, M.W., and Haire-Joshu, D. (2021). 'Implementation science should give higher priority to health equity'. *Implementation Science*, 16(28), 1–16. https://doi.org/10.1186/s13012-021-01097-0.

Brown, T. (2005). 'Strategy by design'. *Fast Company* 95: 52–54.

Brown, T. (2008). 'Design thinking'. *Harvard Business Review* (June): 84–91. https://hbr.org/2008/06/design-thinking.

Brown, T. (2009). *Change by Design: How Design Thinking Transforms Organizations and Inspires Innovation*. Harper Collins.

Bruker. (2009). 'Bruker announces Avance 1000, the world's first 1 gigahertz NMR spectrometer'. https://ir.bruker.com/press-releases/press-release-details/2009/Bruker-Announces-Avance-1000-the-Worlds-First-1-Gigahertz-NMR-Spectrometer/default.aspx.

Bruker. (2019). https://www.bruker.com/pt/news-and-events/news/2019/bruker-announces-worlds-first-1-2-ghz-high-resolution-protein-nmr-data.html.

Burget, M., Bardone, E., and Pedaste, M. (2017). 'Definitions and conceptual dimensions of responsible research and innovation: A literature review'. *Science and Engineering Ethics* 23: 1–17. https://doi.org/10.1007/s11948-016-9782-1.

Burns, J.M. (1978). *Leadership*. Harper & Row.

Bush, V. (1945). *Science, the Endless Frontier*. Washington D.C: United States Government Printing Office. https://www.nsf.gov/od/lpa/nsf50/vbush1945.htm.

Butler, J.N. (2017). 'Highlights and perspectives from the CMS experiment', in *Proceedings of the Fifth Annual LHCP CMS-CR-2017/226 FERMILAB-CONF-17-366-CMS*: 1–10. https://arxiv.org/pdf/1709.03006.pdf.

Buytaert J. et al. (2020). 'The HEV ventilator proposal'. *CERN-EP-TECH-NOTE-2020-0*. https://doi.org/10.48550/arXiv.2004.00534.

Calviani, M. (2021). 'Intercepting the beams. *CERN Courier*, May. https://cerncourier.com/a/intercepting-the-beams/.

Campbell, K., Aspray, W., Ensmenger, N., and Yost, J. (2014). *Computer: A History of the Information Machine*. London: Routledge. https://doi.org/10.4324/9780429495373.

Canals, A., Ortoll, E., and Nordberg, M. (2017). 'Collaboration networks in big science: The ATLAS experiment at CERN'. *El profesional de la información* 26(5): 961–971. https://doi.org/10.3145/EPI.

Caraça, J., Lundvall, B-E., and Mendonça, S. (2009). 'The changing role of science in the innovation process: From Queen to Cinderella?'. *Technological Forecasting and Social Change* 76(6): 861–867. https://doi.org/10.1016/j.techfore.2008.08.003.

Cardinal, L.B., Alessandri, T.M., and Turner, S.F. (2001). 'Knowledge codifiability, resources, and science-based innovation'. *Journal of Knowledge Management* 5(2): 195–204. https://doi.org/10.1108/13673270110393266.

Carlier, E., Ducimetière, L., Filippini, R., Goddard, B., and Uythoven, J. (2005). *LHC Project Report* 811, CERN. https://cds.cern.ch/record/841092/files/lhc-project-report-811.pdf.

Caryotakis, G. (1998). 'The Klystron: A microwave source of surprising range and endurance'. *Physics of Plasmas* 5: 1590–1598. https://doi.org/10.1063/1.872826.

Castelvecchi, D. (2019). 'Black hole pictured for first time—in spectacular detail'. *Nature* (June). 568(7752): 284–285. doi: https://doi.org/10.1038/d41586-019-01155.

Castelvecchi, D., Gibney, E. (2020). 'CERN makes bold push to build€ 21-billion super-collider'. *Nature* (June). https://doi.org/10.1038/d41586-020-01866-9.

Cavaglià M., Staats K., and Gill, T. (2018). 'Finding the origin of noise transients in LIGO data with machine learning'. *Commun. Comput. Phys.* 25(4): 963–987. https://doi.org/10.4208/cicp.OA-2018-0092.

CBI. (2020). https://www.cbi-course.com/.

CBI A3. (2018). https://www.cbi.dfm.org.au/.

CEPC-SPPC Study Group. (2015). *CEPC-SPPC Preliminary Conceptual Design Report Volume I—Physics and Detector*. http://cepc.ihep.ac.cn/preCDR/main_preCDR.pdf.

CERN. (1984). 'Large Hadron Collider in the LEP Tunnel'. *Proceedings of ECFA-CERN Workshop March 1984. CERN* 84–10. *CERN.* https://inis.iaea.org/collection/NCLCollectionStore/_Public/16/034/16034963.pdf?r=1&r=1.

CERN. (1987). *Report of the Long-Range Planning Committee to the CERN Council*, 19 June. Geneva: CERN. https://cds.cern.ch/record/30710/files/CM-P00082677-e.pdf.

CERN. (1992). 'Towards the LHC experimental programme', in *General Meeting on LHC Physics and Detectors CERN/ECFA, 5–8 Mar 1992, Evian-les-Bains, France. CERN*. http://cds.cern.ch/record/236265?ln=en.

CERN. (1992b). 'CMS: letter of intent by the CMS Collaboration for a general purpose detector at LHC'.

CERN. (1993). 'The Large Hadron Collider and the long-term scientific programme of CERN: Executive summary'. CERN/SPC/679, CERN/CC/2016. CERN.

CERN. (1994). 'The LHC technological challenge'. https://home.cern/news/press-release/cern/lhc-technological-challenge.

CERN. (1994b). 'Hundredth session of the council, First Part', June. CERN/2052. CERN. https://cds.cern.ch/record/33419/files/CM-P00079664-e.pdf.

CERN. (1994c). 'Hundredth session of the CERN Council Second Part, resolution approval of the Large Hadron Collider (LHC) Project'. CERN/2075. CERN. https://cds.cern.ch/record/33574/files/CM-P00079681-e.pdf.

CERN. (1994d). 'ATLAS: letter of intent for a general-purpose PP experiment at the Large Hadron Collider at CERN'. *ATLAS Collaboration, Technical Proposal CERN-LHCC-92-004; LHCC-I-2. CERN.* https://cds.cern.ch/record/291061?ln=en.

CERN. (2001). *LHC Power Converters: A Precision Game.* CERN. https://cds.cern.ch/journal/CERNBulletin/2001/34.

CERN. (2004). *The LHC Design Report 1: The LHC Main Ring.* CERN-2004-003. CERN.

CERN. (2008). 'CERN releases analysis of LHC incident'. https://home.cern/news/press-release/cern/cern-releases-analysis-lhc-incident.

CERN. (2013). 'The European strategy for particle physics: Update 2013'. CERN-Council-S/106:1–2. https://cds.cern.ch/record/2690131?ln=en.

CERN. (2014). 'CERN's Japanese pied-à-terre (and vice versa)'. https://home.cern/news/news/accelerators/cerns-japanese-pied-terre-and-vice-versa.

CERN. (2017). 'A new compact accelerator for cultural heritage'. https://home.cern/news/news/knowledge-sharing/new-compact-accelerator-cultural-heritage.

CERN. (2017b). 'India becomes Associate Member State of CERN'. https://home.cern/news/news/cern/india-becomes-associate-member-state-cern.

CERN. (2019). *IdeaSquare Progress Report 2017–2018.* CERN. https://doi.org/10.17181/CERN.vxjZ.I72N.

CERN. (2019b). https://home.cern/resources/video/computing/brief-history-world-wide-web.

CERN. (2020). 'CERN and the LHC experiments' computing resources in the global research effort against COVID-19'. https://kt.cern/article/cern-and-lhc-experiments-computing-rezzs-global-research-effort-against-covid-19.

CERN. (2021). *Facts and Figures about the LHC.* Geneva: CERN. https://home.cern/resources/faqs/facts-and-figures-about-lhc.

CERN. (2021b). 'Data preservation'. https://home.cern/science/computing/data-preservation.

CERN. (2021c). 'Diversity office' and 'Diversity and Inclusion programme'. https://diversity-and-inclusion.web.cern.ch/ https://bit.ly/398hSmI.

CERN. (2021d). 'ATTRACT Phase 2'. https://home.cern/news/news/knowledge-sharing/phase-2-attract-launches-new-call-proposals.

CERN. (2021e). 'Dark matter and dark energy'. https://home.cern/science/physics/dark-matter.

CERN. (2022). 'CERN Council declares its intention to terminate cooperation agreements with Russia and Belarus at their expiration dates in 2024'. https://home.cern/news/news/cern/cern-council-cooperation-agreements-russia-belarus.

CERN. (2023). 'A vacuum as empty as interstellar space'. https://home.cern/science/engineering/vacuum-empty-interstellar-space.

CERN Courier. (2007). 'LHC magnet tests: The Indian connection'. *CERN Courier*, June. https://cerncourier.com/a/lhc-magnet-tests-the-indian-connection/.

CERN Courier. (2013). 'ATLAS AND CMS How it all began'. *CERN Courier*, Volume 53 Issue 5, pg. 22. http://cds.cern.ch/record/1550751/files/CERN%20Courier%20June%202013.pdf.

CERN Courier. (2021). 'Farewell to Daya Bay, hello Juno'. https://cerncourier.com/a/farewell-daya-bay-hello-juno/.

CERN Courier (2021b). 'Intercepting the beams'. *CERN Courier*, May. https://cerncourier.com/a/intercepting-the-beams/.

CERN Courier (2021c). 'Making a difference'. *CERN Courier*. https://cerncourier.com/a/making-a-difference/.

CERN Courier (2021d). 'World's most powerful MRI unveiled'. *CERN Courier*, November. https://cerncourier.com/a/worlds-most-powerful-mri-unveiled/.

CERN openlab. (2017). *A Public–Private Partnership to Drive ICT Innovation in Science*. CERN.

CERN openlab. (2017b). *White Paper. Future ICT Challenges in Scientific Research*. CERN.

CERN Personnel Statistics. (2021). Human Resource Department, March 2022. https://cds.cern.ch/record/2809746/files/CERN-HR-STAFF-STAT-2021-RESTR.pdf.

CERN & Society Foundation. (2021). *A Gateway to CERN- the CERN Science Gateway* by *rinevati*. https://cernandsocietyfoundation.cern/news/gateway-cern-cern-science-gateway

Chandrasekaran, K. (2015). *Essentials of Cloud Computing*. Taylor & Francis.

Chang, J.H., Lim Joon, D., Lee, S.T., Gong, S.J., Anderson, N.J., Scott, A.M., Davis, I.D., Clouston, D., Bolton, D., Hamilton, C.S., and Khoo, V. (2012). 'Intensity modulated radiation therapy dose painting for localized prostate cancer using (11)C-choline positron emission tomography scans'. *International Journal of Radiation Oncology, Biology, Physics*. 83(5): e691–6. https://doi.org/10.1016/j.ijrobp.2012.01.087.

Charpak, G. (1992). *Nobel Prize Press Release. CERN Courier* 32. https://www.nobelprize.org/prizes/physics/1992/press-release.

Chernyaev, A.P., and Varzar, S.M. (2014). 'Particle accelerators in modern world'. *Phys. Atomic Nuclear*, 77(10): 1203–1215. https://doi.org/10.1134/S1063778814100032.

Chesbrough, H. (2003). *Open Innovation: The New Imperative for Creating and Profiting from Technology*. Harvard Business Press.

Chesbrough, H. (2017). 'The future of open innovation'. *Research-Technology Management*, 60(1): 35–38. https://doi.org/10.1080/08956308.2017.1255054.

Chesbrough, H. (2020). 'To recover faster from Covid-19, open up: Managerial implications from an open innovation perspective'. *Industrial Marketing Management* 88: 410–413. https://doi.org/10.1016/j.indmarman.2020.04.010.

Child, J., Ihrig, M., and Merali, Y. (2014). 'Organization as information—a space odyssey'. *Organization Studies* 35(6): 801–824. https://doi.org/10.1177/0170840613515472.

Chubin, D.E. (1985). 'Open Science and Closed Science: Tradeoffs in a Democracy'. *Science, Technology and Human Value* 10(2): 73–81. https://doi.org/10.1177/016224398501000211.

Ciborowski, R.W., and Skrodzka, I. (2020). 'International technology transfer and innovative changes adjustment in EU'. *Empir. Econ.* 59: 1351–1371. https://doi.org/10.1007/s00181-019-01683-8.

Cirilli, M. (2021). 'CERN 's impact on medical technology'. *CERN Courier* 24 June 2021.

Clarkson, C., Jacobs, Z., and Marwick, B. et al. (2017). 'Human occupation of northern Australia by 65,000 years ago'. *Nature* 547 (7663): 306–310. https://doi.org/10.1038/nature22968.

CMS Collaboration (1994). *CMS, the Compact Muon Solenoid: Technical Proposal.* CERN-LHCC-92-003; CERN-LHCC-92-3; LHCC-I-1. CERN. https://cds.cern.ch/record/290808?ln=enalsohttps://inspirehep.net/literature/390839, access on 11-12-2022.

CMS Collaboration. (2012). 'Observation of a new boson at a mass of 125 GeV with the CMS experiment at the LHC'. *Physics Letters B* 716(1): 30–61. https://doi.org/10.1016/j.physletb.2012.08.021.

CMS Collaboration. (2012b). *CMS Technical Design Report for the Pixel Detector Upgrade.* CERN-LHCC2012-016; CMS-TDR-11. CERN.

CMS Collaboration (2020). https://home.cern/science/experiments/cms.

CMS Collaboration (2022). 'A portrait of the Higgs boson by the CMS experiment ten years after the discovery'. *Nature* 607, 60–68. https://doi.org/10.1038/s41586-022-04892-x.

Cohen, L. R., and Noll, R. G. (2002). *The technology pork barrel.* Brookings Institution Press.

Cohen, L., Scully, M., and Scully, R. (2009). 'Willis E. Lamb, Jr 1913–2008: A biographical memoir'. National Academy of Sciences, 6. http://www.nasonline.org/publications/biographical-memoirs/memoir-pdfs/lamb-jr-willis.pdf.

Cohen, W.M., Nelson, R.R., and Walsh, J.P. (2002). 'Links and impacts: the influence of public research on industrial R&D'. *Management Science* 48(1): 1–23.

Collins, H.M. (2003). 'LIGO becomes big science'. *Historical Studies in the Physical and Biological Sciences* 33(2): 261–297. https://doi.org/10.1525/hsps.2003.33.2.261.

Comroe, Jr. J.H. (1978). 'Retrospectroscope: Insights into medical discovery'. Book review: 451. *University of Chicago Press Journals.* https://www.journals.uchicago.edu/doi/pdf/10.1086/352087.

Cooper, C. (1981). *Aboriginal Australia.* Sydney: National Gallery of Victoria, Australian Gallery Directors' Council. ISBN 0642896895. OCLC 8487510.

Cooper, R. (2019). 'Design research—its 50-year transformation'. *Design Studies* 65: 6–17. https://doi.org/10.1016/j.destud.2019.10.002.

Cornet, A., and Bonnivert, S. (2008). 'De l'importance d'introduire les contextes culturels et nationaux pour comprendre la relation leadership et genre'. https://www.agrh.fr/assets/actes/2008bonnivert-cornet.pdf.

Cramer, K.C., and Hallonsten, O. (eds). (2020). 'Big Science and research infrastructures in Europe'. ElgarOnline. https://doi.org/10.4337/9781839100017.

Cramer, K.C., Hallonsten, O., Bolliger, I.K., and Griffiths, A. (2020b). 'Big Science and research infrastructures in Europe: History and current trends', in Cramer, K.C. and Hallonsten, O. (eds), *Big Science and Research Infrastructures in Europe.* Cheltenham: Elgar: 1–26. https://doi.org/10.4337/9781839100017.00007.

Creager, A.N.H. (2006). 'Nuclear energy in the service of biomedicine: The U.S. Atomic Energy Commission's Radioisotope Program, 1946–1950'. *J Hist Biol* 39: 649–684. https://doi.org/10.1007/s10739-006-9108-2.

Cross, N. (1993). 'Science and design methodology: A review'. *Research in Engineering Design* 5: 63–69. https://doi.org/10.1007/BF02032575.

CROWD4SDG (2020). https://crowd4sdg.eu/.

Cucinotta, F.A., Kim, M.H.Y., Chappell, L.J., and Huff, J.L. (2013). 'How safe is safe enough? Radiation risk for a human mission to Mars'. *PLOS ONE* 8(10): e74988. https://doi.org/10.1371/journal.pone.0074988.

Daniyal, D., and Hassan Kazmi, S.J. (2019). 'Optimal site selection for an optical-astronomical observatory in Pakistan using Multicriteria Decision Analysis'. *Res. Astron. Astrophys.* 19(9): 129. https://doi.org/10.1088/1674-4527/19/9/129.

David, P. (1998). 'Common agency contracting and the emergence of 'Open Science' institutions'. *The American Economic Review* 88(2): 15–21.

David, P. and Foray, D. (2001). *An Introduction to the Economy of the Knowledge Society.* Discussion Paper. Oxford: Department of Economics. ISSN 1471–0498.

Davies, K. (2002). *Cracking the Genome: Inside the Race to Unlock Human DNA.* JHU Press. https://www.press.jhu.edu/books/title/1380/cracking-genome.

Dayton, L. (2020). 'How South Korea made itself a global innovation leader'. *Nature* 581: S54–S56. https://doi.org/10.1038/d41586-020-01466-7.

de Blas, J., Durieux, G., Grojean, C., et al. (2019). 'On the future of Higgs, electroweak and diboson measurements at lepton colliders'. *J. High Energy Phys.* 117. https://doi.org/10.1007/JHEP12(2019)117.

Design Council (2018). *The Design Economy 2018: The State of Design in the UK.* Design Council. https://www.designcouncil.org.uk/fileadmin/uploads/dc/Documents/Design_Economy_2018_exec_summary.pdf.

De Solla Price, D.J. (1963, 1986). *Little Science, Big Science—and Beyond.* Columbia University Press.

De Solla Price, D.J. (1984). 'The science/technology relationship, the craft of experimental science, and policy for the improvement of high technology innovation'. *Research Policy*, 13: 3–20.

De Winter J.C.F., and Hancock, P.A. (2021). 'Why human factors science is demonstrably necessary: historical and evolutionary foundations'. *Ergonomics* 64(9): 1115–1131. https://doi.org/10.1080/00140139.2021.1905882.

Degiovanni, A., and Amaldi, U. (2015). 'History of hadron therapy accelerators'. *Physica Medica* 31(4): 322–332. https://doi.org/10.1016/j.ejmp.2015.03.002.

deGrasse Tyson, N. (2010). 'Neil deGrasse Tyson explains why the sky is blue'. https://www.youtube.com/watch?v=UvmWxm3nR6E.

Del Rosso, A. (2014). 'World-record current in the MGB2 superconductor'. *CERN Bulletin* 16–17. http://cds.cern.ch/journal/CERNBulletin/2014/16/News%20Articles/1693853.

Del Rosso, A. (2014b). 'ICTP: theorists in the developing world'. *CERN Courier*, November. https://cerncourier.com/a/ictp-theorists-in-the-developing-world/.

Della Negra, M., Jenni, P., and Virdee, T.S. (2012). 'Journey in the search for the Higgs boson: The ATLAS and CMS experiments at the Large Hadron Collider'. *Science* 338(6114): 1560–1568. https://doi.org/10.1126/science.1230827.

Della Negra, M., Jenni, P., and Virdee, T.S. (2018). 'The Construction of ATLAS and CMS'. *Annual Review of Nuclear and Particle Science* 68: 183–209. https://doi.org/10.1146/annurev-nucl-101917-021038.

DFGN. (2020). https://dfgn.org/.

Di Meglio, A. (2021). 'The CERN Quantum Technology Initiative: A hub for collaboration R&D in quantum information science and technology'. http://quanthep-seminar.org/wp-content/uploads/2021/01/QuantHEP-Seminar-2021-01-Alberto-Di-Meglio.pdf.

Di Meglio, A., Doser, M., Frisch, B., Grabowoska, D.M., Pierini, M., Vallecorsa, S. (2020) 'CERN Quantum Technology Initiative Strategy and Roadmap'. CERN Quantum Technology Initiative Strategy and Roadmap. Version 1.0_Rev2. Published 15 October 2021. https://doi.org/10.5281/zenodo5571809.

Di Meglio, A., Girone, M., Purcell, A., and Rademakers, F. (2017). 'CERN openlab white paper on future ICT challenges in scientific research'. *Zenodo.* https://doi.org/10.5281/zenodo.998694.

Diamante, L. (2018). 'Using CERN technology for medical challenges'. https://home.cern/news/news/knowledge-sharing/using-cern-technology-medical-challenges.

Diehl, H.T. (2012). 'The Dark Energy Survey Camera (DECam)'. *Physics Procedia* 37: 1332–1340. https://doi.org/10.1016/j.phpro.2012.02.472.

DOE. (1996). *US Assessment of the Large Hadron Collider,* DOE/ER-0677 (Washington DC: Department of Energy). https://inis.iaea.org/collection/NCLCollectionStore/_Public/27/068/27068372.pdf.

Dran, J.C. (2002). 'Accelerators in art and archaeology', in *Proceedings of EPAC 2002, Paris, France.* https://accelconf.web.cern.ch/e02/papers/frygb001.pdf.

Drell, S.G. (1994). *High Energy Physics Advisory Panel's Subpanel on Vision for the Future of High Energy Physics.* DOE/ER-0614P. US Department of Energy Office of Energy Research Division of High Energy Physics. https://inspirehep.net/literature/374485.

Dretske, F. I. (1981). *Knowledge and the Flow of Information.* MIT Press.

Drupp, M.A., Freeman, M.C., Groom, B., and Nesje, E. (2018). 'Discounting disentangled'. *American Economic Journal: Economic Policy* 10(4): 109–134. https://doi.org/10.1257/pol.20160240.

Du, N., Force, N., Khatiwada, R., Lentz, E., Ottens, R., Rosenberg, L.J., et al. (2018). 'Search for invisible axion dark matter with the axion dark matter experiment'. *Phys. Rev. Lett.* 120(15). https://doi.org/10.1103/PhysRevLett.120.151301.

Duffy, A.R. (2014). 'Probing the nature of dark energy through galaxy redshift surveys with radio telescopes'. *Ann. der Phys.* 526(7–8): 283–293. https://doi.org/10.1002/andp.201400059.

Duffy, A.R., Battye, R.A., Davies, R.D., Moss A., and Wilkinson P.N. (2008). 'Galaxy redshift surveys selected by neutral hydrogen using the Five-hundred metre Aperture Spherical Telescope'. *MNRAS* 383(1): 150–160. https://doi.org/10.1111/j.1365-2966.2007.12537.x.

Duffy, A.R., Moss, A., and Staveley-Smith, L. (2012a). 'Cosmological surveys with the Australian Square Kilometre Array Pathfinder'. *Pub. Astron. Soc. Aus.* 29(2): 202–211. https://doi.org/10.1071/AS11013.

Duffy, A.R., Meyer, M.J., Staveley-Smith, L., Bernyk, M., Croton, D.J., Koribalski, B.S., Gerstmann, D., et al. (2012b). 'Predictions for ASKAP neutral hydrogen surveys'. *MNRAS*, 426(4): 3385–3402. https://doi.org/10.1111/j.1365-2966.2012.21987.x.

Dulov, E. (2020). 'Evaluation of decision-making chains and their fractal dimensions'. *Integrative Psychological and Behavioural Science* 55: 386–429. https://doi.org/10.1007/s12124-020-09566-9.

DUNE. (2020). https://www.dunescience.org/.

Dunne, A., and Raby, F. (2013). *Speculative Everything: Design, Fiction, and Social Dreaming.* MIT Press.

Durante, M., and Paganetti, H. (2016). 'Nuclear physics in particle therapy: a review'. *Reports on Progress in Physics* 79. https://doi.org/10.1088/0034-4885/79/9/096702.

Durante, M., Debus, J., and Loeffler, J.S. (2021). 'Physics and biomedical challenges of cancer therapy with accelerated heavy ions'. *Nat Rev Phys.* 3(12): 777–790. doi: 10.1038/s42254-021-00368-5. Epub 2021 Sep 17. PMID: 34870097; PMCID: PMC7612063.

Edwards, P. (2012). 'Fifty years in fifteen minutes: The impact of the Parkes Observatory'. https://www.atnf.csiro.au/research/conferences/Parkes50th/ProcPapers/edwards.pdf.

Edwards, W. (1954). 'The theory of decision making'. *Psychological Bulletin* 51(4): 380–417. https://doi.org/10.1037/h0053870.

Ekers, R. (2012). 'The history of the Square Kilometre Array (SKA)—born global'. https://doi.org/10.48550/arXiv.1212.3497.

Eller, J.D. (2010). 'Cultural anthropology: Global forces, local lives'. *Anthropos: International Review of Anthropology and Linguistics* 105(2): 633–634. https://doi.org/10.5771/0257-9774-2010-2-633-1.

Elliott, K. and Resnik, D.B. (2014). 'Science, Policy, and the Transparency of Values'. *Environmental Health Perspectives* 122: 647–650.

Elliott, K. and Resnik, D.B. (2019). 'Making Open Science Work for Science and Society'. *Environmental Health Perspectives* 127(7): 647–650. https://doi.org/10.1289/EHP4808.

Ellis, N., and Virdee, T.S. (1994). 'Experimental challenges in high luminosity collider physics'. *Annu. Rev. Nucl. Part. Sci.* 44: 609–653. https://www.annualreviews.org/doi/pdf/10.1146/annurev.ns.44.120194.003141.

ELT. (2021). The European Extremely Large Telescope 'ELT' Project. https://www.eso.org/sci/facilities/eelt/.

EMBL. (2020). https://www.embl.org/topics/coronavirus/.

EMBL. (2021). History of EMBL. https://www.embl.org/about/history/.

Enkel, E., Gassmann, O., and Chesbrough, H. (2009). 'Open R&D and open innovation: exploring the phenomenon'. *R&D Management* 39(4): 311–316. https://doi.org/10.1111/j.1467-9310.2009.00570.x.

ESO. (2021). European Southern Observatory. https://www.eso.org/public/about-eso/.

ESO. (2022). 'Astronomers reveal first image of the black hole at the heart of our galaxy'. ESO Press Release, 12 May 2022. https://www.eso.org/public/news/eso2208-eht-mw/.

Esparza, J., and Yamada, T. (2007). 'The discovery value of "Big Science"'. *Journal of Experimental Medicine* 204(4): 701–704. https://doi.org/10.1084/jem.20070073.

ESPPU. 2013. *The European Strategy for Particle Physics Update 2013*. Geneva: CERN. https://cds.cern.ch/record/1567258/files/esc-e-106.pdf.

ESPPU. 2019. *Physics Briefing Book*. European Strategy for Particle Physics Preparatory Group, CERN: Geneva. arxiv.org/abs/1910.11775.

ESPPU. 2020. *Update of the European Strategy for Particle Physics*. European Strategy Group, Geneva: CERN. https://cds.cern.ch/record/2721370/files/CERN-ESU-015-2020%20Update%20European%20Strategy.pdf.

ESRF. (2020). http://www.esrf.eu/home/news/general/content-news/general/covid-19-scientific-research.html.

ESS. (2020). https://europeanspallationsource.se/article/2020/03/27/ess-demax-lab-prioritise-proposals-covid-19-related-research.

Etzkowitz, H., and Kemelgor, C. (1998). 'The role of research centers in the collectivization of academic science', *Minerva* 36(3): 271–288. Springer.

Etzkowitz, H., and Leydesdorff, L. (2000). 'The dynamics of innovation: from National Systems and "Mode 2" to a Triple Helix of university–industry–government relations'. *Research Policy* 29(2): 109–123. https://doi.org/10.1016/S0048-7333(99)00055-4.

EU. (2015). *European Commission: The Role of Science, Technology and Innovation Policies to Foster the Implementation of the Sustainable Development Goals. Report of the Expert Group*. DG-RTD.

EU. (2016). *European Commission: Open Innovation, Open Science, Open to the World*. DG-RTD.

EU. (2018). *European Commission: Mission-oriented R&D Policies: In-depth Case Studies. Case Study Report on War on Cancer*. DG-RTD.

EU. (2020). 'European Commission: EU's open science policy'. https://ec.europa.eu/info/research-and-innovation/strategy/goals-research-and-innovation-policy/open-science_en.

European Commission. (2014). 'Towards a circular economy: A zero waste programme for Europe'. https://www.oecd.org/env/outreach/EC-Circular-economy.pdf.

European Commission. (2018). 'A renewed European Agenda for Research and Innovation—Europe's chance to shape its future'. https://ec.europa.eu/info/sites/default/files/com-2018-306-a-renewed-european-agenda-_for_research-and-innovation_may_2018_en_0.pdf.

European Commission. (2020). *White Paper on Artificial Intelligence—A European Approach to Excellence and Trust*. COM (2020), 65 final (Brussels: EC).

European Monetary Union. (1992). Convergence Criteria. https://www.insee.fr/en/metadonnees/definition/c1348.

European Strategy for Particle Physics (ESPPU 2020). https://cds.cern.ch/record/2721370/files/CERN-ESU-015-2020%20Update%20European%20Strategy.pdf.

Evans, L. (1998). 'LHC accelerator physics and technological challenge', in *Proc. EPAC'98, Stockholm, Sweden, June*. https://accelconf.web.cern.ch/e98/PAPERS/MOX02A.PDF.

Evans, L. (2010). 'The Large Hadron Collider from conception to commissioning: A personal recollection'. *Reviews of Accelerator Science and Technology* 3(1): 261–280. https://doi.org/10.1142/S1793626810000373.

Evans, L. (ed). (2018). *The Large Hadron Collider: A Marvel of Technology*. Second Edition. Lausanne: EPFL Press. https://cds.cern.ch/record/2645935/files/Evans2018.pdf.

Evans L. and Jenni, P. (2021). 'Discovery machines'. *CERN Courier*, January. https://cerncourier.com/a/discovery-machines/.

Event Horizon Telescope Collaboration et al. (2019). 'First M87 Event Horizon Telescope Results. VI the shadow and mass of the central black hole'. *ApJL*, 875L6. DOI: 10.3847/2041-8213/ab1141.

Event Horizon Telescope Collaboration et al. (2022). 'First Sagittarius A* Event Horizon Telescope Results. I. The Shadow of the Supermassive Black Hole in the Center of the Milky Way'. *ApJL* 930 L12. https://doi.org/10.3847/2041-8213/ac6674.

Fabjan, C., Taylor, T., Treille, D., and Wenninger, H. (eds). (2017). *Technology Meets Research: 60 Years of CERN Technology: Selected Highlights*. World Scientific. https://doi.org/10.1142/9921.

Fahrenthold, D.A. (2013). 'The (86,000) budget-cutting ideas that got away'. *The Washington Post*. https://www.washingtonpost.com/politics/the-85000-budget-cutting-ideas-that-got-away/2013/03/27/85a075a0-90a4-11e2-9c4d-798c073d7ec8_story.html.

FCC. (2020). https://home.cern/science/accelerators/future-circular-collider.

FCC Collaboration. (2019). 'FCC-ee: The Lepton Collider'. *Eur. Phys. J. Spec. Top.* 228: 261–623. https://doi.org/10.1140/epjst/e2019-900045-4.

FCC Collaboration. (2019b). 'FCC-hh: The Hadron Collider'. *Eur. Phys. J. Spec. Top.* 228: 755–1107. https://doi.org/10.1140/epjst/e2019-900087-0.

FCC Collaboration. (2019c). 'FCC physics opportunities'. *Eur. Phys. J. C* 79(6): 474. https://doi.org/10.1140/epjc/s10052-019-6904-3.

Fermilab. (1995). 'Physicists discover top quark'. https://news.fnal.gov/1995/03/physicists-discover-top-quark/.

Fermilab. (2010). 'Fermilab experiments narrow allowed mass range for Higgs boson'. https://news.fnal.gov/2010/07/fermilab-experiments-narrow-allowed-mass-range-higgs-boson/.

Fermilab. (2014). 'The Tevatron: 28 years of discovery and innovation'.

Fermilab. (2017). 'IEEE honors game-changing Tevatron technology'. https://news.fnal.gov/2017/12/ieee-honors-game-changing-tevatron-technology/.

Filipovic, A. (2018). 'The biggest challenges for science communication in the digital age'. *The Elephant in the Lab*. https://doi.org/10.5281/zenodo.1400555.

Fleisch, D.A. (2020). *A Student's Guide to the Schrödinger Equation*. Cambridge University Press.

Fleming, M. (1955). 'External economies and the doctrine of balanced growth'. *The Economic Journal* 65(258): 241–256.

Flexner, A. (2017). *The Usefulness of Useless Knowledge*. Princeton University Press. https://www.ias.edu/sites/default/files/library/UsefulnessHarpers.pdf.

Florio, M. (2019). *Investing in Science: Social Cost–Benefit Analysis of Research Infrastructures*. MIT Press. https://doi.org/10.7551/mitpress/11850.001.0001.

Florio, M., and Sirtori, E. (2016). 'Social benefits and costs of large scale research infrastructures'. *Technological Forecasting and Social Change* 112: 65–78. https://doi.org/10.1016/j.techfore.2015.11.024.

Florio, M., Forte, S., and Sirtori, E. (2016). 'Forecasting the socio-economic impact of the Large Hadron Collider: A cost-benefit analysis to 2025 and beyond'. *Technological Forecasting and Social Change* 112: 38–53. https://doi.org/10.1016/j.techfore.2016.03.007.

Florio, M., Forte, S., Pancotti, C., Sirtori, E., and Vignetti, S. (2016). 'Exploring cost-benefit analysis of research, development and innovation infrastructures: an evaluation framework'. https://doi.org/10.48550/arXiv.1603.03654.

Florio, M., Giffoni, F., Giunta, A., and Sirtori, E. (2018). 'Big science, learning, and innovation: evidence from CERN procurement'. *Industrial and Corporate Change* 27(5): 915–936. https://doi.org/10.1093/icc/dty029.

Fogel, A.L., and Kvedar, J.C. (2018). 'Artificial intelligence powers digital medicine'. *NPJ Digital Medicine* 1 Article 3. https://www.nature.com/articles/s41746-017-0012-2.

Forman, P. (1995). 'Swords into ploughshares: Breaking new ground with radar hardware and technique in physical research after World War II'. *Rev. Mod. Phys.*, 67(2): 397–456. https://doi.org/10.1103/RevModPhys.67.397.

Fothergill, B.T., Knight, W., Stahl, B.C., and Ulnicane, I. (2019). 'Responsible data governance of neuroscience big data'. *Frontiers in Neuroinformatics* 13: 28. https://doi.org/10.3389/fninf.2019.00028.

Frieman, J., Turner, M., and Huterer, D. (2008). 'Dark energy and the accelerating universe'. *Ann. Rev. Astron. Astrophys.* 46: 385–432. https://doi.org/10.1146/annurev.astro.46.060407.145243.

Frisoni, G., Redolfi, A., Manset, D. et al. (2011). Virtual imaging laboratories for marker discovery in neurodegenerative diseases. *Nat Rev Neurol* 7: 429–438. https://doi.org/10.1038/nrneurol.2011.99.

Froborg, F., and Duffy, A.R. (2020). 'Annual modulation in direct dark matter searches'. *J. Phys. G Nuc. Phys.* 47(9). https://doi.org/10.1088/1361-6471/ab8e93.

Fukumoto, S. (1995). 'Cyclotron versus Synchrotron for proton beam therapy', in *Proceedings of the 14th International Conference on Cyclotrons and their Applications, Cape Town, South Africa*.

Funk, C., Kennedy, B., and Hefferon, M. (2018). 'Public perspectives on food risks'. Pew Research Center. https://www.pewresearch.org/science/2018/11/19/public-perspectives-on-food-risks.

Furukawa, S., Nagamatsu, A., et.al. (2020). 'Space radiation biology for "living in space"'. *BioMed Research International* 2020 Article ID 4703286: 25. https://doi.org/10.1155/2020/4703286.

Gaillard, M. (2017). 'CERN data-centre passes the 200 petabyte milestone'. https://home.cern/news/news/computing/cern-data-centre-passes-200-petabyte-milestone.

Galison, P. (1997). *Image and Logic: A Material Culture of Microphysics*. Chicago: University of Chicago Press.

Galison, P., and Hevly, B. (eds). (1992). *Big Science: The Growth of Large-Scale Research*. Stanford: Stanford University Press.

Galison, P., and Stump, D. (eds). (1996). *The Disunity of Science. Boundaries, Contexts and Power*. Stanford: Stanford University Press.

Garud, R., Gehman, J., and Giuliani, A.P. (2018). 'Serendipity arrangements for exapting science-based innovations'. *Academy of Management Perspectives* 32(1): 125–140. https://doi.org/10.5465/amp.2016.0138.

Gastrow, M., and Oppelt, T. (2018). 'Big science and human development—what is the connection?'. *South African Journal of Science* 114(11/12). https://doi.org/10.17159/sajs.2018/5182.

Gauvin, P. (1995). 'Russell Varian (1899–1959)'. *Palo Alto Online*. https://www.paloaltoonline.com/weekly/morgue/news/1995_Jan_4.CREATR44.html.

Gazni, A., Sugimoto, C.R., and Didegah, F. (2012). 'Mapping world scientific collaboration: Authors, institutions, and countries'. *Journal of the American Society for Information Science and Technology* 63(2): 323–335. doi/abs/10.1002/asi.21688.

Geles, C., Lindecker, G., Month, M., and Roche, C. (2000). *Managing Science: Management for R&D Laboratories*. Wiley.

Gianotti, F. (2019). *Implementation of the 2013 European Strategy Update, Open Symposium, Grenada, May*. Geneve: CERN. https://indico.cern.ch/event/808335/timetable/?view=standard.

Gibbons, M. (1994). 'Transfer sciences: management of distributed knowledge production'. *Empirica* 21, 259–270.

Gilbert, J., Grigoriev, A., King, S., Mathew, J., Sharp, R., and Vaccarella, A. (2019). 'Linear-mode avalanche photodiode arrays for low-noise near-infrared imaging in space', in *70th International Astronautical Congress, 21–25 October 2019, Washington, DC*. https://doi.org/10.48550/arXiv.1911.04684.

Gillies, J., and Cailliau, R. (2000). *How the Web Was Born: The Story of the World Wide Web*. Oxford University Press.

Gillies, J (2011). 'Luminosity? Why don't we just say collision rate?'. *Quantum Diaries*, https://home.cern/news/opinion/cern/luminosity-why-dont-we-just-say-collision-rate.

Giudice, G.F. (2012). 'Big Science and the Large Hadron Collider'. *Physics in Perspectives* 14: 95–112. https://doi.org/10.1007/s00016-011-0078-1.

Glänzel W., and Schubert A. (2004). 'Analysing scientific networks through co-authorship', in Moed, H.F., Glänzel, W., and Schmoch, U. (eds), *Handbook of Quantitative Science and Technology Research*. Springer. https://doi.org/10.1007/1-4020-2755-9_12.

Goddard, J. (ed.). (2010). *National Geographic Concise History of Science and Invention: An Illustrated Time Line*. National Geographic.

Goitein, M., and Jermann, M. (2003). 'The relative costs of proton and X-ray radiation therapy'. *Clinical Oncology* 15(1): S37–50. https://doi.org/10.1053/clon.2002.0174.

Granstrand, O., and Holgersson, M. (2014). 'The challenge of closing open innovation: The intellectual property disassembly problem'. *Research-Technology Management* 57(5): 19–25. https://doi.org/10.5437/08956308X5705258.

Grant, R.M. (1996). 'Toward a knowledge-based theory of the firm'. *Strategic Management Journal* 17(S2). https://onlinelibrary.wiley.com/doi/10.1002/smj.4250171110.

Green Car Congress. (2021). 'Airbus aims for a breakthrough in electric aircraft with ASCEND; cryogenics and superconductivity'. https://www.greencarcongress.com/2021/03/20210331-airbus.html.

Gregoris, G., Chin Ong, Y., and Wang, B. (2019). 'Curvature invariants and lower dimensional black hole horizons'. *The European Physical Journal C* 79: 925. https://doi.org/10.1140/epjc/s10052-019-7423-y.

Grey, F. (2003). 'The CERN openlab: a novel testbed for the Grid'. *CERN Courier*, October. https://cerncourier.com/a/the-cern-openlab-a-novel-testbed-for-the-grid/.

Gribbin, J. (2002). *Science: A History*. London: Penguin Press.

Guest, D., Cranmer, K., and Whiteson D. (2018). 'Deep learning and its application to LHC physics', *Annual Review of Nuclear and Particle Science* 68(1): 161–181.

Guida, R., Mandelli, B., and Rigoletti, G. (2020). Performance studies of RPC detectors with new environmentally friendly gas mixtures in presence of LHC-like radiation background. *Nuclear Instruments and Methods in Physics Research Section A: Accelerators, Spectrometers, Detectors and Associated Equipment* 958: 162073.

Gutleber, J. (2021). 'Rethinking the socio-economic value of big science: Lessons from the FCC study', in Beck, H.P. and Charitos, P. (eds), *The Economics of Big Science*. Science Policy Reports. Springer. https://doi.org/10.1007/978-3-030-52391-6_7.

GWIC. (2022). 'Gravitational Wave International Committee'. https://gwic.ligo.org.

Hallonsten, O. (2016). *Big Science Transformed*. Cham: Springer. https://doi.org/10.1007/978-3-319-32738-9.

Hallonsten, O. (2021). 'Stop evaluating science: A historical-sociological argument'. *Social Science Information* 60 (1): 7–26. https://doi.org/10.1177/0539018421992204.

Hameri, A.P. (1997). 'Innovating from Big Science Research'. *Journal of Technology Transfer* 22(3): 27–36.

Hameri, A.P., and Nordberg, M. (1998). 'From experience: Linking available resources and technologies to create a solution for document sharing—the early years of the WWW'. *Journal of Product Development* 15(4): 322–334. https://doi.org/10.1111/1540-5885.1540322.

Hardin, G. (1981). 'Who cares for posterity?', in Partridge, E. (ed.), *Responsibilities to Future Generations*. Prometheus Books.

Harling, O.K. (2009). 'Fission reactor based epithermal neutron irradiation facilities for routine clinical application in BNCT—Hatanaka memorial lecture'. *Applied Radiation and Isotopes* 67(7–8): S7–11. https://doi.org/10.1016/j.apradiso.2009.03.095.

Harmon, B., Ardishvili, A., Cardozo, R., Elder, T., Leuthold, J., Parshall, J., Raghian, M., and Smith, D. (1997). 'Mapping the university technology transfer process'. *Journal of Business Venturing* 12(6): 423–434. https://doi.org/10.1016/S0883-9026(96)00064-X.

Harris, E. (2004). 'Building scientific capacity in developing countries'. *EMBO Reports* 5(1): 8–11.

Harris, A., and Spillane, J. (2018). 'Distributed leadership through the looking glass'. *Management in Education* 22(1): 31–34. https://doi.org/10.1177/0892020607085623.

Harwit, M. (1981). *Cosmic Discovery—The Search, Scope and Heritage of Astronomy*. Basic Books.

Hashimy, L., Treiblmaier, H., and Jain, G. (2021). 'Distributed ledger technology as a catalyst for open innovation adoption among small and medium-sized enterprises'. *The Journal of High Technology Management Research* 32(1) May. https://doi.org/10.1016/j.hitech.2021.100405.

Hatanaka, H. (1991). 'Boron-neutron capture therapy for tumours', in Karim, A.B.M.F. and Laws, E.R. (eds), *Glioma*. Berlin: Springer: 233–249. https://doi.org/10.1007/978-3-642-84127-9_18.

HBR. (2020). 'How the Japanese government's new "Sandbox" program is testing innovations in mobility and technology'. *Government-Sponsored Content for 'Innovating the Future'*. https://hbr.org/sponsored/2020/02/how-the-japanese-governments-new-sandbox-program-is-testing-innovations-in-mobility-and-technology.

Heidler, R. (2017). 'Epistemic cultures in conflict: The case of astronomy and high energy physics'. *Minerva* 55: 249–277. https://doi.org/10.1007/s11024-017-9315-3.

Heiss, A. (2019). 'Big data challenges in Big Science'. *Computer Software for Big Science* 3(15). https://doi.org/10.1007/s41781-019-0030-7.

Hemmer, F. (2018). 'CERN and information technology', in Streit-Bianchi, M. (ed.), *CERN—Science Bridging Cultures*: 31–34. DOI:10.52817/zenodo.1193238.

Henkel, J. (2006). 'Selective revealing in open innovation processes: The case of embedded Linux'. *Research Policy* 35(7): 953–969. https://doi.org/10.1016/j.respol.2006.04.010.

Herfeld, C., and Doehne, M. (2019). 'The diffusion of scientific innovations: A role typology'. *Studies in History and Philosophy of Science Part A* 77: 64–80.

Herschel, W. (1800). 'Experiments on the refrangibility of the invisible rays of the Sun'. *Phil. Trans. Roy. Soc. London* 90: 284–292.

Hess, C., and Ostrom, E. (eds). (2007). *Understanding Knowledge as a Commons: From Theory to Practice*. MIT Press.

Heuer, R. (2018). 'Science policy and society in CERN', in Streit-Bianchi, M. (ed.), *Science Bridging Cultures*: 41. DOI:10.52817/zenodo.1193238.

Hey, T., Butler, K., Jackson, S., and Thiyagalingam, J. (2020). 'Machine learning and big scientific data'. *Philosophical Transactions Royal Society A*. 378. dx.https://doi.org/10.1098/rsta.2019.0054.

Hey, T., Tansley, S., and Tolle, K. (eds). (2009). *The Fourth Paradigm. Data-intensive Scientific Discovery*. Microsoft Research.

HIAF. (2020). https://physics.anu.edu.au/nuclear/hiaf.php.

Hicks, D., and Katz, J.S. (1996). 'Where is science going?' *Science, Technology, & Human Values* 21(4): 379–406. https://doi.org/10.1177/016224399602100401.

Hicks, D., and Katz, J.S. (1997). 'How much is a collaboration worth? A calibrated bibliometric model'. *Scientometrics* 40(3): 541–554. https://doi.org/10.1007/BF02459299.

Hill, S., Yagi, T., and Yamah'ta S. (2022). *The Kyoto Post-COVID Manifesto for Global Economics—Confronting Our Shattered Society*. Springer. https://doi.org/10.1007/978-981-16-8566-8.

Ho, S.L., Cha, H., Oh, I.T., Jung, K.H., Kim, M.H., Lee, Y.J., Miao, X., Tegafaw, T., Ahmad, M.Y., Chae, K.S., Chang, Y., and Lee, G.H. (2018). 'Magnetic resonance imaging, gadolinium neutron capture therapy, and tumour cell detection using ultrasmall Gd2O3 nanoparticles coated with polyacrylic acid-rhodamine B as a multifunctional tumour theragnostic agent'. *RSC Advances* 8: 12653–12665. https://doi.org/10.1039/c8ra00553b.

Hobbs, D., Leitz, C., Bartlett, J., Hepburn, I., Kawata, D., Cropper, M., et al. (2019). 'All-sky near infrared space astrometry'. *Bull. Amer. Astron. Soc.* 51(7). https://baas.aas.org/pub/2020n7i019.

Hofmann, B., Fischer, C.O., Lawaczeck, R., Platzek, J., and Semmler, W. (1999). 'Gadolinium Neutron Capture Therapy (GdNCT) of melanoma cells and solid tumours with the magnetic resonance imaging contrast agent Gadobutrol'. *Investigative Radiology* 34: 126–133. https://doi.org/10.1097/00004424-199902000-00005.

Hofstede, G. (2011). 'Dimensionalizing cultures: The Hofstede model in context'. *Online Readings in Psychology and Culture* 2(1). https://doi.org/10.9707/2307-0919.1014.

Hofstede, G., Hofstede, G.J., and Minkov, M. (2010). *Cultures and Organizations: Software of the Mind: Intercultural Cooperation and Its Importance for Survival*. McGraw-Hill.

Hogan, R., and Kaiser, R.B. (2005). 'What we know about leadership'. *Review of General Psychology* 9(2), 1 June: 169–180. https://doi.org/10.1037/1089-2680.9.2.1.

Holden, C. (1985). 'Global cooperation in big science'. *Science* 229(4708): 32–33. https://doi.org/10.1126/science.229.4708.32.

Horizon (2020). 'Horizon 2020 Details of the EU funding programme which ended in 2020 and links to further information'. https://research-and-innovation.ec.europa.eu/funding/funding-opportunities/funding-programmes-and-open-calls/horizon-2020_en.

Hortala, T. (2021). 'CLEAR study paves the way for novel electron-based cancer therapy'. https://home.cern/news/news/knowledge-sharing/clear-study-paves-way-novel-electron-based-cancer-therapy.

Howell, E. (2018). 'Cosmic microwave backgound: Remnant of the Big Bang'. https://www.space.com/33892-cosmic-microwave-background.html.

Hsu, J., and Huang, D. (2011). 'Correlation between impact and collaboration'. *Scientometrics* 86: 317–324. https://doi.org/10.1007/s11192-010-0265-x.

IAEA. (2021). *International Atomic Energy Agency. Current Status of Neutron Capture Therapy.* IAEA-TECDOC-1223. IAEA.

IAEA. (2021b). *Directory of Radiotherapy Centres.* IAEA.

IceCube. (2020). https://icecube.wisc.edu/.

ICFA. (2019). Taylor, G. ICFA Report. *LP2019V3*, 10th August 2019, Toronto, CA. http://icfa.fnal. gov/ and https://indico.cern.ch/event/688643/contributions/3410173/attachments/1892292/3120953/ICFA_Report_LP2019V3.pdf.

Ichikawa, H., Watanabe, T., Tokumitsu, H., and Fukumori, Y. (2007). 'Formulation considerations of Gadolinium lipid nanoemulsion for intravenous delivery to tumours in neutron-capture therapy'. *Current Drug Delivery* 4: 131–140. https://doi.org/10.2174/156720107780362294.

ICTP. (2021). 'Dirac medallists 2021 announced'. https://www.ictp.it.

Ihrig, M., and Child, J. (2013). Max Boisot and the dynamic evolution of knowledge. *Knowledge, Organization, & Management–Building on the Work of Max Boisot*, 3–18.

ILL. (2020). https://www.ill.eu/neutrons-for-society/science-at-the-ill/biology-and-health/how-neutron-science-supports-research-on-viruses-such-as-covid-19.

Innovationnewsnetwork. (2021). Discussing the Future Circular Collider feasibility study. https://www.innovationnewsnetwork.com/discussing-the-future-circular-collider-feasibility-study/14108/.

The International Human Genome Mapping Consortium. (2001). 'Initial sequencing and analysis of the human genome'. *Nature* 409: 860–921. https://doi.org/10.1038/35057062.

Inverardi, P. (2019). 'The European perspective on responsible computing'. *Communications of the ACM* 62(4): 64. https://doi.org/10.1145/3311783.

IPPOG (2022). Hands on Particle Physics Masterclasses. Published papers. https://physicsmasterclasses.org/index.php?cat=papers.

Irastorza, I.G. (2021). 'BabyIAXO submits for publication its Conceptual Design Report'. https://ep-news.web.cern.ch/content/babyiaxo-submits-publication-its-conceptual-design-report.

Jarlskog, G., and Fernández, E. (eds). (1989). 'ECFA study week on "Instrumentation Technology for High-luminosity Hadron Colliders"'. *CERN-89-10-V-1; ECFA-89-124-V-1, Barcelona, Spain.* https://cds.cern.ch/record/203070?ln=en.

Jarlskog, G., and Rein, D. (1990). *Large Hadron Collider Workshop, European Committee for Future Accelerators. CERN 90–10 ECFA 90–133* 1 (I). Geneva: CERN. https://inis.iaea.org/collection/NCLCollectionStore/_Public/22/030/22030490.pdf.

Jenni, P. (2017). 'The history of ATLAS collaboration'. https://www.youtube.com/watch?v=dZfsCncSB9E.

JETRO. (2018). *Japan External Trade Organization. Detailed Overview: New Regulatory Sandbox Framework in Japan.* https://www.jetro.go.jp/ext_images/en/invest/incentive_programs/pdf/Detailed_overview.pdf.

Jia, H., and Liu, L. (2014). 'Unbalanced progress: The hard road from science popularisation to public engagement with science in China'. *Public Understanding of Science* 23: 32–37. https://doi.org/10.1177/0963662513476404.

Johannessen, S.O. (2009). 'The complexity turn in studies of organizations and leadership: relevance and implications'. *International Journal of Learning and Change* 3(3): 214–229. https://doi.org/10.1504/IJLC.2009.024689.

Johnson, M. (2008). *The Meaning of the Body: Aesthetics of Human Understanding.* University of Chicago Press.

Johnson, S. (2010). *Where Good Ideas Come From: The Natural History of Innovation.* Riverhead Books.

Jongen, Y. (2010). 'Review on cyclotrons for cancer therapy'. In *Proceedings of CYCLOTRONS 2010 Lanzhou Cina,* 398–403. Joint Accelerator Conferences Website (JACoW), CERN.

Jung, J. (2012). 'A conversation with CERN's Rolf Heuer'. *Forbes,* January. https://www.forbes.com/sites/jaynejung/2012/01/13/a-conversation-with-cerns-rolf-dieter-heuer/?sh=482b47ab23a8.

Kadhum, M., Smock, E., Khan, A., and Fleming, A. (2017). 'Radiotherapy in Dupuytren's disease: a systematic review of the evidence'. *The Journal of Hand Surgery (European Volume)* 42(7): 689–692. https://doi.org/10.1177/1753193417695996.

Kahneman D., Sibony O., and Sunstein C.R. (2021). *Noise: A Flaw in Human Judgement.* Little Brown Spark.

Kamada, T., Tsujii, H., Blakely, E.A, Debus, J., De Neve, W., Durante, M., Jäkel, O., Mayer, R., Orecchia, R., Pötter, R., Vatnitsky, S., and Chu, W.T. (2015). 'Review carbon ion radiotherapy in Japan: An assessment of 20 years of clinical experience'. *Lancet Oncol.* Feb. 16(2): e93–e100. https://www.academia.edu/25923050/Carbon_ion_radiotherapy_in_Japan_an_assessment_of_20_years_of_clinical_experience.

Kauffman, S.A. (2000). *Investigations.* Oxford University Press.

Kennefick, D. (2005). 'Einstein versus the Physical Review'. *Physics Today* 58(9): 43–48. https://doi.org/10.1063/1.2117822.

Kersting, K., and Meyer, U. (2018). 'From big data to big artificial intelligence?'. *Künstl Intell* 32: 3–8. https://doi.org/10.1007/s13218-017-0523-7.

Khachatryan, V., Sirunyan, A.M., Tumasyan, A., et al. (2015). 'Precise determination of the mass of the Higgs boson and tests of compatibility of its couplings with the standard model predictions using proton collisions at 7 and 8TeV'. *Eur. Phys. J. C* 75: 212. https://doi.org/10.1140/epjc/s10052-015-3351-7.

Kim, Y. (2009). 'Around the world in 180 minutes: Differences and similarities among women physicists', in *APS Meeting, 17 March 2009, Fermilab, Chicago.*

Kimbell, L. (2011). 'Designing for service as one way of designing services'. *International Journal of Design* 5(2): 41–52.

Kimmitt, R., and Vieira, M. (2020). 'Research synthesis: Time and success rates of pharmaceutical R&D'. knowledgeportalia.org/_files/ugd/356854_9dd6f18e2b114015b253736b1c666cfb.pdf.

Kinsella, W.J. (1999). 'Discourse, power, and knowledge in the management of Big Science: The production of consensus in a nuclear fusion research laboratory'. *Management Communication Quarterly* 13(2): 171–208. https://doi.org/10.1177/0893318999132001.

Kitcher, P. (1993). *The Advancement of Science. Science without Legend, Objectivity without Illusions.* Oxford University Press.

Kitcher, P. (2001). *Science, Truth and Democracy.* Oxford University Press.

Klitsie, J.B., Price, R.A., Stefanie, C, and De Lille, H. (2019). 'Overcoming the valley of death: A design innovation perspective'. *Design Management Journal* 14(1): 28–41. https://doi.org/10.1111/dmj.12052.

Knapp, J., Zeratsky, J., and Kowitz, B. (2016). *Sprint: How to Solve Big Problems and Test New Ideas in Just Five Days.* New York: Simon & Schuster.

Knorr-Cetina, K. (1999). *Epistemic Cultures: How the Sciences Make Knowledge,* Harvard University Press.

Koch, C., and Jones, A. (2016). 'Big Science, Team Science, and Open Science for Neuroscience'. *Neuron* 92: 612–616. https://https://doi.org/10.1016/j.neuron.2016.10.019.

Kogut, B., and Zander, U. (1992). 'Knowledge of the firm, combinative capabilities, and the replication of technology'. *Organization Science* 3(3). https://doi.org/10.1287/orsc.3.3.383.

Kokko, A., Gustavsson Tingvall, P. and Videnord, J. (2015). The Growth Effects of R&D Spending in the EU: A Meta-Analysis. Economics Discussion Papers, No. 2015-29, Kiel Institute for the World Economy. http://www.economics-ejournal.org/economics/discussionpapers/2015-29.

Kokurewicz, K., Schüller, A., Brunetti, E., Subiel, A. Kranzer, R. Hackel, T., Meier, M., Kapsch, R., and Jaroszynski, D.A. (2020). 'Dosimetry for New Radiation Therapy Approaches Using High Energy Electron Accelerators'. *Frontiers in Physics* 8: 1–12. https://doi.org/10.3389/fphy.2020.56830.

Komamiya, S. (2001). 'The e+e- Linear Collider JLC', in *Proceedings of the Second Particle Accelerator Conference, Beijing, China*: 130–134. https://accelconf.web.cern.ch/a01/PDF/MOB01.pdf.

Kowarski, L. (1972). 'Computers: Why?' *CERN Courier* 12 (3): 59. https://cerncourier.com/a/computers-why/.

Krige, J. (1993). 'Some socio-historical aspects of multinational collaborations in high energy physics at CERN between 1975 and 1985', in Crawford, E., Shinn, T., and Sörlin, S. (eds), *Denationalizing Science. Sociology of the Sciences: A Yearbook*, 16: 233–262. Springer. https://doi.org/10.1007/978-94-017-1221-7_9.

Kronegger, L., Ferligoj, A., and Doreian, P. (2011). 'On the dynamics of national scientific systems'. *Quality & Quantity* 45: 989–1015. https://doi.org/10.1007/s11135-011-9484-3.

Kuhn, T. (1962, 2012). *The Structure of Scientific Revolutions*. University of Chicago Press.

Kurakawa, S. (2005). 'Asian view point as ACFA chair and global view point as new ILCSC Chair', in *Second ILC workshop, August Snowmass Colorado*. https://www.slac.stanford.edu/econf/C0508141/proc/pres/PLEN0054_TALK.PDF.

Kurki-Suonio, R., and Kurki-Suonio, K. (1994). *Fysiikan Merkitykset ja Rakenteet* (*The Meanings and Structures of Physics*).

Kurokawa, S. I. (2005). 'Asian View Point as ACFA Chair and Global View Point as New ILCSC Chair'. slac.stanford.edu.

Laine, A. (2016). 'International Symposium on Ion Therapy: Planning the first hospital-based heavy ion therapy center in the United States'. *Int J Part Ther*. Winter 2(3): 468–471.

Lamont, M. (2013). 'The first years of LHC operation for luminosity production', in *4th International Particle Accelerator Conference, Shanghai*. http://cds.cern.ch/record/2010134.

Lampe, N. (2017). 'Understanding low radiation background biology through controlled evolution experiments'. *Evolutionary Applications* 10(7): 658–666. https://doi.org/10.1111/eva.12491.

Langley, S.P. (1881). 'The bolometer and radiant energy'. *Proc. Am. Academy of Arts and Sciences (May 1880–Jun. 1881)* 16: 342–358. https://doi.org/10.2307/25138616.

Larkin, J., McDermott, J., Simon, D. P., and Simon, H. A. (1980). 'Expert and novice performance in solving physics problems'. *Science* 208(4450): 1335–1342.

Latour, B. (1987). *Science in Action*. Harvard University Press.

Lave, J., and Wenger, E. (1991). *Situated Learning: Legitimate Peripheral Participation*. Cambridge University Press.

Lawrence, E.O., and Livingston, M.S. (1932). 'The production of high speed light ions without the use of high voltages'. *Physical Review* 40(19). https://doi.org/10.1103/PhysRev.40.19.

Lawrence, J.H., Tobias, C.A., Born, J.L., McCombs, K., Roberts, J.E., Anger, H.O., Low-Beer, B.V., and Huggins, C.B. (1958). 'Pituitary irradiation with high energy proton beams: a preliminary report'. *Cancer Res*. 18(2): 121–134. PMID: 13511365.

Lawson, W.D., Nielson, S., Putley, E.H., and Young A.S. (1959). 'Preparation and properties of HgTe and mixed crystals of HgTe–CdTe', *J. Phys. Chem. Solids* 9(3–4): 325–329. https://doi.org/10.1016/0022-3697(59)90110-6.

Lazzarotti, V., Manzini, R. (2009). 'Different modes of open innovation: a theoretical framework and an empirical study'. *International Journal of Innovation Management* 13(04): 615–636. https://doi.org/10.1142/S1363919609002443.

Le, U.M., and Cui, Z. (2006). 'Long-circulating gadolinium-encapsulated liposomes for potential application in tumour neutron capture therapy'. *International Journal of Pharmaceutics* 312: 105–112. https://doi.org/10.1016/j.ijpharm.2006.01.002.

Lehman, J., and Stanley, K.O. (2015). *Why Greatness Cannot Be Planned*. Cham: Springer.

Lekka-Kowalik, A. (2010). 'Why science cannot be value-free understanding the rationality and responsibility of science'. *Science Engineering Ethics* 16: 33–41. https://doi.org/10.1007/s11948-009-9128-3.

Leogrande, E., and Nicassio, R. (2021). 'Collaborative processes in science and literature: An in depth look at the cases of CERN and SIC'. *Frontiers in Research and Metrics Analytics* 5, 1–11. https://doi.org/10.3389/frma.2020.592819.

Levy, R.P., and Blakely, E.A. (2009). 'The current status and future directions of heavy charged particle therapy in medicine'. *AIP Conference Proceedings* 1099(410): 410–425. https://doi.org/10.1063/1.3120064.

LHC Study Group (1991). *Design Study of the Large Hadron Collider—a Multiparticle Hadron Collider of the LEP Tunnel.* CERN, Geneva. https://cds.cern.ch/record/220493/files/CERN-91-03. pdf.

LHCb Collaboration. (2008). 'The LHCb Detector at the LHC'. Journal of Instrumentation 3: 1–205. https://iopscience.iop.org/article/10.1088/1748-0221/3/08/S08005/pdf.

LHCb Collaboration. (2021). 'Test of lepton universality in beauty-quark decays'. arXiv:2103.11769 [hep-ex].

Li Vigni, F. (2021). 'Regimes of evidence in complexity sciences'. *Perspectives on Science* 29(1): 62–103.

LIGO. (2017). 'LIGO Data Management Plan', June 2017. dcc.ligo.org/LIGO-M1000066/public/.

LIGO-Virgo. (2007). 'Memorandum of Understanding'. dcc.ligo.org/LIGO-M060038-x0/public.

LIGO and Virgo Collaborations. (2016). 'Observation of gravitational waves from a binary black hole merger'. *Phys. Rev. Lett.* 116: 1–17. https://doi.org/10.1103/PhysRevLett.116.061102.

Liyanage, S. (1995). 'Breeding innovation clusters through collaborative research networks'. *Technovation* 15(9): 553–567. https://doi.org/10.1016/0166-4972(95)96585-H.

Liyanage, S., and Boisot, M. (2011). 'Leadership in the ATLAS collaboration', in Boisot, M., Nordberg, M., Yami, S., and Nicquevert, B. (eds), *Collisions and Collaboration.* Oxford: University Press. https://doi.org/10.1093/acprof:oso/9780199567928.003.0012.

Liyanage S., Greenfield P.F., and Don R. (1999). 'Towards a fourth generation R&D management model-research networks in knowledge management'. *International Journal of Technology Management* 18(3–4): 372–393. https://doi.org/10.1504/IJTM.1999.002770.

Liyanage, S., Wink, R., and Nordberg, M. (2007). *Managing Path-breaking Innovations: CERN-ATLAS, Airbus, and Stem Cell Research.* Westport: Praeger.

Lockspeiser, B. (1954). Quoted in CERN celebrates 40th Anniversary. https://home.cern/news/press-release/cern/cern-celebrates-40th-anniversary.

Long, L (2011). 'More "hands-on" particle physics: Learning with ATLAS'. *Phys.Edu.* 46(3): 270–280. https://iopscience.iop.org/article/10.1088/0031-9120/46/3/003/meta.

Long, G. (2019). 'Fundamental Research Security'. MITRE Corporation JSR-19-2I. https://www.nsf.gov/news/special_reports/jasonsecurity/JSR-19-2IFundamentalResearchSecurity_12062019FINAL.pdf.

Lorimer, D.R., Bailes, M., McLaughlin, M.A., Narkevic, D.J., and Crawford, F. (2007). 'A bright millisecond radio burst of extragalactic origin'. *Science* 318(5851): 777–780. https://doi.org/10.1126/science.1147532.

Lovell, B. (1977). 'The effects of defence science on the advance of astronomy'. *J. Hist. Astron.* 8: 151–173.

Luderer, M.J., de la Puente, P., and Azab, A.K. (2015). 'Advancements in tumour targeting strategies for boron neutron capture therapy'. *Pharmaceutical Research* 32: 2824–2836. https://doi.org/10.1007/s11095-015-1718-y.

Lupton, E. (1986). 'Reading isotype'. *Design Issues* 3(2): 47–58.

Lutz, S.T., Jones J., and Chow E. (2014). 'Role of radiation therapy in palliative care of the patient with cancer'. *J Clin Oncol.* 32(26): 2913–2919. https://doi.org/10.1200/JCO.2014.55.1143.

MacArthur & IDEO. (2017). 'The Circular Design Guide'. https://www.circulardesignguide.com/.

MacKee, G.M. (1921). *X-rays and Radium in the Treatment of Diseases of the Skin.* New York: Nabu Press. ISBN-13 978–1145802988.

MacMillan, I.C., van Putten, A.B., Gunther McGrath, R., and Thompson, J.D. (2015). 'Using real options discipline for highly uncertain technology investments'. *Research-Technology Management* 49(1): 29–37. https://doi.org/10.1080/08956308.2006.11657356.

Maeda, J. (2019). *How to Speak Machine: Laws of Design for a Computational Age.* London: Portfolio Penguin.

Maher, B. and Noorden, R.V. (2021). 'How the COVID pandemic is changing global science collaborations'. *Nature,* June. https://www.nature.com/articles/d41586-021-01570-2.

Mak, D., Corry, J., Lau, E., Rischin, D., and Hicks, R.J. (2011). 'Role of FDG-PET/CT in staging and follow-up of head and neck squamous cell carcinoma'. *Quarterly Journal of Nuclear Medicine Molecular Imaging* October 55(5): 487–499. https://europepmc.org/article/med/22019706.

Makarova, E. et al. (2019). 'The gender gap in STEM Fields: The impact of the gender stereotype of math and science on secondary students' career aspirations'. *Frontiers in Education* 4 (July). https://doi.org/10.3389/feduc.2019.00060.

Malina, R.F. (1997). 'Leonardo: Our fourth decade'. *Leonardo* 30(1): 1–2. muse.jhu.edu/article/607399.

Malina, R.F. (2008). 'A call for new Leonardos'. *Leonardo* 41(1): 2. https://doi.org/10.1162/leon.2008.41.1.2.

Malina, R.F. (2011). 'Alt.Art-Sci: We need new ways of linking arts and sciences'. Leonardo 44(1): 2–2. muse.jhu.edu/article/412940.

Manset, D. (2019). Annual report covering CERN openlab's activities, Slide 66, In Purcell, A.R., Di Meglio, A., Gunne, K., Carminati F. Rademaker, F., Girone, M., Baechle, H., and Lazuka, A. (eds), CERN openlab final annual report_2019.pdf. https://doi.org/10.5281/zenodo.4040603 and 'be-studys rejoint le CERN openlab'. https://www.brefeco.com/actualite/logiciels-services-numeriques/be-studys-rejoint-le-cern-openlab.

Manset, D. (2021). Personal interviews on Big Science operation, May.

Manset, D., Hikkerova, L., and Sahut, J-M. (2017). 'Rethinking the humanitarian model from efficiency to resilience'. *Review Gestion & Management Public* 5(4): 83–106. http://www.airmap.fr/wp-content/uploads/2017/12/vol5n4_eng_6_Manset_etal.pdf.

Manzini, E. (2015). *Design, When Everybody Designs: An Introduction to Design for Social Innovation*. MIT Press.

Marchand, D.A., and Margery, P. (2009). 'Leadership through collaboration and harmony: how to lead without formal authority'. *Perspectives for Managers* 180. Lausanne: IMD: 4. https://www.proquest.com/docview/2367788216?parentSessionId=j3Yt1fQ8KYV7TVcrRBCwEJNl8t7Fe5DCHJNIISapqG8%3D.

Marfavi, A., Kavianpour, P., and Rendina, L.M. (2022). 'Carboranes in drug discovery, chemical biology and molecular imaging'. *Nat Rev Chem* 6: 486–504. https://doi.org/10.1038/s41570-022-00400-x.

Mark, H., and Levine, A. (1984). *The Management of Research Institutions: A Look at Government Laboratories*. NASA SP-481. Washington, DC: NASA. https://archivesdev.lib.purdue.edu/repositories/2/archival_objects/21884.

Maroto, A., Gallego, J., and Rubalcaba, L. (2016). 'Publicly funded R&D for public sector performance and efficiency: evidence from Europe'. *R D Manag.* 46(S2): 564–578.

Martin, B.R. (2012). 'Statistics for physical sciences', in Martin, B.R. (ed.), *Statistics for Physical Sciences*. London: Elsevier. https://doi.org/10.1016/C2010-0-68810-9.

Martin, I., and Child, J. (2013). 'Max Boisot and the dynamic evolution of knowledge', in J. Child and M. Ihrig (eds), *Knowledge, Organization, & Management: Building on the Work of Max Boisot*, 216. Oxford: Oxford University Press.

Massimi, M. (2019). 'Two kinds of exploratory model'. *Philosophy of Science* 86(5) 869–881. https://doi.org/10.1086/705494.

Massimi, M. (2021). 'Realism, perspectivism, and disagreement in science'. *Synthese* 198(S25): 6115–6141. https://doi.org/10.1007/s11229-019-02500-6.

Matsubara, H., Karasawa, K., Furuichi, W., Wakaisami, M., Shiba, S., Wakatsuki, M., Omatsu, T., Inaniwa, T., Fukuda, S., and Kamada, T. (2018). 'Comparison of passive and scanning irradiation methods for carbon-ion radiotherapy for breast cancer'. *Journal of Radiation Research* 59: 625–631. https://doi.org/10.1093/jrr/rry052.

Mabey, C. Kulich, C., Lorenzi-Cioldi, F. (2012). 'Knowledge leadership in global scientific research'. *International Journal of Human Resources Development and Management* 23(12): 2450–2467. https://doi.org/10.1080/09585192.2012.668386.

Mazzuccato, M. (2013) *The entrepreneurial state: debunking public vs. private sector myths*. Anthem Press. ISBN 978-0857282521.

Mazzuccato, M. (2013). *The Entrepreneurial State: Debunking Public vs. Private Sector Myths: 1*. Anthem Press.

McGrath Gunther, R., and MacMillan, I.C. (2009). *Discovery-Driven Growth: A Breakthrough Process to Reduce Risk and Seize Opportunities.* Harvard University Press.

McPhee, C., Bliemel, M., and van der Bijl-brouwer, M. (2018). 'Transdisciplinary innovation'. *Technology Innovation Management Review* 8(8): 3–6. https://timreview.ca/article/1173.

Melchor, S. (2021). 'The coevolution of particle physics and computing'. https://www.symmetrymagazine.org/article/the-coevolution-of-particle-physics-and-computing.

Mentink, M., Curé, B., Deelen, N., et al. (2023) 'Superconducting detector magnets for high energy physics'. *Journal of Instrumentation* 18(6): T06013. https://doi.org/10.1088/1748-0221/18/06/T06013.

Merton, R.K. (1942). 'The normative structure of science', in N.W. Storer (ed.), *The Sociology of Science: Theoretical and Empirical Investigations*, 267–278. The University of Chicago Press.

Merton, R. (1957). 'Priorities in scientific discovery: A chapter in the sociology of science'. *American Sociological Review* 22(6): 635–659.

Merton, R.K., and Barber, E. (2004). *The Travels and Adventure of Serendipity.* Princeton University Press.

Merz, M., and Sorgner, H. (2022). 'Organizational complexity in big science: Strategies and practices', *Synthese* 200: 211, https://doi.org/10.1007/s11229-022-03649-3.

Meyer, E.T. (2009, 2012). 'Moving from small science to big science: Social and organizational impediments to large scale data sharing', in N. Jankowski (ed.), *e-Research: Transformation in Scholarly Practice*, 147–159. Routledge.

Michlewski, K. (2015). *Design Attitude.* Routledge. https://doi.org/10.4324/9781315576589.

Minkov, M., Dutt, P., Schachner, M., Morales, O., Sanchez, C., Jandosova, J., Khassenbekov, Y., and Mudd, B. (2017). 'A revision of Hofstede's individualism-collectivism dimension'. *Cross Cultural & Strategic Management* 24(3): 386–404. https://doi.org/10.1108/CCSM-11-2016-0197.

Mishima, Y., Honda, C., Ichihashi, M., Obara, H., Hiratsuka, J., Fukuda, H., Karashima, H., Kobayashi, T., Kanda, K., and Yoshino, K. (1989). 'Treatment of malignant melanoma by single thermal neutron capture therapy with melanoma-seeking 10b-compound'. *The Lancet* 334: 388–389. https://doi.org/10.1016/s0140-6736(89)90567-9.

Mitchell, N., Devred A., Lubeyre P., Lim B., and Savary F. (2012). 'The ITER magnet: Design and construction status'. *IEEE Trans. on Applied. Superconductivity* 22 (3. https://doi.org/10.1109/TASC.2011.2174560.

Mitchell, S. D. (2009). *Unsimple truths: Science, complexity, and policy.* University of Chi-cago Press.

Mitnick, B.M. (2015). 'Agency theory', in Cooper, C.L (ed.), *Wiley Encyclopedia of Management.* https://doi.org/10.1002/9781118785317.weom020097.

Mitsou, V.A. (2011). 'Dark matter searches at LHC'. *J. Phys.: Conf. Ser.* 335, 012003. https://doi.org/10.1088/1742-6596/335/1/012003.

Morrison, D.E., Aitken, J.B., de Jonge, M.D., Ioppolo, J.A., Harris, H.H., and Rendina, L.M. (2014). 'High mitochondrial accumulation of new gadolinium(iii) agents within tumour cells'. *Chemistry Communications* 50: 2252–2254. https://doi.org/10.1039/c3cc46903d.

Morton, L.W., Eigenbrode, S.D., and Martin, T.A. (2015). 'Architectures of adaptive integration in large collaborative projects'. *Ecology and Society* 20(4): 5.

Murayama, K., Nirei. M., and Shimizu, H. (2015). 'Management of science, serendipity, and research performance: Evidence from a survey of scientists in Japan and the U.S'. *Research Policy* 44(4): 862–873. https://doi.org/10.1016/j.respol.2015.01.018.

Murillo, L.F.R., and Kauttu, P. (2017). 'Open hardware as an experimental innovation platform: preliminary research questions and findings'. *CERN IdeaSquare Journal of Experimental Innovation* 1(1): 26–34. https://doi.org/10.23726/cij.2017.462.

Murphy, K.M., Shleifer, A., and Vishny, R.W. (1989). 'Industrialization and the big push'. *Journal of Political Economy* 97(5): 1003–1026.

Murray, I., Skene, K., and Haynes, K. (2017). 'The circular economy: An interdisciplinary exploration of the concept and application in a global context'. *Journal of Business Ethics* 140(3): 369–380. https://doi.org/10.1007/s10551-015-2693-2.

Myrdal, G. (1968). *Asian Drama: An Inquiry into the Poverty of Nations*, 3 Volumes. Pantheon.

Nahapiet, J., and Ghoshal, S. (1998). 'Social capital, knowledge capital and the organizational advantage'. *Academy of Management Review* 3(2): 242–266.

Naim, K., Basaglia, T., Brankovic, J., Fokianos, P., Gonzalez Lopez, J.B., Pia, M.G., et al. (2020). 'Pushing the Boundaries of Open Science at CERN: Submission to the UNESCO Open Science Consultation'. https://en.unesco.org/sites/default/files/cern-unesco_consultation_jul_15.pdf.

NASA. (2018). 'CCDs provide clearer and less painful biopsies'. https://www.nasa.gov/feature/goddard/ccds-provide-clearer-and-less-painful-biopsies/.

NASA. (2022). 'Technology transfer'. https://technology.nasa.gov.

Natarajan, P. (2021). 'Cosmos, chaos and order: Mapping as knowing'. *Leonardo* 54(1): 107–114.

Nature (2004). 'Time for Japan to shine?' *Nature* 427, 763. https://doi.org/10.1038/427763a.

Nature (2021). 'The world's largest radio telescope should open its skies to all'. Nature, 590: 527. https://doi.org/10.1038/d41586-021-00468-3.

Nature Index. (2019). 'Collaboration and big science'. 575(7783).

Nayyar, D. (ed). (2019). *Asian Transformations: An Inquiry into the Development of Nations*. Oxford University Press. https://doi.org/10.1093/oso/9780198844938.001.0001.

Nedunchezhian, K. (2016). 'Boron neutron capture therapy—a literature review'. *Journal of Clinical and Diagnostic Research* 10(12). https://doi.org/10.7860/JCDR/2016/19890.9024.

Negroponte, N. (2018). 'Big idea famine'. *Journal of Design and Science*. https://doi.org/10.21428/7174a671.

Nelson, F.A., and Weaver, H.E. (1964). 'Nuclear magnetic resonance spectroscopy in superconducting magnetic fields'. *Science* 146 (3641): 223–232. https://doi.org/10.1126/science.146.3641.223.

Nelson, R.R. (1977). *The Moon and the Ghetto*. Norton.

Nelson, R.R. (2011). 'The moon and the ghetto revisited'. *Science and Public Policy* 38(9): 681–690. https://doi.org/10.1093/scipol/38.9.681.

Nelson, R.R., and Romer, P.M. (1996). 'Science, economic growth, and public policy'. *Challenge* 39(1): 9–21. https://doi.org/10.1080/05775132.1996.11471873.

Nemoto, I., and Kubono, M. (1996). 'Complex associative memory'. *Neural Networks* 9(2): 253–261. https://doi.org/10.1016/0893-6080(95)00004-6.

Nilsen V., and Anelli G. (2016). 'Technology Transfer at CERN'. *Technol. Forecast. Soc. Change* 112(2016): 113–120.

Nobel. (1935). Irène Joliot-Curie—Nobel Lecture. https://www.nobelprize.org/prizes/chemistry/1935/joliot-curie/lecture/.

Nobel Prize (1943). George de Hevesy – Facts. NobelPrize.org. https://www.nobelprize.org/prizes/chemistry/1943/hevesy/facts/.

Nobili, L. (1830). 'Description d'un thermo-multiplicateur ou thermoscope électrique'. *Bibliothèque Universelle* (in French), 44: 225–234.

Nonaka, I., and Takeuchi, H. (1995). *The Knowledge Creating Company: How Japanese Companies Create the Dynamics of Innovation*. Oxford University Press.

Nordberg, M., and Nessi M. (2013). 'ATTRACT Initiative'. Mimeo. https://indico.cern.ch/event/281585/contributions/1629911/attachments/517051/713383/ATTRACT-Mimeo.pdf.

Nordberg, M., and Verbeke, A. (1999). *The Strategic Management of High Technology Contracts: The Case of CERN*. Emerald Publishing.

NQIT. (2019). 'Responsible research and innovation in networked quantum IT'. *Networked Quantum Information Technologies*. https://www.cs.ox.ac.uk/projects/NQITRRI/index.html.

NRC. (2013). *Nuclear Physics: Exploring the Heart of Matter*. National Academies Press.

Nunlist, T. (2016). 'Big China, Big Science'. *CKGSB Knowledge, #Science*. https://english.ckgsb.edu.cn/knowledges/big-china-big-science/.

Nurske, R. (1953). '*Problemas de formación de capital en los países insuficientemente desarrollados*'. México: Fund of Economic Culture.

Nurske, R. (1953b). *Problems of Capital Formation in Underdeveloped Countries*, Oxford University Press.

Obaldo, J.M. and Hertz, B.E. (2021), 'The early years of nuclear medicine: A retelling', *Asia Ocean J Nucl Med Biol.* Spring; 9(2): 207–219. https://doi.org/10.22038/aojnmb.2021.55514.1385.

OECD. (2008). *Report on Roadmapping of Large Research Infrastructures.* OECD.

OECD. (2010). *Large Research Infrastructures. OECD Global Science Forum:* 19. OECD.

OECD. (2014). *The Impacts of Large Research Infrastructures on Economic Innovation and on Society: Case Studies at CERN.* OECD.

OECD. (2014b). *PISA 2012 Results: What Students Know and Can Do—Student Performance in Mathematics, Reading and Science I,* Revised edition. OECD. https://doi.org/10.1787/19963777.

OECD (2015), *Frascati Manual 2015: Guidelines for Collecting and Reporting Data on Research and Experimental Development, The Measurement of Scientific, Technological and Innovation Activities.* OECD. DOI: http://dx.https://doi.org/10.1787/9789264239012-en.

OECD. (2016). 'G20 Innovation Report'. Figure 14. https://www.oecd.org/china/G20-innovation-report-2016.pdf.

OECD. (2020). 'Addressing societal challenges using transdisciplinary research'. *OECD Policy Paper* 88. https://www.oecd-ilibrary.org/docserver/0ca0ca45-en.pdf?expires =1669915761&id=id&accname=guest&checksum=59427FC9867AA7A10DE5FD25816111EE.

Olan, F., Jayawickrama, U., Arakpogun, E.O., et al. (2022). 'Fake news on social media: The impact on society'. *Inf Syst Front.* https://doi.org/10.1007/s10796-022-10242-z.

Oliveira, M., and Teixeira, A. (2010). *The Determinants of Technology Transfer Efficiency and the Role of Innovation Policies: A Survey.* FEP Working Paper (Porto: Universidade do Porto, Faculdade de Economia do Porto). https://www.researchgate.net/publication/46448511_The_ determinants_of_technology_transfer_efficiency_and_the_role_of_innovation_policies_a_ survey.

Ortoll, E., Canals, A., Garcia, and M. Cobarsí J. (2014). 'Principales parámetros para el estudio de la colaboración científica en big science'. *Revista Española de Documentación Científica* 37(4): e069. https://doi.org/10.3989/redc.2014.4.1142.

Orton J.W. (2004). *The Story of Semiconductors.* Oxford University Press. http://ndl.ethernet.edu. et/bitstream/123456789/16909/1/429.pdf.

Oxford English Dictionary. (2020). https://www.oed.com/.

Oxman, N. (2016). 'The age of entanglement'. *Journal of Design and Science.* https://doi.org/10. 21428/7e0583ad.

Palmieri, L., and Jensen, H.J. (2020). 'Investigating critical systems via the distribution of correlation lengths'. *Phys. Rev. Research*2(1): 013199. https://doi.org/10.1103/PhysRevResearch.2.013199.

Panofsky, W.K.H. (1997). 'The evolution of particle collider and accelerators, *Beam Line,* Spring: 36–44. https://www.slac.stanford.edu/pubs/beamline/27/1/27-1-panofsky.pdf.

Panofsky, W.K.H. (2007). *Panofsky on Physics, Politics, and Peace.* Springer. https://link.springer. com/content/pdf/bbm:978-0-387-69732-1/1.pdf.

Paraderie, F. (1996). 'Big Science, why? where? and how?'. *Mem Sa It.* 67(4): 889–900. https:// adsabs.harvard.edu/full/1996MmSAI.67.889P.

Particle Therapy Co-Operative Group. (2021). 'Particle therapy facilities in clinical operation'. https://www.ptcog.ch/index.php/facilities-in-operation.

Particle Data Group, Zyla, P.A., Barnett, R.M., Beringer, J., Dahl, O., D Dwyer, A.D., Groom, E.D., ... and Profumo, S. (2020). 'Review of Particle Physics (RPP 2020)'. *Progress of Theoretical and Experimental Physics,* 2020(8): 1–2093.

Patel, Z.S., Brunstetter, T.J., and Tarver, W.J. (2020). 'Red risks for a journey to the red planet: The highest priority human health risks for a mission to Mars'. *Microgravity* 6 33. https://doi.org/10. 1038/s41526-020-00124-6.

Pavitt, K. (1998). 'The social shaping of the national science base'. *Research Policy* 27(8): 793–805.

Peacock, M. S. (2009). 'Path dependence in the production of scientific knowledge'. *Social Epistemology* 23(2): 105–124. https://doi.org/10.1080/02691720902962813.

Pearce, D., Groom, B., Hepburn, C., and Koundouri, P. (2003). 'Valuing the future'. *World Economics* 4(2): 121–141.

Pentland, A. (2014). *Social Physics—How Good Ideas Spread—Lessons from a New Science*. Penguin Press.

Petrenko, V., Smith, A., Schaefer, H., et al. (2017). 'Minimal geological methane emissions during the Younger Dryas–Preboreal abrupt warming event'. *Nature* 548: 443–446. https://doi.org/10.1038/nature23316.

Pickering, A. (1988). 'Big Science as a form of life', in DeMaria, M., Grilli M., and Sebastian, F. (eds), *Restructuring of the Physical Sciences in Europe and the United States: 1945–1960*, 42–54. World Scientific Publishing.

Pietarinen, A.V. (2011). 'Principles and practices of Neurath's picture language', in Symons, J., Pombo, O., and Torres, J. (eds), *Otto Neurath and the Unity of Science. Logic, Epistemology, and the Unity of Science* 18: 71–82. Dordrecht: Springer. https://doi.org/10.1007/978-94-007-0143-4_6.

Planck Collaboration (2019). 'Planck finds no new evidence for cosmic anomalies'. https://sci.esa.int/web/planck/-/61396-planck-finds-no-new-evidence-for-cosmic-anomalies.

Plank Collaboration: Aghanim, N., Akrami, Y., Arojjo, F., Ashdown, M., Aumont, J., Baccigalupi, C., Ballardini, M., et al. (2020a). 'Planck 2018 results. I. Overview and the cosmological legacy of Planck'. *Astron. & Astrophys.* 641(A1). https://doi.org/10.1051/0004-6361/201833880.

Planck Collaboration: Aghanim N., Akrami, Y., Ashdown, M. et al. (2020b). 'Planck 2018 Results V. CMB power spectra and likelihoods'. *Astron. & Astrophys.* 641.A5 1-92

Planck Collaboration: Aghanim, N., Akrami, Y., Ashdown, M., Aumont, J., Baccigalupi, C., Ballardini, M., et al. (2020c). 'Planck 2018 results. VI Cosmological Parameters'. *Astron. & Astrophys.* 641 A6. https://doi.org/10.1051/0004-6361/201833910.

Planel, H., Soleilhavoup J.P., Tixador, R., Richoilley, G., Conter, A., Croute, F., and Gaubin, Y. (1987). 'Influence on cell proliferation of background radiation or exposure to very low, chronic gamma radiation'. *Health Physics* 52(5): 571–578. https://doi.org/10.1097/00004032-198705000-00007.

Polanyi, M. (1962). 'The republic of science—its political and economic theory'. *Minerva* 1: 54–73. https://doi.org/10.1007/BF01101453.

Praderie, F. (1996). 'Big Science: Why? where? and how?'. *Memorie della Società Astronomia Italiana* 67: 889.

Praderie, F. (1996b). 'A new opportunity for science in Europe'. *Nature* 384: 108.

Pralavorio, C. (2020), Electricity transmission reaches even higher intensities. https://home.cern/news/news/accelerators/electricity-transmission-reaches-even-higher-intensities.

Price, D.J. (1963). *Little Science, Big Science*. Columbia University Press.

Pujol Priego, L. (2020). 'At the crossroads of big science, open science, and technology transfer'. PhD Thesis. Barcelona; ESADE ADE. http://hdl.handle.net/10803/669220.

Punturo, M., Abernathy, M., Acernese, F., Allen, B., Andersson, E., Arun, K., et al. (2010), 'The Einstein Telescope: A third-generation gravitational wave observatory'. *Classical Quantum Grav.* 27(19). https://doi.org/10.1088/0264-9381/27/19/194002.

Qi Dong, J., McCarthy, K.J., and Schoenmakers, W.W.M.E. (2017). 'How central is too central? Organizing interorganizational collaboration networks for breakthrough innovation'. *Journal of Product Innovation Management* 34(4): 526–542. https://doi.org/10.1111/jpim.12384.

Quijada. J.A.M. et al. (2017). 'Recent upgrades and results from the CMS experiment'. *J. Phys.: Conf. Ser.* 912. https://iopscience.iop.org/article/10.1088/1742-6596/912/1/012005.

Radhe, M. (2017). 'Proton Therapy—Present and Future'. *Advanced Drug Delivery Reviews* 109: 26–44. https://doi.org/10.1016/j.addr.2016.11.006.

Raeburn, P. (2017). 'Wise buy? Proton beam therapy'. https://www.medpagetoday.com/radiology/therapeuticradiology/65422.

Rawsthorn, A. (2018). *Design as an Attitude*. JRP Ringier.

Reardon, S. (2011). 'High energy physics experiments in Japan weather the crises'. *ScienceInsider* (March). https://www.science.org/content/article/highenergy-physics-experiments-japan-weather-crises.

Reas, C., and Maeda, J. (2004). *Creative Code: Aesthetics+ Computation*. Thames & Hudson.

Rees, M. (2003). 'Albert Einstein World Award of Science 2003'. https://www.consejoculturalmundial.org/winners/winners-of-the-world-award-of-science/prof-martin-rees/.

Rees, M. (2021). Interview by Hans Ulrich Obist, Purple Magazine, Cosmos Issue #32. https://purple.fr/magazine/the-cosmos-issue-32/an-interview-with-martin-rees/.

Reitze, D. (2017). 'Gravitational waves: An entirely new window on to the cosmos', CERN Seminar 2017, LIGO-G1701533-v1. https://indico.cern.ch/event/656490/attachments/1500024/2365345/CERNColloquium_95MB.pdf.

Reitze, D., Adhikari, R. X., Ballmer, S., Barish, B., Barsotti, L., Billingsley, G., et al. (2020). 'Cosmic Explorer: the US contribution to gravitational-wave astronomy beyond LIGO'. *Bull. Am. Astron. Soc.* 51(7). https://baas.aas.org/pub/2020n7i035/release/1.

Rendina, H.J., Talan, A.J., Cienfuegos-Szalay, J., Carter, J.A., and Shalhav, O. (2020). 'Treatment is more than prevention: Perceived personal and social benefits of messaging among sexual minority men living with HIV'. *Aids Patient Care and Stds.* 34: 444–451. https://doi.org/10.1089/apc.2020.0137.

Resnik, D.B., and Elliott, K.C. (2016). 'The ethical challenge of socially responsible science'. *Account Res.* 23 (1): 31–46.

Reynaud, E. G. (2005). Scientists for better world, *EMBO Reports* 6(2): 103–107. https://doi.org/10.1038/sj.embor.7400338.

Riken. (2020). http://xfel.riken.jp/eng/.

Rinck, P.A. (2008). 'A short history of magnetic resonance imaging'. *Spectroscopy Europe* 20(1): 7–9. https://www.spectroscopyeurope.com/system/files/pdf/NMR_20_1.pdf.

Robertson, D.A., and Caldart, A.A. (2008). 'Natural science models in management'. *Opportunities and Challenges E: CO* 10(2): 61–75. https://www.researchgate.net/publication/241311601_Natural_Science_Models_in_Management_Opportunities_and_Challenges.

Robinson, M. (2021). 'Big Science collaborations; Lessons for global governance and leadership'. *Global Policy.* https://doi.org/10.1111/1758-5899.12861.

Rochat, S., Masdonati, J., and Dauwalder, J-P. (2017). 'Determining career resilience', in Maree, K. (ed.), *Psychology of Career Adaptability, Employability and Resilience.* Springer: 125–141. https://doi.org/10.1007/978-3-319-66954-0_8.

Rogers, E.M. (1962). *Diffusion of Innovations.* Free Press of Glencoe. https://doi.org/10.1002/jps.2600520633.

Rogers, E.M. (2003). *Diffusion of Innovations*, 5th Edition. Simon & Schuster.

Rogers, E.M., and Kincaid, D.L. (1981). *Communication Networks: Toward a New Paradigm for Research.* Free Press of Glencoe.

Romasanta, A.K., Wareham, J., Pujol Priego, L., Garcia Tello, P., and Nordberg, M. (2021). 'Risky business: How to capitalize on the success of Big Science'. *Issues in Science and Technology*, 23 July. https://issues.org/risky-business-big-science-deep-tech-transfer-commercialization/.

Romer, P.M. (1990). 'Endogenous technological change'. *Journal of Political Economy* 98(5) Part 2: S71–S102.

Rosa, H. (2010). *High-speed Society: Social Acceleration, Power, and Modernity.* Penn State Press.

Rossi, L., and Bottura, L. (2012). 'Superconducting magnets for particle accelerators'. *Reviews of Accelerator Science and Technology* 5: 51–89. https://doi.org/10.1142/S1793626812300034.

Rossi, I., and Brüning, O. (2015). 'Introduction to the HL-LHC Project'. *Advanced Series on Directions in High Energy Physics* 24: 1–17. https://doi.org/10.1142/9789814675475_0001.

Rossi, L. (2021). Personal interview with Lucio Rossi, CERN, May 2021.

Rubin, V.C., and Ford, W.K. Jr. (1970). 'Rotation of the Andromeda nebula from a spectroscopic survey of emission regions'. *The Astrophysical Journal* 159: 379–403. https://doi.org/10.1086/150317.

Rubin, V.C., Ford, W.K. Jr., and Thonnard, N. (1980). 'Rotational properties of 21 SC galaxies with a large range of luminosities and radii, from NGC 4605 (R=4kpc) to UGC 2885 (R=122kpc)'. *Astrophysical Journal Part 1* 238: 471–487. https://doi.org/10.1086/158003.

Rubin, V.C., Burstein. D., Ford, W. K. Jr., and Thonnard, N. (1985). 'Rotation velocities of 16 SA galaxies and a comparison of Sa, Sb, and SC rotation properties'. *Astrophys. J.* 289: 81–104. https://doi.org/10.1086/162866.

Rubin, V.C., Graham, J.A., and Kenney, J.D.P. (1992). 'Cospatial counterrotating stellar disks in the Virgo E7/ S0 galaxy NGC 4550'. *The Astrophysical Journal* 394: L9–L12. https://doi.org/10.1086/186460.

Sabado, M.M. (1991). 'Medical applications of accelerators', in Nonte J. (ed.), *Supercollider 3.* Springer. https://doi.org/10.1007/978-1-4615-3746-5_60.

Sánchez, E. (2016). 'The Dark Energy Survey: status and first results'. *Nuclear and Particle Physics Proceedings* 273–275: 302–308. https://doi.org/10.1016/j.nuclphysbps.2015.09.042.

Sapienza, A.M. (2004). *Managing Scientists: Leadership Strategies in Scientific Research.* John Wiley & Sons.

Sarkissian J.M. (2001), 'On eagle's wings: The Parkes Observatory's support of the Apollo 11 mission'. *Pub. Astron. Soc. Aus.* 18(3): 287–310. https://doi.org/10.1071/AS01038.

Schinle, M., Erler, C., and Stork, W. (2020). 'Distributed ledger technology for the systematic investigation and reduction of information asymmetry in collaborative networks', in *Proceedings of the 53rd Hawaii International Conference on System Sciences.* https://scholarspace.manoa.hawaii.edu/server/api/core/bitstreams/e49a6322-c61d-478f-9f72-af3f5d890573/content.

Schmidt, R. (2016), 'Introduction to machine protection', *Proceedings of the 2014 Joint International Accelerator School: Beam Loss and Accelerator Protection,* https://e-publishing.cern.ch/index.php/CYR/article/view/227 Vol 2, https://doi.org/10.5170/CERN-2016-002.1.

Schmied, H. (1975). 'A study of economic utility resulting from CERN Contracts'. CERN Yellow Report CERN 75–05 (English) CERN 75-06 (French) CERN 79-10 (German) and 1977 *IEEE Trans. Eng. Mgt.* EM–24 N°4: 125–138.

Science Business. (2015). *Big Science—What's It's Worth?.* 25 February. SB Publishing. https://sciencebusiness.net/report/big-science-whats-it-worth.

Science Business. (2019). *Why Open Science Is the Future—and How to Make It Happen.* 16 July. SB Publishing. https://sciencebusiness.net/report/why-open-science-future-and-how-make-it-happen.

Sciolino, E. (2018). 'A celebrated physicist with a passion for music', New York Times, 7 Marck. https://www.nytimes.com/2018/03/07/science/fabiola-gianotti-physics-cern.html.

Scitovsky, T. (1954). 'Two concepts of external economies'. *Journal of Political Economy* 62(2): 143–151.

Seebeck, T.J. (1822). 'Magnetische polarisation der metalle und erze durch temperatur–differenz'. *Abh. Deutsch. Akad. Wiss.* Berlin: 265–373.

Seegenschmiedt, M.H., Wielpütz, M., Hanslian, E., and Fehlauer, F. (2012). 'Long-term outcome of radiotherapy for primary and recurrent Ledderhose disease', in Eaton Ch. et al. (eds), *Dupuytren's Disease and Related Hyperproliferative Disorders.* Berlin: Springer: 349–371. https://doi.org/10.1007/978-3-642-22697-7_44.

Seeman, J., Schulte, D., Delahaye, J.P., Ross, M., Stapnes, S., Grudiev, A., Yamamoto, A., Latina, A., Seryi, A., Thoma Gracia, R., Guiducci, Y., Papahilippou, Y., Bogacz, S.A., and Krafft, G.A. (2020). 'Design and principles of linear accelerators and colliders', in Myers, S. and Schopper, H. (eds), *Particle Physics Reference Library.* Springer. https://doi.org/10.1007/978-3-030-34245-6_7.

Segedy, J.R., Kinnebrew, J.S., and Biswas, G. (2015). 'Using coherence analysis to characterize self-regulated learning behaviours in open-ended learning environments'. *Journal of Learning Analytics* 2(1): 13–48. https://learning-analytics.info/index.php/JLA/article/view/4129.

Seltzer, E., and Mahmoudi, D. (2013). 'Citizen participation, open innovation, and crowdsourcing: Challenges and opportunities for planning'. *Journal of Planning Literature* 28(1): 3–18. https://doi.org/10.1177/0885412212469112.

Sheehy, S. (2022). *The Matter of Everything—Twelve Experiments that Changed Our World.* Bloomsbury Publishing.

Shihong, Xu. (2017). 'Benefits and drawbacks of China's top-down innovation campaign in tertiary science education in creative education'. *Scientific Research* 8(1): 2151–4771. https://doi.org/10.4236/ce.2017.81005.

Shipsey, I. (2020). Lessons in leadership. *CERN Courier*, September. https://cerncourier.com/a/lessons-in-leadership/.

Sikivie, P. (1983). 'Experimental tests of the "invisible" axion'. *Phys. Rev. Lett.* 51(16): 1415–1417. https://doi.org/10.1103/PhysRevLett.51.1415.

Sime, J. (1900). 'William Herschel and his work'. *Wikisource*. https://en.wikisource.org/wiki/William_Herschel_and_his_work.

Simon, H.A. (1962). 'The architecture of complexity'. *Proceedings of the American Philosophical Society* 106(6): 467–482.

Skaburskis, A. (2008). 'The origin of "wicked problems"'. *Planning Theory & Practice* 9(2): 77–280. https://doi.org/10.1080/14649350802041654.

SLAC. (2020). https://www6.slac.stanford.edu/.

Sloane, P. (2011). 'The brave new world of open innovation'. *Strategic Direction* 27(5): 3–4. https://doi.org/10.1108/02580541111125725.

Smart, J., Scott, M., McCarthy, J.B., et al. (2012). 'Big science and big administration'. *Eur. Phys. J. Spec. Top.* 214: 635–666. https://doi.org/10.1140/epjst/e2012-01708-x.

Smith, A.R. (2009). 'Vision 20/20: Proton therapy'. *Medical Physics* 36(2): 556–568. https://doi.org/10.1118/1.3058485.

Smith, C.L. (2014). 'Genesis of the Large Hadron Collider', *Phyl. Trans Roy. Soc. London A* 373 (2014) 2032, 20140037. https://doi.org/10.1098/rsta.2014.0037.

Snyder, H.R., Reedy, A.J., and Lennarz, W.J. (1958). 'Synthesis of aromatic boronic acids. Aldehydo boronic acids and a boronic acid analog of Tyrosine'. *Journal of the American Chemical Society* 80(4): 835–838. https://doi.org/10.1021/ja01537a021.

Sollerman, J., Motsell, E., David, T.M., et al. (2009). 'First-year Sloan Digital Sky Survey-II(SDSS-II) supernova results: Constraints on nonstandard cosmological models'. *Astrophys. J.* 703(2): 1374–1385. https://doi.org/10.1088/0004-637X/703/2/1374.

Spender, J.C. (1996). 'Making knowledge the basis of a dynamic theory of the firm'. *Strategic Management Journal* 17(S2). https://doi.org/10.1002/smj.4250171106.

Spinney, L. (2022). 'Are we witnessing the dawn of post-theory science?'. *The Guardian*, 9 January. https://www.theguardian.com/technology/2022/jan/09/are-we-witnessing-the-dawn-of-post-theory-science.

Spitzberg, B.H. (2014). 'Toward a model of meme diffusion (M3D)'. *Communication Theory* 24(3): 311–339. https://doi.org/10.1111/comt.12042.

Stahl, B.C. (2013). 'Responsible research and innovation: The role of privacy in an emerging framework'. *Science and Public Policy* 40(6): 708–716. https://doi.org/10.1093/scipol/sct067.

Starovoitova, V.N., Tchelidze, L., and Wells, D.P. (2014). 'Production of medical radioisotopes with linear accelerators'. *Applied Radiation and Isotopes* 85: 39–44. https://doi.org/10.1016/j.apradiso.2013.11.122.

Stefik, M., and B. Stefik. (2004). *Breakthrough Stories and Strategies of Radical Innovation*. MIT Press.

Steinwachs, C.F. (2019). 'Higgs field in cosmology'. https://arxiv.org/abs/1909.10528.

Streit-Bianchi, M., Cimadevila, M., and Trettnak, W. (2020). *Mare Plasticum—The Plastic Sea*. Berlin: Springer. https://doi.org/10.1007/978-3-030-38945-1.

Streit-Bianchi, M., Michelini, M., Bonivento, W., and Tuveri, M. (eds) (2023). *Challenges and Opportunities in digital education*. Springer. https://www.springer.com/series/16575.

Strom, R. (2017). 'Spectral imaging: From CERN to medical technologies'. https://home.cern/news/news/knowledge-sharing/spectral-imaging-cern-medical-technologies.

SWAFS (2020). European Commission, Directorate-General for Research and Innovation, Iagher, R., Monachello, R., Warin, C. et al., *Science with and for society in Horizon 2020 – Achievements and Recommendations for Horizon Europe*, Delaney, N. and Tornasi, Z. (eds), Publications Office. https://data.europa.eu/doi/10.2777/32018.

Swinburne Astronomy Productions. (2021). https://astronomy.swin.edu.au/production/.

Tanabashi, M., Hagiwara, K., Hikasa, K., Nakamura, K., Sumino, Y., Takahashi, F., et al. (2018). 'Review of particle physics', *Phys. Rev. D* 98(3): 1–1896. https://doi.org/10.1103/physrevd.98.030001.

Tavernelli, I., and Barkoutsos, P. (2021). 'IBM welcomes CERN as a new hub in the IBM quantum network'. https://research.ibm.com/blog/cern-lhc-qml?lnk=ushpv18nf1.

Taylor, A. (2012). 'The dark universe', in Majid, S. (ed.), *On Space and Time*. Cambridge University Press: 1–55. https://doi.org/10.1017/CBO9781139197069.002.

Thelen, E., and Smith, L.B. (1994). *A Dynamic Systems Approach to the Development of Cognition and Action*. MIT Press.

Thom, R. (1977). 'High density infrared detector arrays'. U.S. Patent No. 4, 039,833.

Thomas, H., Adela, M., Chervyakov, A., Rubbia, C., and Salmieri, D. (2016). 'Efficiency of superconducting transmission lines: An analysis with respect to the load factor and capacity rating'. *Electric Power Systems Research* 14: 381–391. https://doi.org/10.1016/j.epsr.2016. 07.007.

Thong, C., Cotoranu, A., Down, A., Kohler, and K., Batista, C. (2021). 'Design innovation integrating deep technology, societal needs, radical innovation, and future thinking: a case study of the CBI A3 program'. Experimenting *with Challenge-Based Innovation* 5(1). https://doi.org/10. 23726/cij.2021.1291.

Thursby, J., and Thursby, M. (2007). 'Knowledge creation and diffusion of public science with intellectual property rights', in Maskus, K.E. (ed.) *Intellectual Property, Growth and Trade (Frontiers of Economics and Globalization*, Vol. 2), Ch. 6: 199–232, Emerald Group Publishing Ltd. https://doi.org/10.1016/S1574-8715(07)00006-1.

Thwaites, D.I., and Tuohy, J.B. (2006). 'Back to the future: the history and development of the clinical linear accelerator'. *Physics in Medicine and Biology* 51(13): R343–R362. https://doi.org/10.1088/0031-9155/51/13/R20.

Tommasini, D. et. al. (2018). 'Status of the 16 T Dipole Development Program for a future Hadron Collider'. *IEEE Trans. On Appl. Supercon.* 28 (2018) 3: 4001305. https://doi.org/10.1109/TASC.2017.2780045.

Tsujii, H. (2017). 'Overview of carbon-ion radiotherapy'. *Journal of Physics: Conference Series* 777 (1): 012032. https://doi.org/10.1088/1742-6596/777/1/012032.

Tuertscher, P., Garud, R., and Kumaraswamy, A. (2014). 'Justification and interlaced knowledge at ATLAS, CERN'. *Organiation Science* 25(6): 1573–1877. https://doi.org/10.1287/orsc.2013.0894.

Tuertscher, P., Garud, R., and Nordberg, M. (2008). 'The emergence of architecture: Coordination across boundaries at ATLAS, CERN'. Conference Paper, Academy of Management.

Tufte, E. (1983). *The Visual Display of Quantitative Information*. Cheshire: Graphics Press. https://www.visualizingsociety.com/class/05/notes/vdqch2.pdf.

Uhl-Bien, M., and Marion, R. (2009). « Complexity leadership in bureaucratic forms of organizing: A meso model'. *The Leadership Quarterly* 20: 631–650. https://digitalcommons.unl.edu/cgi/viewcontent.cgi?referer=&httpsredir=1&article=1037&context=managementfacpub.

UK Design Council. (2018). 'Designing a future economy'. https://www.designcouncil.org.uk/sites/default/files/asset/document/Designing_a_future_economy18.pdf.

UK Design Council. (2019). 'What is the Framework for Innovation?' https://www.designcouncil.org.uk/news-opinion/what-framework-innovation-design-councils-evolved-double-diamond.

UK Design Council. (2021). 'Beyond net zero: A systemic design approach'. https://www.designcouncil.org.uk/resources/guide/beyond-net-zero-systemic-design-approach.

Ulanowicz, R.U., Goerner S.J., Ljetaer and Gomez, R. (2009). 'Quantifying sustainability: Resilience, efficiency and the return of information theory'. *Ecological Complexity* 6(1): 27–36. https://doi.org/10.1016/j.ecocom.2008.10.005.

Ulnicane, I. (2020). 'Beyond "good old CERN": data and digital research infrastructures'. *Ethics Dialogues*. https://www.ethicsdialogues.eu/2020/11/06/beyond-good-old-cern-data-and-digital-research-infrastructures/.

Ulnicane, I. (2020a). 'Ever-changing big science and research infrastructures: Evolving European Union policy', in Kramer, K.C. and Hallonsten, O. (eds), *Big Science and Research Infrastructures in Europe*, 76–100. Elgar.

UN. (2020). The 17 Goals. https://sdgs.un.org/goals.

UNESCO. (2021). *UNESCO Science Report 2021: Towards 2030, Regional Overview: Asia and the Pacific*. Paris: UNESCO. https://unesdoc.unesco.org/ark:/48223/pf0000235406.

Unnikrishnan, C. S. (2013). 'IndIGO and LIGO-India: scope and plans for gravitational wave research and precision metrology in India'. *International J. Mod. Phys. D* 22(01). https://doi.org/10.1142/S0218271813410101.

USAID. (2018). 'Embracing enterprise-driven development'. https://medium.com/usaid-2030/embracing-enterprise-driven-development-e290d0729fab.

Utting, P. (2007). 'CSR and equality'. *Third World Quarterly* 28(4): 697–712. http://www.jstor.org/stable/20454957.

Vallisneri, M., Kanner, J., Williams, R., Weinstein, A., and Stephens, B. (2015). 'The LIGO Open Science Center'. https://doi.org/10.1088/1742-6596/610/1/012021.

Van den Bosch, F.A., Volberda, H.W., and de Boer, M. (1999). 'Coevolution of firm absorptive capacity and knowledge environment: organizational forms and combinative capabilities'. *Organization Science* 10(5): 551–568. https://doi.org/10.1287/orsc.10.5.551.

Van der Bijl-brouwer, M., and Dorst, K. (2017). 'Advancing the strategic impact of human-centred design'. *Design Studies* 53: 1–23. https://doi.org/10.1016/j.destud.2017.06.003.

Van Noorden, R. (2015). Interdisciplinary research by the numbers'. *Nature*, 525(7569): 306–307. https://doi.org/10.1038/525306a.

Van Putten, A.B., and MacMillan, I.C. (2004). 'Making real options really works'. *Harvard Business Review* 82(12): 134–141.

Venter, C.J. et al. (2001). 'The sequence of the human genome'. *Science* 291(5507): 1304–1351. https://doi.org/10.1126/science.1058040.

Verganti, R. (2008). 'Design, meanings, and radical innovation: A metamodel and a research agenda'. *Journal of Product Innovation Management*, 25(5): 436–456.

Verganti, R. (2009). *Design Driven Innovation: Changing the Rules of Competition by Radically Innovating What Things Mean.* Boston: Harvard Business Press.

Voegtlin, C., Frisch, C., Walther, A., and Schwab, P. (2020). 'Theoretical development and empirical examination of a three-roles model of responsible leadership'. *Journal of Business Ethics* 167(3): 411–431. https://doi.org/10.1007/s10551-019-04155-2.

Von Hippel, E. (1988). *Sources of Innovation.* Oxford University Press. https://papers.ssrn.com/sol3/papers.cfm?abstract_id=2877276.

Von Schomberg, R. (2013). 'A vision of responsible innovation', in Owen, R., Heintz, M., and Bessant, J. (eds), *Responsible Innovation. Managing the Responsible Emergence of Science and Innovation in Society*, 51–74. Wiley. https://doi.org/10.1002/9781118551424.ch3.

Vuola, O. (2006). 'Innovation and new business through mutually beneficial collaboration and proactive procurement'. Doctoral thesis. Ecole des hautes études commerciales.

Vuola, O., and Hameri, A.P. (2006). 'Mutually benefiting joint innovation process between industry and big-science'. *Technovation* 26(1): 3–12. https://doi.org/10.1016/j.technovation.2005.03.003.

Wadia, S.R. (2009). 'Homi Jehangir Bhabha and the Tata Institute of Fundamental Research', *Current Science* 96(5): 725–733. https://www.currentscience.ac.in/Volumes/96/05/0725.pdf.

Wagner, C.S., Park, H.W., and Leydesdorff, L. (2015). 'The continuing growth of global cooperation networks in research: A conundrum for national governments'. *PLoS ONE* 10(7). https://doi.org/10.1371/journal.pone.0131816.

Wagner, K., Boehle, A., Pathak, P., et al. (2021). 'Imaging low-mass planets within the habitable zone of α Centauri'. *Nat Commun* 12: 922 (2021). https://doi.org/10.1038/s41467-021-21176-6.

Wareham, J. D., Pujol Priego, L., Romasanta, A. K., Mathiassen, T.W., Nordberg, M., and Garcia Tello, P. (2022). 'Systematising serendipity for big science infrastructures: the ATTRACT project'. *Technovation* 116: 1–16. https://doi.org/10.1016/j.technovation.2021.102374.

Watanabe, T., Ichikawa, H., and Fukumori, Y. (2002). 'Tumour accumulation of gadolinium in lipid-nanoparticles intravenously injected for neutron-capture therapy of cancer'. *European Journal of Pharmaceutics and Biopharmaceutics* 54: 119–124. https://doi.org/10.1016/s0939-6411(02)00085-1.

Wei, W., Huerta, E.A., Yun, M., Loutrel N., Shaikh, M.A., Kumar, P., Haas R., and Kindratenko, V. (2021). 'Deep learning with quantized neural networks for gravitational-wave forecasting of eccentric compact binary coalescence'. *The Astrophysical Journal*, 919(2) :1–10. https://iopscience.iop.org/article/10.3847/1538-4357/ac1121.

Weiglein, G. (2021). 'Linear collider physics', in *LCWS 2021 International Linear Collider Conference, CERN*. https://indico.cern.ch/event/995633/contributions/4240291/attachments/2206930/3736132/lcws_21_physlect.pdf.

Weinberg, A.M. (1961). 'Impact of large-scale science on the United States'. *Science*. 134 (3473): 161–164. Bibcode:1961Sci...134..161W. https://doi.org/10.1126/science.134.3473.161. JSTOR 1708292. PMID 17818712.

Weinberg, A.M. (1968). *Reflections on Big Science*. MIT Press.

Weinberg, S. (1978). 'A new light boson?' *Phys. Rev. Lett.* 40(4): 223–226. https://doi.org/10.1103/PhysRevLett.40.223.

Weinberg, S. (2011). 'Particle physics from Rutherford to LHC'. *Physics Today* 64(8) 29. https://doi.org/10.1063/PT.3.1216.

Weisberg, J.W., Nice, D.J., and Taylor, J.H. (2010). 'Timing measurements of the relativistic binary pulsar PSR B1913+16'. *Astrophy. J.* 722(2): 1030–1034. https://doi.org/10.1088/0004-637X/722/2/1030.

Weisskopf, V.F. (1977). 'The Development of the Concept of an Elementary Particle,' in: Karimaki V. (ed) *Proceedings of the symposium on the foundation of modern physics*, LomaKoli, Finland, August 11–18, 1977.

Wheeler, J.A. (1998). *Geons, Black Holes, and Quantum Foam: A Life in Physics*. Norton.

WHO. (2021). Public Health Surveillance. https://www.who.int/emergencies/surveillance.

Wideröe, R. (1928). 'Über ein neues Prinzip zur Herstellung hoher Spannungen'. *Archiv för Elektrotechnik (in German)*. 21(4):387–406. https://doi.org/10.1007/BF01656341.

Wilczek, F. (1978). 'Problem of strong P and T invariance in the presence of instantons'. *Phys. Rev. Lett.* 40(5):279–282. https://doi.org/10.1103/PhysRevLett.40.279.

Wilkinson, P. (1991), 'The hydrogen array, in radio interferometry: Theory, techniques, and applications', in T.J. Cornwell and R.A. Perley (eds), *Proceedings of the 131st IAU Colloquium, Socorro, NM, Oct. 8–12, 1990*. San Francisco, CA, Astronomical Society of the Pacific, 19, 428–432.

Williams, D. (2004). '50 years of computing at CERN. Personal reflections and photographs', in *CHEP04, 24 September–1 October, Interlaken, Switzerland*. https://slideplayer.com/slide/12414425/.

Wilson, R.R. (1946). 'Radiological use of fast protons'. *Radiology* 487–491. https://doi.org/10.1148/47.5.487.

Wilson, S. (2002). *Information Arts: Intersections of Art, Science, and Technology*. MIT Press.

Wittig A., and Sauerwein, W.A.G. (2012). 'Neutron capture therapy: Principles and applications', in W.A.G. Sauerwein, A. Wittig, R. Moss, and N. Nakagawa (eds), *Neutron Capture Therapy: Principles and Applications*. Springer Science & Business Media.

Wojcicki, S. (2008). 'The Supercollider: The pre-Texas days—a personal recollection of its birth and Berkeley years'. *RAST* 1(1): 259–302. https://doi.org/10.1142/S1793626808000113.

Wojcicki, S. (2009). 'The Supercollider. The Texas days—a personal recollection of its short life and demise'. *RAST* 2: 265–301. https://doi.org/10.1142/9789814299350_0013.

Womack, J.P., and Jones, D.T. (1996). *Lean Thinking*. Simon & Schuster.

World Bank Group. (2018). 'Blockchain & Distributed Ledger Technology (DLT)'. https://www.worldbank.org/en/topic/financialsector/brief/blockchain-dlt.

Wright, D. (2020). 'Open Science is good science'. In ARC News. https://www.esri.com/about/newsroom/arcnews/open-science-is-good-science/.

XFEL. (2020). https://www.xfel.eu/news_and_events/news/index_eng.html?openDirect Anchor=1777.

Xu. (2017). 'Benefits and drawbacks of China's top-down innovation campaign in tertiary science education in creative education'. *Scientific Research* 8(1): 2151–4771. https://doi.org/10.4236/ce.2017.81005.

Yamamoto, H. (2021). 'The International Linear Collider Project—its physics and status'. *Symmetry* 13: 674–689. https://doi.org/10.3390/sym13040674.

Yammarino, F. J., and Dansereau, F. (2002). 'Individualized leadership'. *Journal of Leadership & Organizational Studies* 9(1): 90–99.

Yaqub, O. (2018). 'Serendipity: Towards a taxonomy and a theory'. *Research Policy* 47(1): 169–179. https://doi.org/10.1016/j.respol.2017.10.007.

Yeong, C.H., Cheng, Mh., and Ng, K.H. (2014). 'Therapeutic radionuclides in nuclear medicine: current and future prospects'. *J. Zhejiang Univ. Sci. B* 15: 845–863. https://doi.org/10.1631/jzus. B1400131.

Yin, Y., Dong, Y., Wang, K. Wang, D and Jones, B.F. (2022). 'Public use and public funding of science'. *Nature Human Behaviour* 6: 1344–1350. https://doi.org/10.1038/s41562-022-01397-5.

Yli-Renko, H., Autio, E., and Sapienza, H.G. (2001). 'Social capital, knowledge acquisition, and competitive advantage in technology-based young firms'. *Strategic Management Journal*, 22 (6–7). https://doi.org/10.1002/smj.183.

Yunus, M., and Weber, C. (2007). *Creating a World Without Poverty: Social Business and the Future of Capitalism.* Public Affairs.

Yunus, M. (2009). 'Economic Security for a World in Crisis'. *World Policy Journal* 26(2): 5–12. http://www.jstor.org/stable/40468628.

Yunus, M. (2011). *Building Social Business: The New Kind of Capitalism that Serves Humanity's Most Pressing Needs.* Public Affairs.

Yunus, M., and Weber, C. (2017). *A World of Three Zeros: The New Economics of Zero Poverty, Zero Unemployment, and Zero Net Carbon Emissions.* Hachette.

Yves, J. (2010). *Proceedings of CYCLOTRONS, Lanzhou, China.* https://accelconf.web.cern.ch/Cyclotrons2010/papers/frm1cio01.pdf.

Zanella, P. (1990). *30 Years of Computing at CERN. Parts 1–3.* Geneva: CERN. http://cdsweb.cern.ch/record/205360/files/p1.pdf.

Zanella, P. (2014). 'Technology fallout in Bioinformatics', in Bressan, B. (ed.), *From Physics to Daily Life*, Part 2, Ch. 7: 153–178. Wiley. https://doi.org/10.1002/9783527687077.ch07.

Zareravasan, A., Ashrafi, A., and Krčál, M. (2020). 'The implications of blockchain for knowledge sharing'. *IFKAD 2020, Matera, Italy 9–11 September.* ISBN 978-88-96687-13-0. ISSN 2280-787X.

Zhang, M.Y., Gann, D., and Dodgson, M. (2022). 'China's "Innovation Machine": How it works, how it's changing and why it matters'. *The Conversation*, May. https://theconversation.com/chinas-innovation-machine-how-it-works-how-its-changing-and-why-it-matters–180615.

Zheng, D. (2018). 'Design thinking is ambidextrous'. *Management Decision* 56(4): 736–756.

Zhu, X., and Qian, S. (2012). 'The Quality Management of the R&D in High Energy Physics Detector', in Akyar, I. (ed.), *Latest Research into Quality Control.* IntechOpen. Chapter 15. https://doi.org/10.5772/51434.

Ziman, J. (2000). *Real Science—What It Is, and What It Means.* Cambridge: Cambridge University Press.

Zimmermann, F. (2015). 'High energy physics strategies and future large-scale projects'. *Nuclear Instruments and Methods in Physics Research* B 355(2015): 4–10, https://doi.org/10.1016/j.nimb.2015.03.090.

Zimmermann, F. (2018). 'Future colliders for particle physics—big and small'. *Nuclear Inst. and Methods in Physics Research A* 909: 33–37. https://doi.org/10.1016/j.nima.2018.01.034.

Zuckerman, H. (1988). The sociology of science. In N. J. Smelser (ed.), *Handbook of sociology* (pp. 511–574). Sage Publications, Inc.

Zweig, G. (2013). 'Concrete quarks'. *EP Newsletter*, December. Geneva: CERN. https://ep-news.web.cern.ch/sites/default/files/issues/Concrete%20Quarks%2012-12-14.pdf.

Zweigzig, J., and Lazzarini A. (2007). 'Best strain sensitivities for the LIGO interferometers—comparisons among S1–S5 runs'. Public LIGO doc. https://dcc.ligo.org/LIGO-G060009/public.

Zwicky, F. (1933). 'The redshift of extragalactic nebulae'. *Helvetica Physica Acta* 6: 110–127. https://authors.library.caltech.edu/92130/1/Fritz_redshift%20of%20extragalactic%20nebulae.pdf.

Zwicky, F. (1937). 'On the masses of nebulae and clusters of nebulae'. *The Astrophysical Journal* 86 (3): 217–246. https://doi.org/10.1086/143864.

# Index

*For the benefit of digital users, indexed terms that span two pages (e.g. 52–53) may, on occasion, appear on only one of those pages.*